I0053183

# Water Pollution Sources and Purification: Challenges and Scope

## Edited by

### R. M. Belekar

*Department of Physics*
*Institute of Science*
*Ravindranath Tagore Road*
*Nagpur, M.S.*
*India-440 001*

### Renu Nayar

*Department of Chemistry*
*D. P. Vipra College*
*Bilaspur C.G.,*
*India-495001*

### Pratibha Agrawal

*Department of Applied Chemistry*
*Laxminarayan Institute of Technology*
*R.T.M. Nagpur University*
*Nagpur, India–440010*

&

### S. J. Dhoble

*Department of Physics*
*R.T.M. Nagpur University*
*Nagpur*
*India-440033*

# Water Pollution Sources and Purification: Challenges and Scope

Editors: R. M. Belekar, Renu Nayar, Pratibha Agrawal & S. J. Dhoble

ISBN (Online): 978-981-5050-68-4

ISBN (Print): 978-981-5050-69-1

ISBN (Paperback): 978-981-5050-70-7

© 2022, Bentham Books imprint.

Published by Bentham Science Publishers Pte. Ltd. Singapore. All Rights Reserved.

First published in 2022.

## BENTHAM SCIENCE PUBLISHERS LTD.
### End User License Agreement (for non-institutional, personal use)

This is an agreement between you and Bentham Science Publishers Ltd. Please read this License Agreement carefully before using the ebook/echapter/ejournal (**"Work"**). Your use of the Work constitutes your agreement to the terms and conditions set forth in this License Agreement. If you do not agree to these terms and conditions then you should not use the Work.

Bentham Science Publishers agrees to grant you a non-exclusive, non-transferable limited license to use the Work subject to and in accordance with the following terms and conditions. This License Agreement is for non-library, personal use only. For a library / institutional / multi user license in respect of the Work, please contact: permission@benthamscience.net.

### Usage Rules:

1. All rights reserved: The Work is the subject of copyright and Bentham Science Publishers either owns the Work (and the copyright in it) or is licensed to distribute the Work. You shall not copy, reproduce, modify, remove, delete, augment, add to, publish, transmit, sell, resell, create derivative works from, or in any way exploit the Work or make the Work available for others to do any of the same, in any form or by any means, in whole or in part, in each case without the prior written permission of Bentham Science Publishers, unless stated otherwise in this License Agreement.
2. You may download a copy of the Work on one occasion to one personal computer (including tablet, laptop, desktop, or other such devices). You may make one back-up copy of the Work to avoid losing it.
3. The unauthorised use or distribution of copyrighted or other proprietary content is illegal and could subject you to liability for substantial money damages. You will be liable for any damage resulting from your misuse of the Work or any violation of this License Agreement, including any infringement by you of copyrights or proprietary rights.

### Disclaimer:

Bentham Science Publishers does not guarantee that the information in the Work is error-free, or warrant that it will meet your requirements or that access to the Work will be uninterrupted or error-free. The Work is provided "as is" without warranty of any kind, either express or implied or statutory, including, without limitation, implied warranties of merchantability and fitness for a particular purpose. The entire risk as to the results and performance of the Work is assumed by you. No responsibility is assumed by Bentham Science Publishers, its staff, editors and/or authors for any injury and/or damage to persons or property as a matter of products liability, negligence or otherwise, or from any use or operation of any methods, products instruction, advertisements or ideas contained in the Work.

### Limitation of Liability:

In no event will Bentham Science Publishers, its staff, editors and/or authors, be liable for any damages, including, without limitation, special, incidental and/or consequential damages and/or damages for lost data and/or profits arising out of (whether directly or indirectly) the use or inability to use the Work. The entire liability of Bentham Science Publishers shall be limited to the amount actually paid by you for the Work.

### General:

1. Any dispute or claim arising out of or in connection with this License Agreement or the Work (including non-contractual disputes or claims) will be governed by and construed in accordance with the laws of Singapore. Each party agrees that the courts of the state of Singapore shall have exclusive jurisdiction to settle any dispute or claim arising out of or in connection with this License Agreement or the Work (including non-contractual disputes or claims).
2. Your rights under this License Agreement will automatically terminate without notice and without the

need for a court order if at any point you breach any terms of this License Agreement. In no event will any delay or failure by Bentham Science Publishers in enforcing your compliance with this License Agreement constitute a waiver of any of its rights.

3. You acknowledge that you have read this License Agreement, and agree to be bound by its terms and conditions. To the extent that any other terms and conditions presented on any website of Bentham Science Publishers conflict with, or are inconsistent with, the terms and conditions set out in this License Agreement, you acknowledge that the terms and conditions set out in this License Agreement shall prevail.

**Bentham Science Publishers Pte. Ltd.**
80 Robinson Road #02-00
Singapore 068898
Singapore
Email: subscriptions@benthamscience.net

# CONTENTS

# FOREWORD

The authors of this edited book are renowned researchers in the field of water research, photocatalytic degradation, and solar cells. With their research experience, they have edited this book by referring to and analyzing a wide range of literature. This edited book not only covers a critical review of some water research problems but also contains some original research work regarding catalytic degradation. In short, this book reveals the causes of water pollution and its impact on animal life. Moreover, it also discusses various water purification methods developed so far. This edited book further focuses on fluoride contamination in drinking water, its effect on human life, and fluoride removal by using activated alumina modified with different materials. The authors of this book have included original research like that on the degradation of substituted benzoic acid by nanoparticles and analysis of seasonal and spatial variations of water quality of Dulhara and Vedponds in Ratnapur, Chhattisgarh, India. Few chapters of this book cover wastewater treatment using modern methods, the impact of water mismanagement on the environment, and suggestions on the preventive measures for proper water utilization. The authors have made all possible efforts to enhance the usefulness of the book for the research community.

**K. G. Rewatkar**
Professor in Physics
Ambedkar College
Nagpur-440010, India

# PREFACE

Water pollution and its impact on human health are frequently discussed in this age. Several research studies and projects are undertaken and accomplished year after year. However, sustainable water resource management is still a serious concern in most developing countries. Moreover, the problem of water contamination and its purification are greater challenges and a lot of the research tends to be futile. There is, therefore, a need to design and develop appropriate methodologies in order to improve the quality of water. By keeping this view in mind, the present book is written with clear objectives: to develop a basic idea about various challenges in water pollution and their removal.

Regarding the organization, the book consists of seven chapters, well-arranged in a coherent manner.

- Chapter one deals with the different water purification techniques used for safe drinking water production, potential threats, and challenges.

- Chapter two focuses exclusively on fluoride removal by adsorption method using activated alumina modified with different materials and isothermal studies.

- Chapter three is the result-oriented chapter that discusses different parameters affecting photocatalytic degradation of substituted benzoic acids.

- Chapter four covers the analysis of seasonal and spatial variations of water quality of Dulhara and Ved ponds in Ratnapur, Chhattisgarh, India.

- Chapter five examines the degradation of benzoic acid by iron nanoparticles as a photo-catalyst using an advanced oxidation process (AOP). This chapter also discusses the synthesis of Fe nanoparticles via hydrothermal process at ordinary and elevated temperatures.

- Chapter six deals with wastewater treatment using modern methods supported by nanoscale materials.

- Chapter seven discusses the impact of water mismanagement on the environment and suggests preventive measures for proper water utilization.

This book is meant for postgraduate and research scholars in the field of physical sciences, chemistry, and material sciences interested in water treatment, photocatalytic degradation, advanced oxidation process, and solar cell. The book will explore and help readers understand fundamental as well as advanced studies on these processes. Chapters in the book also provide future scope and challenges in both the phenomena, which allow readers to understand basic and current status in the fields. It is hoped that the book shall provide guidelines to all interested in research studies of one sort or the other.

We are highly indebted to our students and learned colleagues for providing the necessary stimulus for writing the book. We are grateful to all those persons whose writings and works have helped us in the preparation of this book. We are equally thankful to the reviewers of this edited book who made extremely valuable suggestions and have thus contributed to improving its quality.

We will feel highly rewarded if the book proves helpful in the development of genuine research studies. We look forward to suggestions from all readers, researchers, and scholars for further improving the content of the book.

**R. M. Belekar**
Department of Physics
Institute of Science
Ravindranath Tagore Road
Nagpur, M.S.
India

**Renu Nayar**
Department of Chemistry
D. P. Vipra College
Bilaspur C.G.,
India

**Pratibha Agrawal**
Department of Applied Chemistry
Laxminarayan Institute of Technology
R.T.M. Nagpur University
Nagpur, India

&

**S. J. Dhoble**
Department of Physics
R.T.M. Nagpur University
Nagpur
India

# List of Contributors

| | |
|---|---|
| **B.D. Deshpande** | Department of Applied Chemistry, Laxminarayan Institute of Technology, RTM Nagpur University, Nagpur, Maharashtra, India-440010 |
| **Bhavna D. Deshpande** | Department of Applied Chemistry, Laxminarayan Institute of Technology, R.T.M. Nagpur University, Nagpur, India–440010 |
| **G.D. Sharma** | Vice Chancellor, AtalBihari Vajpayee University, Bilaspur, India-495 001 |
| **M.G. Bhotmange** | Department of Applied Chemistry, Laxminarayan Institute of Technology, RTM Nagpur University, Nagpur, Maharashtra, India-440010 |
| **M. K. N. Yenkie** | Department of Applied Chemistry, Laxminarayan Institute of Technology, RTM Nagpur University, Nagpur, Maharashtra, India-440010 |
| **Pratibha S. Agrawal** | Department of Applied Chemistry, Laxminarayan Institute of Technology, RTM Nagpur University, Nagpur, Maharashtra, India-440010 |
| **R. M. Belekar** | Department of Physics, Institute of Science, Rabindranath Tagore Road, Nagpur-440 001, M.S., India |
| **Renu Nayar** | Department of Chemistry, D.P.Vipra College, Bilaspur,C.G., India-495001 |
| **Ritesh Kohale** | Department of Physics, Sant Gadge Maharaj Mahavidyalaya, Hingna, Nagpur,India, 441110 |
| **S.J. Dhoble** | Department of Physics, R.T.M. Nagpur University, Nagpur-440033, India |

# CHAPTER 1

# Review on Water Purifications Techniques and Challenges

**R. M. Belekar**[1,*] and **S.J. Dhoble**[2]

[1] *Department of Physics, Institute of Science, Rabindranath Tagore Road, Nagpur-440 001, M.S., India*

[2] *Department of Physics, R.T.M. Nagpur University, Nagpur-440033, India*

**Abstract:** Nowadays, the whole world is facing water containment issues caused by anthropogenic sources, including household waste, agricultural waste, and industrial waste. There is a huge impact of wastewater on the environment; hence, the public concern over it has been increased. This led researchers to be motivated and find radical and cheap solutions to overcome this problem. Several conventional techniques, including boiling, filtration, sedimentation, and chlorination, are used for wastewater treatment; however, they have limited scope. Some other methods like coagulation, flocculation, biological treatment, Fenton processes, advanced oxidation, membrane-based processes, ion exchange, electrochemical, adsorption, and UV-based processes have been applied to remove pollutants, but there are still some limitations. This review chapter sheds some light on these traditional and modern methods applied for water treatment, along with their advantages and disadvantages. These methods have the potential to remove pollutants from wastewater, such as natural organic matter, heavy metals, inorganic metallic matter, disinfection byproducts, and microbial chemicals. The potential threats and challenges of using water treatment methods for safe water production have also been discussed in this chapter.

**Keywords:** Adsorption, Biological treatment, Chemical methods, Electrodialysis, Fenton process, Membrane treatment, Purification methods, UV treatment, Water pollution.

## INTRODUCTION

Water is an essential element in natural resources required for the survival of all living organisms, cultivation, and food production. Today, many cities around the world face severe water shortages. About 40 percent of the global food supply requires irrigation, and the industrial process depends on the extensive use of water [1]. Environment and economic development are severely affected by the

---

* **Corresponding author R. M. Belekar:** Department of Physics, Institute of Science, Rabindranath Tagore Road, Nagpur-440 001, M.S., India; Tel: +91-9822292336; E-mail: rajubelekar@gmail.com

**R. M. Belekar, Renu Nayar, Pratibha Agrawal and S. J. Dhoble (Eds)**
**All rights reserved-© 2022 Bentham Science Publishers**

seasonal availability of water and its quality. Water quality is affected by human activity and is being reduced due to urbanization, population growth, industrial production, and climate change. As a result, water pollution has a severe impact on the earth and its inhabitants [2]. Water treatment produces drinkable water that is chemically, biologically, aesthetically, pure and healthy. The treatment cost for clean raw water is less as it requires fewer purification steps. In rural areas, water usually comes from commonly shared wells, ponds, or hand pumps, whereas in urban areas, it is supplied by municipal corporation water supply [3]. The purification of water involves many steps that are different in different regions and depend upon the quality of water and contaminants. More than 70% of the earth's surface is covered with water, but only around 1% of water is drinkable as per standards. There are many contaminants that make the water unhealthy for drinking purposes, like aluminum, ammonia, arsenic, fluoride, barium, cadmium, copper, *etc* [4 - 6]. There are common treatment methods that include coagulation, sedimentation, biological oxidation, photo-Fenton treatment, advanced oxidation processes (AOPs), oxidation with chemical oxidants, photocatalytic oxidation, membrane processes, electrochemical oxidation/degradation, adsorption, and combined methods [7].

## TRADITIONAL WATER PURIFICATION METHODS

The rural communities have adopted simple and traditional methods for removing visible impurities present in the water collected from various sources. Though these methods are not sufficient to provide quality water in urban areas as per international standards, they are more useful in rural areas where the degree of harmful contamination is almost negligible [8]. These methods can easily remove certain bacteria, pathogens, undissolved matter, dust, *etc*.

### Filtration

This is the most simple and convenient technique for removing wind-borne impurities like plant debris, insects, dust particles, or coarse mud particles. The raw water collected from various sources passes through a cotton cloth or winnowing sieves, and the impurities get filtered. However, this method cannot be used effectively when water is highly turbid or muddy as cotton cloth or sieve cannot filter fine suspended particles. This method of filtration is popular in many villages of India and other parts of the world, where water is collected from wells or clean ponds [9]. To filter highly turbid water, clay vessels with suitable pore sizes can be used. The turbid water is collected in a clay vessel and allowed to settle. The water in the clay vessel trickles through it, and clear water is collected in another jar. This method of filtration is common in Egypt. In the southern part of India, water purification is carried out using plant parts. The turbid water is allowed to settle and coalesce out using nuts and roots of some locally available plants. It was found that nuts excrete coagulation chemicals upon soaking, which settles most fine suspended particles. Besides that, the wiry roots of some plants are placed in a clay jar that has tiny holes at the bottom. In some artificial ponds in Indonesia, Jempeng stone filters are used for the filtration of water. This Jempeng stone is porous in nature and capable of filtering even highly turbid water [10].

## Boiling

Boiling with fuel is the oldest and most commonly practiced water treatment method that kills many bacteria, parasites, cysts, worms, and viruses. It is the simplest and easiest method to remove waterborne pathogens from water. This method of water purification can be implemented anywhere and at anytime as it does not require many accessories. According to WHO, water must be heated until the first big bubble appears in it, which ensures that water is pathogen-free [11]. In an emergency situation such as a flood, pandemic, or war, it is advised to drink boiled water. Besides these advantages, there are certain disadvantages as well. The boiling of water can only kill pathogens and does not remove chemical pollutants like fluoride, arsenic, *etc*. It also cannot remove the turbidity of the water; therefore, pretreatment is required for highly turbid water. Moreover, it consumes traditional fuels (wood, gas, kerosene), which are costly, contributing to deforestation and indoor air pollution. The boiling of water also alters the taste of natural water as it drives out dissolved gases.

## Chlorination

Chlorination involves adding a measured amount of chlorine into the water to kill bacteria, viruses, and cysts. Besides, chlorination can also be used for taste and odor control and to remove some gases such as ammonia and hydrogen sulfide. Chlorine is an effective disinfectant widely used in rural common wells to kill most of the bacteria which are responsible for many diseases. Chorine is added to the water as a final stage of water treatment. Chlorine is widely used in many developing countries to prevent waterborne diseases like typhoid and dysentery. The chlorine is added to the water resources in the form of sodium hypochlorite, bleaching powder, or chlorinated lime in a measured amount. The chlorine is also available commercially in tablet form as halazone, Chlor-dechlor, and hydrochlonazone. Depending upon the water quality, the appropriate amount of freshly prepared chlorine is added to water by trained personnel. Chlorine can produce some harmful effects in some cases [12]. The halogen chlorine can easily react with organic compounds present in the water producing trihalomethanes and haloacetic acids. These materials are hazardous to human health and shows symptoms like sleepiness, and slower brain activity. Chronic exposure to trihalomethanes can be responsible for kidney cancer, heart disease, and unconsciousness [13]. However, WHO states that the risk to health from these by-products is negligible than the risk associated with drinking water without disinfectant. The enhanced filtration method to remove organic matter should be employed to prevent producing hazardous compounds in the treated water.

## Sedimentation

In rural areas, most of the regions are underprivileged and there is no availability of filters, disinfectant chemicals, and trained workers. Sedimentation is the only method to treat turbid water. In the sedimentation process, the suspended particles in water are allowed to settle down under the effect of gravity [14]. The sedimentation is mostly implemented before coagulation as it reduces the concentration of the particles in suspension and fewer coagulation chemicals are required. In the sedimentation technique, the turbid water is filled in the tank and left for a longer time to settle the particles, and decant off the clear water. There are many types of sedimentation techniques like horizontal flow tanks, radial flow tanks, inclined settling, ballasted sedimentation, floc blanket sedimentation, *etc.* The efficiency of sedimentation is depended upon the nature of the suspended particles, size, and characteristics of suspended matter. There are few chemicals that assist sedimentation, but in rural areas use of such chemicals is not feasible [15].

## COAGULATION AND FLOCCULATION

The ground water, soil water, and surface water contain suspended or dissolved particles. These suspended particles vary in shape, size, source, charge, and density. The suspended particles in water possess a negative charge; therefore when coming closer, they repel each other. The result is these small particles cant clump together to form larger structures (flocs) and settle down hence proper coagulation and flocculation are required. In the coagulation process, the repulsive potential of electrical double layers of colloids is reduced and microparticles are produced. The coagulation process removes turbidity, color, and pathogens. In the coagulation process, coagulant chemicals with charges opposite to that of suspended particles are added, which neutralizes negatively charged particles [16]. Such chemicals are usually used for non-settlable solids like clay and organic substances. After neutralizing the charge, the suspended particles stick together and micro flocs are formed which are not visible to the naked eye. The water gets clear after the formation of complete flocs. The rapid mixing of coagulants is required to promote particle collisions and achieve good coagulation. The flocculation is the next step after coagulation which increases the size of submicroscopic micro flocs particles to visible suspended particles. Tiny and neutral micofloc particles collide and bond together to form larger visible floc particles called pin flocs. The coagulant chemical interacts with these flocs and their size continues to grow with collision. The coagulant chemical is usually high molecular weight polymers that help to bind, add weight and settle the flocs. The general process of the coagulation-oriented filtration mechanism is shown in Fig. (**1**).

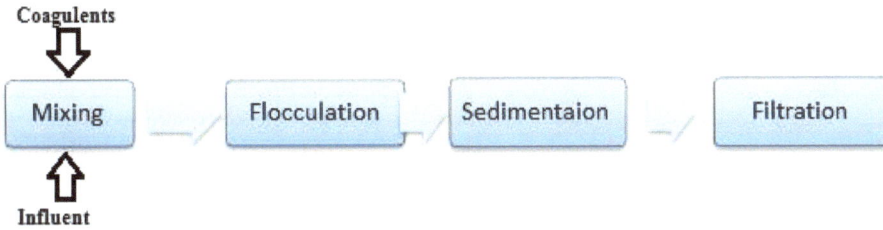

**Fig. (1).** General Coagulation and filtration mechanism.

Besides polymers, there are many inorganics coagulants such as aluminum and iron salts [17]. In water, these salts dissociate into trivalent ions $Al_3^+$ and $Al_3^+$. These ions get hydrolyzed and form positively charged soluble complexes on the surface of negatively charged suspended particles [18]. When the pH of the water is higher than the minimum solubility of the coagulants, the hydrolysis products are HMM polymers whereas when the pH of the water is lower than the minimum solubility of coagulants the hydrolysis products are monomers or medium polymers [19]. The most commonly used coagulant is an alum (aluminum sulfate) and some ferric salts. The leaching of aluminum in drinking water may pose a risk of Alzheimer, and hence the use of ferric salts has become more popular now a day [20]. Table **1** describes the features and properties of some coagulants being available in the market.

**Table 1. Properties of some commercially available coagulants [21].**

| S. No | Coagulants | Features | Target Application |
|---|---|---|---|
| 1 | Melamine, Formaldehyde | Coagulates the suspended particles and produces its own precipitated floc | Organic materials, oil, grease |
| 2 | Polyamine | Charges neutralization | High turbid water |
| 3 | Alum | Promotes coagulation of fine particles | Turbid water and colour |

*(Table 1) cont.....*

| S. No | Coagulants | Features | Target Application |
|---|---|---|---|
| 4 | Polyaluminium chloride | Effective at a lower temperature, less sludge is produced, lower doses required | To remove total phosphorus concentrations and turbidity |
| 5 | Polyaluminiumchloro-sulphate | Enhances cold water performance, strong floc formation, lower chemical solids formation and less alkali demand | Used as a flocculant in water purification. |
| 6 | Polyaluminium silicate sulphate | *in-situ* hydrolysis, complexation and charge neutralization | To remove mono-phosphate ions from aqueous solutions |
| 7 | Ferric sulphate | Less dependent on temperature change | Odour control agent and will aid phosphorus removal, removes fats, oil, grease |
| 8 | Ferric chloride | High efficiency, effectiveness in clarification, and utility as a sludge dewatering agent | Arsenic and turbidity removal |
| 9 | Polymeric ferric sulphate | No corrosion problem, low price and high flocculation ratio | Removal of chemical oxygen demand, color and turbidity of wastewater |

## BIOLOGICAL TREATMENT

The biological treatment employs natural processes to decompose organic contaminants present in wastewater. Biological treatments use bacteria, nematodes, and other small organisms to break down organic waste using a normal cellular process [22]. Organic waste usually consists of vegetables, waste foods, garbage, and pathogenic organisms. There are two types of biological treatment aerobic and anaerobic. The aerobic treatment involves the oxidation of organic material (termed biochemical oxygen demand, BOD) and the oxidation of ammonium ($NH_4^+$) in the presence of oxygen. The organic materials present in the water mineralized to $H_2O$, $CO_2$, and $NH_4^+$. The biological treatment fosters the accumulation of large biomass to affect rapid and complete oxidation in a relatively short liquid detention time. Many water scientists are trying to control and refine biological processes to achieve optimum removal of an organic substance from water. If we use an activated sludge process, the microorganisms usually accumulate into larger particles called flocs as discussed in the previous section. These flocs can settle out in quiescent settlers as they are larger than normal bacteria cells. The settled cell mass can proceed to the aeration tank to build up activated sludge. In another method called the trickling filter system, the cell mass retained in the filter is attached to a fixed and solid surface. In this type organic and $NH_4^+$ ions could be removed and new cell mass (called a biofilm) growth occurs. The wastewater moves from filter to settler for improving the quality of effluents. In aerobic biological plants, bacteria responsible for the oxidation of organic contaminants alone and with $NH_4^+$ are physiologically different. The oxidizers are heterotrophs and nitrifiers are autotrophs. The heterotrophs employ organic molecules as a source of carbon to acquire electrons and energy to synthesize new cell mass. Whereas, the autotroph reduces carbon from $CO_2$ and can oxidize $NH_4^+$ or $NO_2^-$ to acquire energy and electrons. The $CO_2$ reduction demands huge energy and electrons from the autotroph hence the yield of new cell material per unit of oxidized electron donor substrate is lower for autotrophs than heterotrophs. Hence, specific growth rate for autotrophs is much lower than heterotrophs under the same favorable conditions for both microorganisms [23].

Activated sludge process widely used in secondary treatment of domestic and industrial wastewater employs aerobic biological treatment. This method is suitable for treating wastewater streams generated from municipal sewage, pulp and paper mills, meat processing, and other industrial waste streams, which contains carbon molecule. Another process is called membrane aerated biofilm reactor (MABR), which uses 90% less energy than another biological reactor [24]. In the MABR reactor, oxygen diffuses through the gas-permeable membrane. This oxygen is supplied into the biofilm side of the membrane where oxidation of pollutants takes place. This method is suitable for high rate organic carbonaceous pollutant oxidation, organic compound biodegradation, nitrification, and denitrification. High oxygen concentration on the biofilms membrane supports nitrification and an anoxic layer close to liquid-biofilm interface allows denitrification. An aerobic heterotrophic layer supports carbonaceous pollutant removal. It is important to study the location of individual layers of microbial activity in membrane aerated biofilms.

In the anaerobic treatment of wastewater, degradation of organic material into gaseous products and biomass occurs as shown in Fig. (**2**). These gaseous products are usually methane and carbon dioxide. This treatment is remarkably useful for the treatment of highly polluted wastewater [25]. The anaerobic biological water treatment has low energy input hence no energy is required for oxygenation. Besides that, it has lower sludge production and lower nutrient requirement due to lower biological synthesis. The degradation of waste organic materials also produces biogas which is also a valuable source of energy. The anaerobic digestion is used for the stabilization of sludge from sedimentation tanks in the closed digester or open lagoons. The anaerobic open lagoons are generally employed for the treatment of industrial wastewater.

Biogas

Wastewater ⟹ | Balancing Tank | ⟹ | Aeration Tank | ⟹ | Membrane Tank | ⟹ | Permiate Tank | ⟹ Effluent

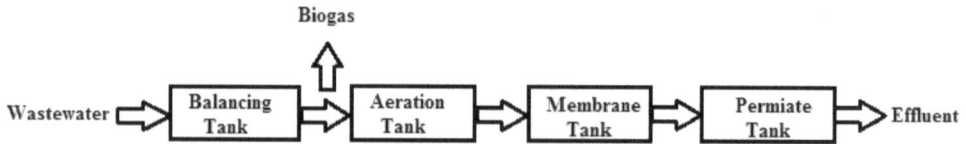

**Fig. (2).** Anaerobic biological treatment plants.

## FENTON OXIDATION PROCESS

Fenton oxidation process employed for direct mineralization of organic matter present in wastewater or improvement of biodegradability of organic pollutants through oxidation. There are many reactions that represent the Fenton process but the following is the general core reaction [26]:

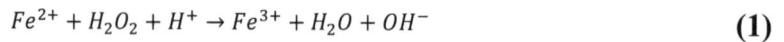

$$Fe^{2+} + H_2O_2 + H^+ \rightarrow Fe^{3+} + H_2O + OH^- \tag{1}$$

The $H_2O_2$ and homogeneous solution of iron ions are called Fenton reagents. They are chemically unstable and concentrated $H_2O_2$ is harmful to humans. Therefore, these reagents increase transformation and storage costs as well as create human health issues [27,88]. The degradation of the organic matter in wastewater is strongly affected by pH, the concentration of Fenton reagent, and the initial concentration of pollutants. The single Fenton optimization process is of three types: heterogeneous Fenton process, photo-Fenton process, and electro-Fenton process. The conventional Fenton process is limited to a narrow pH range and produces a heavy amount of iron sludge. In the case of the heterogeneous Fenton process, $Fe_2^+$ catalyst is replaced by a solid catalytic active component. This prevents the leaching of iron, facilitates a wide pH range, and reduces iron sludge formation. However, the heterogeneous Fenton process is suitable for laboratory scale use due to harsh synthesis conditions, complicated synthesis routes, and high synthesis costs [29]. Therefore, the heterogeneous Fenton process cannot be directly implemented in large-scale industrial applications.

In the case of the photo-Fenton process, ultraviolet or visible light is used in combination with the conventional Fenton process which enhances the catalytic capacity of the catalyst. The use of light also increases the degradation efficiency of the organic pollutants and reduces iron sludge production. The energy provided through light photons reduces $Al_3^+$ ions to $Fe_2^+$ [30, 31]. The photons present in light trigger metal charge transfer excitation from $Fe(OH)_2^+$ and regenerate $Fe_2^+$ which promotes decomposition of $H_2O_2$. The decomposition of $H_2O_2$ produces OH⁻ ions, which finally degrade organic pollutants present in the wastewater.

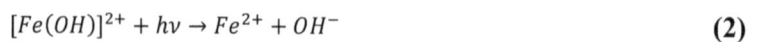

$$[Fe(OH)]^{2+} + h\nu \rightarrow Fe^{2+} + OH^- \tag{2}$$

$$H_2O_2 + hv \rightarrow 2.OH \qquad\qquad (3)$$

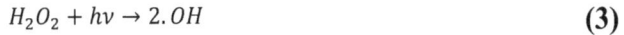

The use of ultraviolet sources shows remarkable increase in degradation but it consumes more energy and has a short life span. Therefore, it is advisable to use a natural light source *i.e.* sunlight, which is renewable and free. Thus, solar photo Fenton processes have gained more attraction for the removal of TOC. However, it has certain disadvantages like utilization of light energy, high operation cost, and design of photo-reactor on large-scale operations.

The electro Fenton process employs *in-situ* generations of $H_2O_2$ by electrochemical reduction of $O_2$ on the cathode. The $Al_3^+$ ions generated in the Fenton process can be reduced to $Fe_2^+$ on the cathode which reduces iron sludge formation [32]. The electro-Fenton process has four types: cathode electro-Fenton process, sacrificial anode electro-Fenton process, $Fe_2^+$ cycling electro-Fenton process, cathode and $Fe_2^+$ cycling electro-Fenton process. In case of the cathode electro-Fenton process, $H_2O_2$ is generated by the electrochemical process on the cathode, and $Fe_2^+$ is added externally. The sacrificial anode electro-Fenton process involves the addition of $H_2O_2$ externally while $Fe_2^+$ is generated electrochemically using a sacrificial anode. In the $Fe_2^+$ cycling electro-Fenton process, both $H_2O_2$ and $Fe_2^+$ are added externally, but $Al_3^+$ generated by the Fenton reaction is reduced to $Fe_2^+$ on the cathode. It reduces the iron sludge production and the requirement of initial $Fe_2^+$ concentration input. The cathode and $Fe_2^+$ cycling electro-Fenton process $H_2O_2$ is generated within the reaction by reduction of $O_2$ and $Fe_2^+$ is regenerated through the reduction of $Al_3^+$ on the cathode, which not only avoids the addition of $H_2O_2$ but also reduces the iron sludge production and the initial $Fe_2^+$ concentration input. Thus, the major challenge in the electro-Fenton process is the development of the electrode material. The electrode must possess good efficiency, high catalytic activity, corrosion resistance, long working life span, and low preparation costs. Finally, it can also be added about the Fenton process that the maximum organic pollutants removal capacity is strongly influenced by optimum pH range, nature of the catalyst and $H_2O_2$ concentration.

## ADVANCED OXIDATION PROCESS (AOP)

The advanced oxidation process (AOP) involves the generation of hydroxyl radicals (OH⁻) in sufficient quantity for water purification. The sulfate radicals ($SO_4^-$) also play a vital role in oxidative processes in AOP [33]. The function of AOP involves the destruction of organic or inorganic pollutants present in wastewater. The radicals like OH⁻ and $SO_4^-$ have a short half-life, hence, they are feebly effective in the inactivation of pathogens. However, these radicals are powerful oxidizing agents which destruct water pollutants and convert them into less toxic products.

The sulfate radical ($SO_4^-$) has a standard oxidation potential ($E°$) of 2.6 V, which is sufficient to initiate a sulfate-based advanced oxidation process. The $SO_4^-$ radicals can be produced from persulfate $S_2O_8^{2-}$ (with $E°=2.01V$) by heat, UV irradiation, or with transition metals as follows-

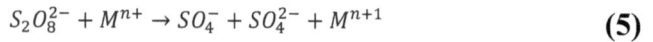

$$S_2O_8^{2-} \rightarrow 2SO_4^- \tag{4}$$

$$S_2O_8^{2-} + M^{n+} \rightarrow SO_4^- + SO_4^{2-} + M^{n+1} \tag{5}$$

There are many ways to activate persulfate: by increasing pH, varying temperature in the range 35°C to 135°C, ultraviolet irradiation process, or transition metal activation. The transition metals used for activation are usually Fe(II), Fe(III), Cu(I), or Ag(I) however, the metal activation process can generate 50% radicals (eq.2.5) therefore it is not an efficient method. The sulfate radicals remove electrons from organic waste material and transformed them into organic radical cations [34]. The hydroxyl radicals can also be generated from sulfate radicals in alkaline conditions. The hydroxyl radical is the most reactive radical with standard oxidizing potential ($E°$) 1.95 V to 2.8 V [35]. Hydroxyl radicals can attack organic pollutants through hydrogen abstraction, radical addition, electron transfer, and radical combination. The hydroxyl radicals usually add to the C=C bond or remove H from the C-H bond when reacted with organic compounds. When reacting with organic compounds, hydroxyl radicals produce R• or R•–OH radicals. These radicals transformed into organic peroxide radicals (ROO•) in the presence of $O_2$. As the lifetime of hydroxyl radicals is very short therefore these radicals should be produced *in-situ* during application in the presence of oxidizing agents. Ozone is a strong oxidizing agent with an oxidation potential of 2.07 V that can react with an ionized and dissociated form of the organic compound directly. The $OH^.$ can also be produced in an indirect mechanism under certain conditions [36].

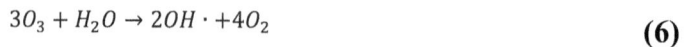

$$3O_3 + H_2O \rightarrow 2OH \cdot + 4O_2 \tag{6}$$

There are many oxidants that can significantly improve hydroxyl radical yield. These radicals can also be generated with ultraviolet photons in the presence of catalysts like $TiO_2$ or RO-type semiconductors. When $TiO_2$ is used as a catalyst, they produce positive holes in the valance band and negative electrons in the conduction band. The holes possess oxidizing property whereas electrons possess reducing property [37].

$$TiO_2 + hv \rightarrow e_{cb}^- + hv_{vb}^+ \qquad (7)$$

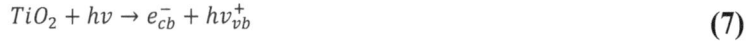

These holes and electrons further reacts with OH⁻ and $H_2O$ adsorbed on the surface of $TiO_2$ produces hydroxyl radicals-

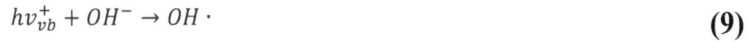

$$hv_{vb}^+ + H_2O \rightarrow OH \cdot + H^+ \qquad (8)$$

$$hv_{vb}^+ + OH^- \rightarrow OH \cdot \qquad (9)$$

The hydroxyl ions can also be produced in Fenton-based AOP by activating $H_2O_2$ using iron metal as discussed in the previous section.

The ultrasound irradiation uses sound waves on the cavities made up of vapor and gas-filled micro-bubbles. This generates high temperature (4200-5000K) and high pressure (200-500 atm) which fragments water molecules in the micro-bubbles and hydroxyl radicals can be generated. The electron beam irradiation also generated hydroxyl radicals or reducing radicals by splitting water-

$$H_2O + e^- \rightarrow 2.7OH \cdot + 2.7H_3O^+ + 2.6e^- + 0.7H_2O_2 + 0.6H \cdot + 0.45H_2 \qquad (10)$$

Many studies have demonstrated that AOPs are viable options for water treatment like leachate treatment, effluent organic matters in biologically treated secondary effluent, water reuse. In the future, an effort should be made on producing cost-effective AOP technology for the treatment of wastewater. The detailed information on this method is discussed in the upcoming chapter.

## MEMBRANE PROCESS

The membrane process is a very popular water purification method that includes reverse osmosis, nanofiltration, ultrafiltration, microfiltration, and electrodialysis. The membrane serves as a selective barrier for unwanted pollutants and allows only certain particles whose size is smaller than membrane pores. There is a driving force between the two sides of the membrane which is capable of moving the constituents across the membrane. Depending on the types of driving force *i.e.* pressure, electrical potential, concentration, or temperature, the membrane processes are classified [38]. The membrane is made up of a number of materials based on mechanical, thermal, chemical stability, and fouling tendency [39]. Polymer-based membrane materials are widely used because they are hydrophobic and are prone to fouling [40]. The membrane fouling is generally caused by deposition of inorganic components, pore blocking, microorganisms, and feed chemistry. The fouling is either reversible (loosely attached of particles) or irreversible (strongly attached particles). In order to overcome these issues,

surface modifications of the polymeric membrane are employed. These modifications include blending, grafting, and incorporation of nanomaterials such as ZnO, carbon nanotubes, grapheme, $Al_2O_3$, and $TiO_2$ [41 - 44]. The graphene oxide membrane has gained more attention due to its hydrophilic properties, flexibility, and high mechanical strength. The graphene oxide membrane is suitable for desalination and wastewater treatment which gives a wide range of pure water flux. The membrane should have high permeability and high selectivity as well as it should possess both hydrophilic and hydrophobic characteristics.

## Reverse Osmosis (RO)

Reverse osmosis is a pressure-driven water purification technique that removes small particles and solids [45]. The RO membrane is permeable to only water molecules. It is well-studied and established technology for various types of water purification. The pressure applied on the membrane must be high enough to overcome the osmotic pressure. The RO technique is capable of removing all particles, bacteria, and organics. It is usually applied in the desalination of brackish water and seawater with less maintenance. It uses a pressure gradient between the water to be treated and permeate side to remove molecules and ions from solutions when it is on one side of a selective membrane. The effective water flow through the membrane is given by the equation-

$$J = A(\Delta P - \Delta \pi) \tag{11}$$

Where, A-membrane permeability coefficient, $\Delta P$-pressure across the membrane, and $\Delta \pi$- osmatic pressure difference across the membrane. The RO membrane allows pure water on one side of the membrane (called permeate stream) and rejects ions and salts on another side of the membrane (reject stream). The membrane is composed of a thin polymeric layer along with porous support that provides mechanical strength to the membrane. Besides so many advantages, there are some disadvantages like the use of high pressure, expensive membrane and are also prone to fouling. The RO membrane also removes useful minerals from water therefore additional mineral cartridge is to be installed which adds cost.

## Ultrafiltration and Microfiltration

The pore size of the microfiltration membrane falls within the range of 0.05-10 µm whereas the pore size of ultrafiltration falls in between nanofiltration (NF) and microfiltration (MF) *i.e.* 0.001-0.05 µm. The UF membranes have an asymmetric structure with a smaller pore size and lower surface porosity than the MF membrane, which produces higher hydrodynamic resistance. The UF operating pressure is low (2-5 bars) due to the larger pore size of the membrane than the NF and RO membrane. The water flux J is given by the equation-

$$J = \frac{\Delta P}{\eta(R_m + R_c)} \qquad (12)$$

Where, $\eta$- fluid viscosity, Rm, and Rc are membrane resistance and cake resistance respectively. The UF membrane consumes low energy and is capable of removing pathogenic microorganisms, macromolecules, and suspended matters [46, 47]. The UF/MF membrane separation is decided by membrane pore size, solute membrane interactions, shape, and size of solutes. The UF/MF process is suitable for potable water treatment, RP pretreatment, tertiary water treatment, and water reclamation. Besides these applications, UF has some disadvantages like its inability to remove any dissolved inorganic substances from water and regular cleaning to maintain high-pressure water flow. The MF cannot remove viruses and dissolved solids with a size less than 1 mm.

## Nanofiltration

Its operation is similar to that of RO, but it operates at a lower pressure than RO. The NF process is better than the RO and UF processes for the treatment of wastewater because it selectively rejects low molecular weight organic compounds and divalent compounds. It is also useful for removing heavy metals and separating dyes and color compounds in the textile industry. Unlike RO, in the case of NF, the rejection of solute depends upon the molecular size and the Donnan exclusion effect [48]. The equilibrium between the solution and the charged membrane is associated with an electric potential called the Donnan potential so that ions smaller than the pore size are rejected because of Donnan exclusion. Pretreatment is required to remove some heavily polluted water using NF as the membranes are sensitive to free chlorine. It was reported by a few researchers that the PMIA/GO composite nanofiltration membrane was found effective for water purification with a greater hydrophilic surface as compared to pure polymer (PMIA). They have also observed high dye rejection and enhanced fouling resistance to bovine serum albumin [49]. The NF method is also useful for textile wastewater treatment, showing excellent removal of heavy metal ions, organic color compounds, and trihalomethane (THM) precursors such as humic acids.

## Electrodialysis

In the dialysis method, the separation of solutes takes place by transport of the solutes, through a membrane instead of using a membrane to retain the solutes whereas water passes through it in reverse diffusion and nano-filtration. Electrodialysis is the most widely used water treatment process that involves the membrane separation process [50]. This process is used to remove salt, acid, and bases from aqueous solutions to separate ionic compounds from neutral; and separate monovalent ions from multivalent ions. In this method, the electric potential is used as a driving force, whereas an ion exchange membrane is used between anode and the cathode. In the presence of electric potential, the negative and positive ions move towards the anode, and cathode, respectively, through the membrane compartment. These membrane compartments are concentrated and diluted alternatively. The electrodialysis process depends upon a number of parameters like pH, cell structure, flow rate, feed water ionic concentrations, and properties of the ion exchange membrane [51]. This process is cost-effective for TDS feed concentrations of less than 3000 parts per million (ppm). When the concentration is above 3000 ppm, RO is more cost-effective than electrodialysis when higher recovery of feed is not required. Membrane fouling is a major problem in electrodialysis, which consumes more energy and declines membrane flux. The fouling problem can be overcome by applying DC polarity to the electrodialysis membrane, which reverses polarity every 15-20 minutes. This prevents the deposition of salts on the membrane surface and eliminates acid and anti-scalant pretreatment. The periodic rinsing of the electrodes also prevents the formation of any gases on the electrodes. The electrodialysis process cannot remove nonionized compounds like silica or other colloids. This method of water purification is useful for the desalination of black water and the treatment of municipal water and wastewater.

## ION EXCHANGE PROCESS

The ion exchange method is the most utilised technique for water treatment as well as in separation processes such as chemical synthesis, medical research, food processing, mining, and agriculture. The ion exchange process involves the removal of dissolved ions from water and replacing them with other similar charged ions [52]. The water causes hardness due to calcium and magnesium ions, which can be softened by the ion exchange process. The replacement of ions involves the replacement of hardness-producing ions with no hardness ions. The water softener usually employs sodium ions supplied from a sodium source called brine. The solid phase in the ion exchange process is a synthetic resin that can selectively adsorb the containments. In the ion-exchange process, the contaminated water is passed through an ion exchange resin until all sites of the resin beads are filled with contaminant ions. The exhausted bed is regenerated by rinsing by suitable regenerant by rinsing the ion exchange column. In order to remove fluoride from water supplies, a strongly basic anion-exchange resin can be used (*e.g.*, chloride-fluoride resin). The fluoride ions replace the chloride ions of the resin until all the sites on the resin are occupied. The supersaturated resin is then backwashed with water. The chloride ions replace halide ions present on the surface, showing a higher replacement tendency towards halide ions. The resins can be cation exchange resins or anion exchange resins based on the type of contaminants present in the water. These resins can be regenerated several times and used for the ion removal process. There are several ion exchange materials, like natural organic/inorganic materials, modified natural materials, synthetic materials, *etc.* These materials may includes: vermiculites, zeolites, clays, polysaccharides, protein, carbonaceous materials, titanates, silicotitanates, transition metal hexacyanoferrate, phenolic and acrylic materials, and many more.

Suppose, the ion exchanger is represented by $M^+X^-$ with $M^+$ being a soluble ion and placed in the salt solution NY [53]. The salt ionises in the solution and gives $N^+$ and $Y^-$ ions and an exchange reaction would take place as-

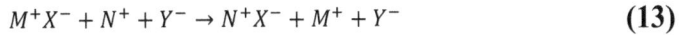

$$M^+X^- + N^+ + Y^- \rightarrow N^+X^- + M^+ + Y^- \tag{13}$$

In a similar manner, we can write the equation for the anion exchange reaction. Hence, ion exchange materials are of two types: cation exchangers and anion exchangers, based on the type of ionic groups attached to the materials.

The mechanism of ion exchange considers the transfer of ions from the interface boundary through a chemical reaction, which is diffusion inside the material and diffusion in the surrounding solution. There are counterions that are exchangeable ions carried by ion exchangers that cause interphase diffusion. These counterions move freely within the framework, and their movement is compensated for by the counter-movement of other ions of the same charge to maintain electroneutrality. The mass action selectivity coefficient determines the thermodynamic equilibrium constants (K) for cation exchange treatment. The activities of adsorbed ions are depends upon mole fractions of the ions adsorbed on the solid, which often varies the ionic composition of the exchanger and the total ionic strength of the solution [54]. According to another model, the variation measured selectivity coefficients with ionic compositions is caused by variation in the activities of adsorbed ions [55]. They have included changes in the chemical potential of water occurring during the ion-exchange reactions. The model proposed by Eriksson is based on a diffuse double-layer theory that estimates selectivity coefficients for heterovalent exchange reactions [56]. This model is based on electrostatic interactions that treat all the cations equally, regardless of their selectivities. In order to represent a mathematical approach to kinetics, diffusion equations have to be considered. The ion exchange interactions are diffusion induced electric forces suggested by Fick's law [57].

$$J_i = -D_i grad \ C_i$$

(14)

Where, Ji-flux of the ion in moles/time, $D$-the diffusion coefficient, and $C_i$-concentration in moles per unit volume. When the species is not subject to any forces besides the concentration inhomogeneity, then the diffusion coefficient is constant and the flux is directly proportional to the concentration gradient $C_i$. The flux does not depend upon concentration itself because diffusion is a purely statistical phenomenon that does not involve any physical force at the molecular level. The diffusion between two counterions without any co-ion transfermeans that the two fluxes are rigorously coupled [58]. Thus, the diffusion of majority ions and thus rates are not much affected by the presence of the electric field. The ions which are going to exchange are present in the phase with different concentrations and the diffusion between them is affected by electro-coupling interactions. The rate of ion exchange depends upon two factors: diffusion of ions inside the material or diffusion of ions through liquid a film. The diffusion inside the material is enhanced by selecting the material with a low density of the gel. The reduction in bead size also increases the rate as it reduces the time of equilibrium achievement for the particles. The increase in temperature also increases the rate, independent of the rate-controlling step.

## ELECTROCHEMICAL METHODS

Electrochemistry is a promising field in cleaner and eco-friendly water treatment. Environmental electrochemistry is developed on the basis of electrochemical techniques used to remove impurities from liquids to minimise environmental pollution. These techniques have certain advantages, like versatility, high energy efficiency, cost-effectiveness, and amenability to automation. The electrochemical process is either employed in pretreatment to increase the biodegradability of pollutants or in advanced steps to reduce COD/color to obtain water standards [59].

In the electrochemical oxidation method, the anodic oxidation of most resistant organic compounds can be broken by both direct and indirect oxidation. Direct oxidation involves the direct transfer of electrons to an anode, and pollutants are broken down after adsorption on the anode surface with no involvement of other substances. This requires more negative potential than is required for water splitting and evolution. The main disadvantage of direct oxidation is the electrode fouling problem, which is due to the accumulation of polymeric layers on the anode surface. This results in poor decontamination performance after a period of time [60]. In an indirect oxidation process, there is no need to add an oxidation catalyst to the solution and it does not produce any byproducts. In this process, oxygen is adsorbed physically or chemically, which leads to decontamination of electro-generated species at the anode. Physio-sorbed oxygen brings complete combustion of organic compounds, whereas chemisorbed oxygen participates in the formation of selective oxidation products. There are active anode electrodes like $IrO_2$, $RuO_2$, or platinum and nonactive anodes like $SnO_2$, $PbO_2$, or boron-doped diamond. The nature of the oxidation electrode influences the efficiency of the process and electrode selectivity. Table **2** represents various electrodes and their properties in an indirect electrochemical oxidation process.

Table 2. Various electrodes and their properties in direct electrochemical oxidation [61].

| Sr. No, | Types of direct oxidation electrode | Features |
|---|---|---|
| 1 | Platinum electrodes | Useful for degradation of phenol but produces residual TOC and aromatic intermediates |
| 2 | Ruthenium- and iridium-based oxide electrodes | Good mechanical resistance and inexpensive. Useful for electrochemical destruction of 4-chlorophenol, organic degradation and electro-oxidation of Reactive Blue 19 |
| 3 | Lead dioxide electrodes | Shorter service life and concerns over the possible release of $Pb^{4+}$ ions into water. Suitable for removal of PFNA,2, 4 dichlorophenoxyacetic acid, and electro-oxidation of cresols |
| 4 | Tin dioxide electrodes | n-type material with wide band gap and poor conductivity. Suitable for oxidation of organic compounds,Electrochemical degradation of phenol,aliphatic acids,dye, drugs, Bisphenol A (BPA),Nitrophenol,2,4-dichlorophenol, 4-chlorophenol, pentachlorophenol, perfluorooctanoicacid,perfluorinated carboxylic acids (PFCAs),industrial wastewater andNaphthylamine |
| 5 | Boron-doped diamond (BDD) electrodes | Useful for the destruction of organic pollutants, such: as phenoliccompounds, synthetic dyes, pesticides and drugs,surfactants and wastewaters |
| 6 | Carbonaceous electrodes | High specific surface area, goodconductivity, excellent adsorption capability and better catalytic and electric capabilities. Suitable for degradation of phenol and other organic pollutants, color and COD removal. |

In the direct oxidation method, there is the problem of deactivation of the anode during oxidation, which can be resolved by using the indirect oxidation method. In the indirect oxidation method, pollutants are destroyed by the electrochemical regeneration of chemical reactants like ozone, chlorine, $H_2O_2$, persulfate, *etc.* Active chlorine in the form of $Cl_2$, HOCl, and OCl⁻ can be electrochemically generated and used in the electrochemical oxidation of organic pollutants both in model solutions and in actual wastewater. The active chlorine is produced in the following way.

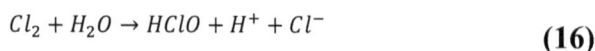

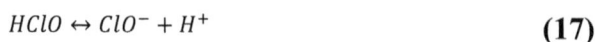

$$2Cl \rightarrow Cl_2 + 2e^-$$

**(15)**

$$Cl_2 + H_2O \rightarrow HClO + H^+ + Cl^-$$

**(16)**

$$HClO \leftrightarrow ClO^- + H^+$$

**(17)**

It is a challenge to develop effective electrodes that do not produce organo-chlorinated intermediates and reduce toxic chemical species in water. Electro-deposition is a metal removal technique based on cathodic deposition, employed in the metallurgical and electroplating industries, printed circuit boards, and the battery manufacturing industry. These heavy metals can be recovered using a chelating agent like EDTA, nitrilotriacetic acid, or citrate. The metal recovery is enhanced by integrating electro-deposition with ultrasonic. The electro-reduction method is an effective method for the dechlorination of chlorinated organic compounds (COC and VOC), polychlorophenols, and polychlorinated hydro-carbons. Cathodic electrochemical denitrification is used for the reduction of nitrate and nitrite ions for the treatment of nitrate-containing groundwater. The rate of reaction on electrodes is determined by the reduction of nitrate to nitrite, the value of the Tafel slope, the effect of co-adsorbing ions, and the kinetic order. Table **3** summarises various advantages and disadvantages of electrochemical reduction methods.

**Table 3. Advantages of electrochemical reduction methods [61].**

| S. No. | Electrochemical reduction method | Advantages | Disadvantage |
|---|---|---|---|
| 1 | Electrodeposition | Deposited metal can easily recycled | Surface of cathode get modified which reduces metal recovery efficiency. |
| 2 | Cathodic electrochemical dechlorination | Attractive technique to destroy COCs due to the mild reaction conditions, and the avoidance of possible secondary pollutants | High energy consumption and requireselectrocatalysts to lower the cell potential and reduce energy demands. |
| 3 | Electrochemical denitrification | Environmentally friendly, safe, selective, and cost effective technique | Need to investigate specific role of the foreign metal, the optimal surface composition, the surface morphology on the process. |

*(Table 3) cont.....*

| S. No. | Electrochemical reduction method | Advantages | Disadvantage |
|--------|----------------------------------|------------|--------------|
| 4 | Electrocoagulation | Small chemical requirement, less sludge production | Anode passivation and sludge deposition onthe electrodes, high concentration of aluminium and iron ions released into effluent |

## ULTRAVIOLET IRRADIATION TECHNOLOGY

The ultraviolet wavelength (250-270 nm) has a germicidal effect and is used for the disinfection of wastewater. In this wavelength region, UV light is lethal to microorganisms like bacteria, protozoa, viruses, molds, yeasts, fungi, nematode eggs, and algae [62]. UV light has the ability to kill faecal coliform and Escherichia coli bacteria at a wavelength of 260 nm. UV light prevents the division of deoxyribonucleic acid (DNA) and the production of enzymes. The nucleotide bases present in the DNA, *i.e.*, adenine, guanine, thymine, and cytosine, absorb UV light. Thymine and cytosine are ten times more sensitive to UV light than adenine and guanine. Two thymine molecules react with UV light and produce a thymine dimer. The dimerization of adjacent pyrimidine molecules causes more photochemical damage. The microorganisms are deactivated by photochemical damage to cellular RNA and DNA. DNA replication is prevented by the formation of a number of thymine dimers in the DNA of bacteria and viruses, which kill the cells. Such damage caused by UV light can be repaired by visible light or sunlight, which is called photo-reactivation [63].

UV light in the germicidal lamps is emitted as a result of electron transmission through ionised vapour between the electrodes. The glass of the germicidal UV lamp is made up of quartz, which transmits 86% of UV light of its total intensity. The slimline instant start lamp is an excellent choice for wastewater treatment because it produces 26.7 watts of UV-C from 100 watts of power at 0.18 watts of UV-C per centimetre of arc length using a conventional core-coil ballast and operates at 40°C. The rapid start lamp is another choice for UV light in which a low voltage (4V) is continuously applied through ballast to the lamp cathodes to operate the arc. The low-pressure flat lamp employs a quartz envelope and a much higher current, which decreases the distance so that ions and electrons quickly reach the wall and increases UV output. Such flat UV lamps are capable of producing 3 times the UV-C output per arc length centimeter. In every UV source, a ballast is used to limit current to the lamp and provide sufficient voltage to start the lamp. In the case of a rapid start circuit, the ballast supplies voltage to heat the lamp cathode continuously [64].

The quality of effluent affects the efficiency of UV disinfection as most of the UV photons are absorbed by pollutants that reduce UV intensity. UV adsorption depends upon suspended solids which typically consist of bacteria-laden particles of varying shapes and sizes. These suspended solids present in wastewater absorbs UV light before it can penetrate the solid to kill microorganisms, which reduces disinfection efficiency. The size of the suspended particles decide the UV light demands. When the size of these particles is less than 10 μm, UV light can easily penetrate and requires less UV demand. As the size of the particles increases (beyond 40 μm), UV light cannot penetrate completely; hence UV demands are higher . The presence of iron in the water also affects UV disinfection. Higher iron concentrations absorb UV light before it can kill any microorganisms, or iron will precipitate out on surface and absorb UV light before it enters into the wastewater. The iron can also be absorbed by suspended solids, clumps of bacteria, and other organic compounds that prevent UV light from penetrating the suspended solids. The hardness of the water caused by calcium and magnesium salts also affects UV disinfection [65]. These salts precipitate and form a layer on the quartz surface, preventing UV radiation from penetrating into the water. Pretreatment is also required to treat industrial waste by UV in order to overcome the effects of factors like the presence of UV absorbing organic materials, iron, hardness, *etc*.

UV radiation is also useful in the inactivation of viruses, including poliovirus, Echo 7, and Coxsackie 9. The degree of UV deactivation of the microorganism is directly related to the UV dose applied to the wastewater (and hence the intensity of UV and exposure time), which is given by-

$$D = It \tag{18}$$

Where, D-Dose (microwatt seconds per $cm^2$, I- irradiation, microwatt per $cm^2$, t- time. Solid material concentration, bacteria in particles, and PSD are the main challenges when a stringent disinfection limit is required [66]. The UV treatment proves better than chlorination and dechlorination because of the reduction of potential chlorinated hydrocarbons in the receiving water and relatively greater safety.

## ADSORPTION

The adsorption method for the removal of heavy metals by the adsorption method has gained considerable importance due to its greater accessibility and lower cost. Even a small amount of some toxic metal ions and dyes can cause serious problems for human health. When such elements are present in small amounts in wastewater, it becomes a difficult and challenging task to remove them. Many techniques have been developed for the elimination of toxic elements like heavy metals, dyes, pesticides, fertilizers, organic acids, *etc* [67]. So far, we have discussed many techniques for the removal of pollutants, but these techniques have some disadvantages. These techniques have low removal efficiency, non-cost and energy efficiency, the generation of toxic by-products which add toxic chemicals into the environment, and high production of sewage sludge. Adsorption is the most effective and economical method, which is less complicated and requires no energy for operation [68]. A lot of research is ongoing on the use of various adsorbents for the purification of wastewater. Several adsorbents can be used for the treatment of wastewater and water, like alumina, activated alumina, carbon, ion exchange resins, carbonates, metal oxides, hydroxides, and clay. The adsorption capacity of any adsorbent is sensitive to temperature, pH, presence of co-existing ions, ionic strength, properties of the adsorbents, initial concentration of the adsorbates, contact time, *etc* [69]. Developing novel, low-cost, and effective adsorbents and improving the efficiency of the existing adsorbents is still a challenge for researchers. Adsorption occurs when pollutant ions diffuse to the external surface of adsorbents from bulk solution across a boundary layer surrounding the adsorbent particle, either through the adsorption of pollutant ions onto the particle surface, or the exchange of pollutant ions with the structural elements inside the adsorbent. In order to study particular adsorbents, we have to consider the adsorption capacity of a dilute solution, pH, time of ion removal, stability of the adsorbent, regeneration, and loading capacity in the presence of other pollutants. There is a wide variety of adsorbents that have been used for the removal of toxic elements from water, including activated, and impregnated alumina, rare earth oxides, activated clay, impregnated silica, red mud, spent catalysts and fly ash, carbonaceous materials, zeolites, and related ion exchangers, bio-adsorbents, alum, alum sludge, and modified chitosan, and many more [70 - 74]. The detailed information about the adsorption process using alumina-based adsorbents for the removal of excess fluoride from drinking water, isotherm studies, and kinetics of adsorption have been discussed in the next chapter.

## CONCLUSION

A number of organic and inorganic substances released from agricultural, domestic, and industrial water pollute the environment. Water pollution is a result of the discharge of these organic and inorganic substances into the environment. There are many local traditional methods that have evolved and propagated from generation to generation amongst rural people. The primary and secondary treatment processes of wastewater were introduced in various places depending on the pollutants in order to remove easily settled materials and oxidize the organic materials present in the wastewater. Simple traditional methods like boiling, filtration, sedimentation, chlorination are still widely used in rural areas for removing visible impurities present in water. Though these methods are not sufficient to provide quality water in urban areas as per international standards, they can easily remove certain bacteria, pathogens, un-dissolved matter, dust, *etc*. It was observed that conventional techniques are not efficient for removing toxic components, heavy metals, nitrogen, phosphorous, *etc*. The coagulation and flocculation processess remove turbidity, color, and pathogens from water and are followed by sedimentation and filtration. The NOM can be removed by coagulation, although, the high mass compounds and hydrophobic fractions of natural organic matter are removed more effectively than hydrophilic and low molar mass compounds. The Fenton process is one of the most rapidly developed techniques used for organic wastewater treatment due to optimization in the conventional Fenton process. There is still scope to improve removal capacity, operation cost reduction, and environmental risk. The other conventional techniques, which include chemical precipitation, ion exchange, carbon adsorption, and membrane processess are potential enough for the treatment of wastewater and sewage water. The biological treatment is gaining popularity to remove toxic and harmful substances and is characterized by the accumulation of large biomass that fosters rapid and complete oxidation in a relatively short liquid detention time. AOP involves the destruction of organic and inorganic pollutants particularly for water treatment like leachate treatment, effluent, an organic matters in biologically treated secondary effluent, water reuse. The membrane process is an efficient water purification method that includes reverse osmosis, nanofiltration, ultrafiltration, microfiltration, and electrodialysis. The ion exchange method is also the most utilised technique for water treatment as well as in separation processes such as chemical synthesis, medical research, food processing, mining, and agriculture. The electrochemical process is either employed in pretreatment to increase the biodegradability of pollutants or in advanced steps to reduce COD/color to obtain water standards. However, it has certain disadvantages like sludge production, anode/cathode modification, and high energy consumption. UV radiation also accelerates the treatment of tannery wastewater using an electrocoagulation process combined with a VUV/UVC photoreactor. It can be concluded that there is no single process that can be used to treat all kinds of wastewater. It can either be treated by combining different methods or by choosing a particular method depending upon the quality of wastewater.

## CONSENT FOR PUBLICATION

Not applicable.

## CONFLICT OF INTEREST

The author declares no conflict of interest, financial or otherwise.

## ACKNOWLEDGEMENTS

Declared none.

## REFERENCES

[1]   Abu Hasan H, Muhammad MH, Ismail NI. A review of biological drinking water treatment technologies for contaminants removal from polluted water resources. J Water Process Eng 2020; 33: 101035.
[http://dx.doi.org/10.1016/j.jwpe.2019.101035]

[2]   Halder J, Islam N. Water pollution and its impact on the human health. Journal of Environment and Human 2015; 2(1): 36-46.
[http://dx.doi.org/10.15764/EH.2015.01005]

[3]   Omarova A, Tussupova K, Hjorth P, Kalishev M, Dosmagambetova R. Water supply challenges in rural areas: A case study from central kazakhstan. Int J Environ Res Public Health 2019; 16(5): 688.
[http://dx.doi.org/10.3390/ijerph16050688] [PMID: 30813591]

[4]   Singh A, Sharma RK, Agrawal M, Marshall FM. Health risk assessment of heavy metals *via* dietary intake of foodstuffs from the wastewater irrigated site of a dry tropical area of India. Food Chem Toxicol 2010; 48(2): 611-9.
[http://dx.doi.org/10.1016/j.fct.2009.11.041] [PMID: 19941927]

[5]   WHO, Fluoride in Drinking-Water. Background Document for Preparation of WHO Guidelines for Drinking-Water Quality. Geneva: World Health Organization 2003.

[6]   Fu F, Wang Q. Removal of heavy metal ions from wastewaters: A review. J Environ Manage 2011; 92(3): 407-18.
[http://dx.doi.org/10.1016/j.jenvman.2010.11.011] [PMID: 21138785]

[7]   Shannon MA, Bohn PW, Elimelech M, Georgiadis JG, Mariñas BJ, Mayes AM. Science and technology for water purification in the coming decades. Nanoscience and Technology 2009; pp. 337-46.
[http://dx.doi.org/10.1142/9789814287005_0035]

[8]   Sadhana Chourasia A K, Tiwari . A review on traditional water purification methods used in rular areas. Indian journal of environmental protection 2016; 36(1): 43-8.

[9]   Tobiason JE, CleasbyJ L, Logsdon GS, O'Melia CR. Granular Media Filtration Chapt10 in Water Quality & Treatment. 6th., McGrawHill 2010.

[10]  Okwadha George, Ahmed AA. Determination of effectiveness of traditional drinking water treatment methods. International Journal of Advanced Engineering Research and Applications 2017; 2

[11]  Guidelines for Drinking-water Quality Recommendations. third edition incorporating the first and second addenda ed. World Health Organization 2008; 1.

[12]  Nakagawara S, Goto T, Nara M, Ozawa Y, Hotta K, Arata Y. Spectroscopic characterization and the ph dependence of bactericidal activity of the aqueous chlorine solution. Anal Sci 1998; 14(4): 691-8.
[http://dx.doi.org/10.2116/analsci.14.691]

[13]  Public health statement: Bromoform&Dibromochloromethane ATSDR 2011.

[14]  Gregory R, Edzwald J. Sedimentation & Flotation. Chapt9 in Water Quality & Treatment. 6th Edtn., McGrawHill 2010.

[15]    Chen Junjie. Dynamics of wet–season turbidity in relation to precipitation, discharge, and land cover in three urbanizing watersheds, oregon, river research and applications 2019; 7(892-904): 10. [http://dx.doi.org/10.1002/rra.3487]

[16]    Jumadi J, Kamari A, Hargreaves JSJ, Yusof N. A review of nano-based materials used as flocculants for water treatment. Int J Environ Sci Technol 2020; 121(7): 3571-94. [http://dx.doi.org/10.1007/s13762-020-02723-y]

[17]    Duan J, Gregory J. Coagulation by hydrolysing metal salts. Adv Colloid Interface Sci 2003; 100-102: 475-502. [http://dx.doi.org/10.1016/S0001-8686(02)00067-2]

[18]    Pham AN, Rose AL, Feitz AJ, Waite TD. Kinetics of Fe(III) precipitation in aqueous solutions at pH 6.0-9.5 and 25 degrees C. Geochim Cosmochim Acta 2006; 70(3): 640-50. [http://dx.doi.org/10.1016/j.gca.2005.10.018]

[19]    Yan M, Wang D, Qu J, Ni J, Chow CWK. Enhanced coagulation for high alkalinity and micro-polluted water: The third way through coagulant optimization. Water Res 2008; 42(8-9): 2278-86. [http://dx.doi.org/10.1016/j.watres.2007.12.006] [PMID: 18206207]

[20]    Duan J, Gregory J. Coagulation by hydrolysing metal salts. Adv Colloid Interface Sci 2003; 100–102: 475-502. [http://dx.doi.org/10.1016/S0001-8686(02)00067-2]

[21]    Bratby J. Coagulation and Flocculation in Water and wastewater Treatment. London, Seattle: IWA Publishing 2006.

[22]    Rani N, Pritam Sangwan, Madhavi Joshi, Anand Sagar, Kiran Bala. Chapter 5 - Microbes: A key player in industrial wastewater treatment, microbial wastewater treatment. Elsevier 2019; pp. 83-102. [http://dx.doi.org/10.1016/B978-0-12-816809-7.00005-1]

[23]    Shaikh & Zia, Dolfing Graham, Jan . Wastewater treatment: Biological. 2013.

[24]    Casey E, Glennon B, Hamer G. Review of membrane aerated biofilm reactors. Resour Conserv Recycling 1999; 27(1-2): 1-2, 203-215. [http://dx.doi.org/10.1016/S0921-3449(99)00007-5]

[25]    Anil Kumar. CHAPTER 28 - Groundwater Decontamination and Treatment, Biotreatment of Industrial Effluents. Butterworth-Heinemann 2005; pp. 285-94. [http://dx.doi.org/10.1016/B978-075067838-4/50029-4]

[26]    Zhang MH, Dong H, Zhao L, Wang DX, Meng D. A review on Fenton process for organic wastewater treatment based on optimization perspective. Sci Total Environ 2019; 670: 110-21. [http://dx.doi.org/10.1016/j.scitotenv.2019.03.180] [PMID: 30903886]

[27]    Liu R, Xu Y, Chen B. Self-assembled nano-FeO(OH)/reduced graphene oxide aerogel as a reusable catalyst for photo-Fenton degradation of phenolic organics. Environ Sci Technol 2018; 52(12): 7043-53. [http://dx.doi.org/10.1021/acs.est.8b01043] [PMID: 29799731]

[28]    Zhao K, Quan X, Chen S, Yu HT, Zhang YB, Zhao HM. Enhanced electroFenton performance by fluorine-doped porous carbon for removal of organic pollutants in wastewater. Chem Eng J 2018; 354: 606-15. [http://dx.doi.org/10.1016/j.cej.2018.08.051]

[29]    Kakde AS, Belekar RM, Wakde GC, Borikar MA, Rewatkar KG, Shingade BA. Evidence of magnetic dilution due to unusual occupancy of zinc on B-site in NiFe2O4 spinel nano-ferrite. J Solid State Chem 2021; 300: 122279. [http://dx.doi.org/10.1016/j.jssc.2021.122279]

[30]    Lee HJ, Lee HS, Lee CH. Degradation of diclofenac and carbamazepine by the copper(II)-catalyzed dark and photo-assisted Fenton-like systems. Chem Eng J 2014; 245: 258-64.

[http://dx.doi.org/10.1016/j.cej.2014.02.037]

[31]   Kalal S, Chauhan NPS, Ameta N, Ameta R, Kumar S, Punjabi PB. Role of copper pyrovanadate as heterogeneous photo-Fenton like catalyst for the degradation of neutral red and azure-B: An eco-friendly approach. Korean J Chem Eng 2014; 31(12): 2183-91.
[http://dx.doi.org/10.1007/s11814-014-0142-z]

[32]   Trellu C, Oturan N, Keita FK, Fourdrin C, Pechaud Y, Oturan MA. Regeneration of activated carbon fiber by the electro-Fenton process. Environ Sci Technol 2018; 52(13): 7450-7.
[http://dx.doi.org/10.1021/acs.est.8b01554] [PMID: 29856620]

[33]   Ikai H, Nakamura K, Shirato M, *et al.* Photolysis of hydrogen peroxide, an effective disinfection system *via* hydroxyl radical formation. Antimicrob Agents Chemother 2010; 54(12): 5086-91.
[http://dx.doi.org/10.1128/AAC.00751-10] [PMID: 20921319]

[34]   Forsey SP. In situ chemical oxidation of creosote/coal tar residuals: Experimental and numerical investigation. University of Waterloo 2004.

[35]   Tchobanoglous G, Burton F, Stensel H. Wastewater engineering. New York: Metcalf & Eddy Inc. 2003.

[36]   Gottschalk C, Libra JA, Saupe A. Ozonation of water and waste water: a practical guide to understanding ozone and its applications. John Wiley & Sons 2009.
[http://dx.doi.org/10.1002/9783527628926]

[37]   Deng Y, Zhao R. Advanced oxidation processes (AOPs) in wastewater treatment. Curr Pollut Rep 2015; 1(3): 167-76.
[http://dx.doi.org/10.1007/s40726-015-0015-z]

[38]   Singh R. Membrane Technology and Engineering for Water Purification Application, Systems Design and Operation. 2nd ed., Oxford, UK: Butterworth-Heinemann 2015.

[39]   Mulder M. Basic Principles of Membrane Technology. Holland: Kluwer Academic Publishers 1997.

[40]   Ahmed F, Lalia BS, Kochkodan V, Hilal N, Hashaikeh R. Electrically conductive polymeric membranes for fouling prevention and detection: A review Desalination 2016; 391: 1-15.
[http://dx.doi.org/10.1016/j.desal.2016.01.030]

[41]   Bet-Moushoul E, Mansourpanah Y, Farhadi K, Tabatabaei M. $TiO_2$ nanocomposite based polymeric membranes: A review on performance improvement for various applications in chemical engineering processes. Chem Eng J 2016; 283: 29-46.
[http://dx.doi.org/10.1016/j.cej.2015.06.124]

[42]   Tan YH, Goh PS, Ismail AF, Ng BC, Lai GS. Decolourization of aerobically treated palm oil mill effluent (AT-POME) using polyvinylidene fluoride (PVDF) ultrafiltration membrane incorporated with coupled zinc-iron oxide nanoparticles. Chem Eng J 2017; 308: 359-69.
[http://dx.doi.org/10.1016/j.cej.2016.09.092]

[43]   Garcia-Ivars J, Iborra-Clar MI, Alcaina-Miranda MI, Mendoza-Roca JA, Pastor-Alcañiz L. Surface photomodification of flat-sheet PES membranes with improved antifouling properties by varying UV irradiation time and additive solution pH. Chem Eng J 2016; 283: 231-42.
[http://dx.doi.org/10.1016/j.cej.2015.07.078]

[44]   Tijing LD, Woo YC, Shim WG, *et al.* Superhydrophobic nanofiber membrane containing carbon nanotubes for high-performance direct contact membrane distillation. J Membr Sci 2016; 502: 158-70.
[http://dx.doi.org/10.1016/j.memsci.2015.12.014]

[45]   Nqombolo A, Mpupa A, Moutloali RM, Nomngongo PN. Wastewater Treatment Using Membrane Technology. Wastewater and Water Quality 2018.
[http://dx.doi.org/10.5772/intechopen.76624]

[46]   Qu F, Liang H, Zhou J, *et al.* Ultrafiltration membrane fouling caused by extracellular organic matter (EOM) from Microcystisaeruginosa: Effects of membrane pore size and surface hydrophobicity. J

Membr Sci 2014; 449: 58-66.
[http://dx.doi.org/10.1016/j.memsci.2013.07.070]

[47]    Krüger R, Vial D, Arifin D, Weber M, Heijnen M. Novel ultrafiltration membranes from low-fouling copolymers for RO pretreatment applications. Desalination Water Treat 2016; 57(48-49): 23185-95.
[http://dx.doi.org/10.1080/19443994.2016.1153906]

[48]    Singh R, Hankins NP. Introduction to Membrane Processes for Water Treatment. Emerging Membrane Technology for Sustainable Water Treatment 2016; pp. 15-52.
[http://dx.doi.org/10.1016/B978-0-444-63312-5.00002-4]

[49]    Yang M, Zhao C, Zhang S, Li P, Hou D. Preparation of graphene oxide modified poly (m-phenyleneisophthalamide) nanofiltration membrane with improved water flux and antifouling property. Appl Surf Sci 2017; 394: 149-59.
[http://dx.doi.org/10.1016/j.apsusc.2016.10.069]

[50]    Oztekin E, Altin S. Wastewater treatment by electrodialysis system and fouling problems. Turkish Online Journal of Science & Technology 2016; p. 6.

[51]    Akhter M, Habib G, Qamar SU. Application of electrodialysis in waste water treatment and impact of fouling on process performance. J Membr Sci Technol 2018; 8(2): 182.
[http://dx.doi.org/10.4172/2155-9589.1000182]

[52]    Zagorodni AA. Ion Exchange Materials Properties and Applications. 1st ed., Oxford, UK: Elseiver 2007.

[53]    Kumar Sanjeev, Jain Sapna. History, introduction, and kinetics of ion exchange materials, journal of chemistry Article ID 957647 2013; 13.
[http://dx.doi.org/10.1155/2013/957647]

[54]    Laudelout H, van Bladel R, Bolt GH, Page AL. Thermodynamics of heterovalentcation exchange reactions in a montmorillonite clay. Transactions of the Faraday Society. 1968; 64: pp. 1477-88.

[55]    Gaines GL Jr, Thomas HC. Adsorption studies on clay minerals. II. A formulation of the thermodynamics of exchange adsorption. J Chem Phys 1953; 21(4): 714-8.
[http://dx.doi.org/10.1063/1.1698996]

[56]    Maes A, Cremers A. Charge density effects in ion exchange: I. Heterovalent exchange equilibria. J Chem Soc, Faraday Trans I 1977; 73(0): 1807-14.
[http://dx.doi.org/10.1039/f19777301807]

[57]    Samsonov GV, Trostyanskaya EB, El'kin GE. Ion Exchange, Sorption of Organic Substances (IonnyiObmen, SorbtsiyaOrganicheskikhVeshchestv). Leningrad, Russia: Nauka 1969.

[58]    Seo J, Paik U. Preparation and characterization of slurry for chemical mechanical planarization (CMP). SuryadevaraBabu, Advances in Chemical Mechanical Planarization (CMP). sWoodhead Publishing 2016; pp. 273-98.
[http://dx.doi.org/10.1016/B978-0-08-100165-3.00011-5]

[59]    Dušan Ž, Mijin Milka L, AvramovIvić Antonije E, Onjia Branimir N. Grgur, Decolorization of textile dye CI Basic Yellow 28 with electrochemically generated active chlorine. Chem Eng J 2012; 204–206: 151-7.
[http://dx.doi.org/10.1016/j.cej.2012.07.091]

[60]    Marcela G, Tavares Lozele VA, da Silva Aline M, *et al.* Zanta, electrochemical oxidation of methyl red using Ti/Ru0.3Ti0.7$O_2$ and Ti/Pt anodes. Chem Eng J 2012; 204–206: 141-50.
[http://dx.doi.org/10.1016/j.cej.2012.07.056]

[61]    Feng Y, Yang L, Liu J, Logan B. Electrochemical Technologies for Wastewater Treatment and Resource ReclamationEnvirontal Science. Water Research & Technology 2016.
[http://dx.doi.org/10.1039/C5EW00289C]

[62]    Xie P, Yue S, Ding J, *et al.* Degradation of organic pollutants by Vacuum-Ultraviolet (VUV): Kinetic

model and efficiency. Water Res 2018; 133: 69-78.
[http://dx.doi.org/10.1016/j.watres.2018.01.019] [PMID: 29367049]

[63]    Das TK. Ultraviolet disinfection application to a wastewater treatment plant. Clean Prod Process 2001; 3(2): 69-80.
[http://dx.doi.org/10.1007/s100980100108]

[64]    Raeiszadeh M, Taghipour F. Microplasma UV lamp as a new source for UV-induced water treatment: Protocols for characterization and kinetic study. Water Res 2019; 164: 114959.
[http://dx.doi.org/10.1016/j.watres.2019.114959] [PMID: 31415967]

[65]    Petersen NB, Madsen T, Glaring MA, Dobbs FC, Jørgensen NOG. Ballast water treatment and bacteria: Analysis of bacterial activity and diversity after treatment of simulated ballast water by electrochlorination and UV exposure. Sci Total Environ 2019; 648: 408-21.
[http://dx.doi.org/10.1016/j.scitotenv.2018.08.080] [PMID: 30121040]

[66]    Juan L, Acero Francisco J, Real F, Javier Benitez, Esther Matamoros. Degradation of neonicotinoids by UV irradiation: Kinetics and effect of real water constituents. Separ Purif Tech 2019; 211: 218-26.
[http://dx.doi.org/10.1016/j.seppur.2018.09.076]

[67]    Edmunds M, Smedley P. Fluoride in natural waters. Essentials of Medical Geology, Impacts of Natural Environment on Public Health. Elsevier Academic Press 2005.

[68]    Belekar RM, Dhoble SJ. Activated Alumina Granules with nanoscale porosity for water defluoridation Nano-Structures & Nano-Objects.,16, 2018; 322-8. Availble: 10.1016/j.nanoso.2018.09.007

[69]    Belekar RM, Athawale SA, Gedekar KA, Dhote AV. Various techniques for water defluoridation by alumina: Development, challenges and future prospects. AIP Conf Proc 2019; 2104: 030004.
[http://dx.doi.org/10.1063/1.5100431]

[70]    Bhatnagar A, Kumar E, Sillanpää M. Fluoride removal from water by adsorption—A review. Chem Eng J 2011; 171(3): 811-40.
[http://dx.doi.org/10.1016/j.cej.2011.05.028]

[71]    Gedekar KA, Wankhede SP, Moharil SV, Belekar RM. d–f luminescence of $Ce_3^+$ and $Eu_2^+$ ions in $BaAl_2O_4$, $SrAl_2O_4$ and $CaAl_2O_4$ phosphors. Journal of Advanced Ceramics 2017; 6(4): 341-50.
[http://dx.doi.org/10.1007/s40145-017-0246-0]

[72]    Gedekar KA, Wankhede SP, Moharil SV, Belekar RM. Synthesis, crystal structure and luminescence in $Ca_3Al_2O_6$. J Mater Sci Mater Electron 2018; 29(8): 6260-5.
[http://dx.doi.org/10.1007/s10854-018-8603-5]

[73]    Gedekar KA, Wankhede SP, Moharil SV, Belekar RM. $Ce_3^+$ and $Eu_2^+$ luminescence in calcium and strontium aluminates. J Mater Sci Mater Electron 2018; 29(6): 4466-77.
[http://dx.doi.org/10.1007/s10854-017-8394-0]

[74]    Wani MA, Dhoble SJ, Belekar RM. Synthesis, characterization and spectroscopic properties of some rare earth activated $LiAlO_2$ phosphor. Optik (Stuttg) 2021; 226(1): 165938.
[http://dx.doi.org/10.1016/j.ijleo.2020.165938]

CHAPTER 2

# The Fluoride Adsorption Isothermal Studies of Activated Alumina Modified with Different Materials: A Critical Review

**R. M. Belekar**[1,*] and **S.J. Dhoble**[2]

[1] *Department of Physics, Institute of Science, Rabindranath Tagore Road, Nagpur, M.S., India-440 001*

[2] *Department of Physics, R.T.M. Nagpur University, Nagpur, India-440033*

**Abstract:** Fluoride in drinking water has become a global problem that has a profound effect on teeth and bones, fostering various health problems. Adsorption is a potential defluoridation technique because of flexibility, cost-effectiveness, environmental friendliness, simplicity in design, relative ease of operation, and capability of producing high water quality. Although activated alumina is an appropriate adsorbent, it has a narrow favorable pH range, a tendency to form toxic aluminum fluoride complexes, and the problem of aluminum metal leaching. This article critically reviews the applicability of activated alumina and its modification by metal oxides, rare earth elements, organic materials, alkaline earth metals, and acid treatment. The effect of process parameters like pH, contact time, adsorbent dose, initial fluoride concentration, and the presence of coexisting ions on the adsorption capacity of fluoride ions is discussed. The adsorption reaction rates were discussed by fitting various rate models into the experimental data and the model equations. The adsorption isotherm models like Langmuir, Freundlich, Temkin, and Dubinin-Radushkevich tested on the adsorption equilibrium data to identify the best fit model for adsorption isotherm are discussed in this chapter. The chapter finally discusses the advantages, disadvantages, and future prospects of all the adsorbents in order to improve their fluoride removal capacity.

**Keywords:** Adsorption, Activated alumina, alumina modification, Defluoridation, Isotherm models, purification, Kinetic models.

## INTRODUCTION

Fluorine is a highly electronegative element in the periodic table, that has the tendency to acquire an electron and form a fluoride ion.

* **Corresponding author R. M. Belekar:** Department of Physics, Institute of Science, Nagpur, M.S., India-440 001; Tel: +91-9822292336; E-mail: rajubelekar@gmail.com

R. M. Belekar, Renu Nayar, Pratibha Agrawal and S. J. Dhoble (Eds)
All rights reserved-© 2022 Bentham Science Publishers

Fluoride is one of the most commonly found elements on the earth's surface due to natural processes like erosion, volcanic activities, hydraulic leaching, *etc.* [1, 2]. Fluoride is an inorganic, monoatomic, negatively charged ion with the chemical formula $F^-$ whose salts are typically white or colorless, and thus being considered a stable form of fluorine. Fluoride can also be found in natural water resources, animal food, or in the rain, with its concentration increasing significantly upon exposure to volcanic activity or atmospheric pollution derived from burning fossil fuels or industry waste [3]. Many studies have found that fluoride in drinking water prevents at least 25 percent of tooth decay in children and adults, even in an era where there is widespread availability of fluoride from other sources. The optimum concentration of fluoride is essential for the human body. However, a higher concentration that is beyond the safe limit affects plants, humans, and animal life. The safe limit of fluoride in drinking water is 0.5-1.0 mg/L as suggested by WHO standards [4]. Drinking water with a fluoride concentration of 2.5-3.0 mg/L may cause dental fluorosis. The concentration of 3.0-4.0 results in stiffened, brittle bone, whereas a concentration higher than 4.0 may result in crippling fluorosis [5, 6]. The concentrations of fluoride in groundwater are reported to be beyond WHO's safe limit in different parts of the world, such as China, India, Mongolia, Japan, Pakistan, Sri Lanka, Iran, Turkey, Southern Algeria, Argentina, Korea, Mexico, Italy, Brazil, Malawi, North Jordan, Ethiopia, Canada, Norway, Ghana, Kenya, in the states of South Carolina, *etc.* [7].

The duration of continuous intake of fluoride contaminated water and the concentration of fluoride within it determines its impact on human health. Generally, fluoride gets deposited in the joints of the knee, pelvic region, neck, and shoulder bone, making it difficult for a person to walk or move. In severe cases, it may lead to rare bone cancer, spondylitis or arthritis osteosarcoma, spine, major joints, muscles, and damage to nervous system [8]. Therefore, maintaining desirable optimum fluoride levels in drinking water has become a need.

There are several techniques that have been studied for the removal of excess amounts of fluoride from drinking water which includes precipitation, coagulation, dialysis, ion exchange, electro-coagulation, adsorption, membrane filtration, *etc.* However, there are certain limitations of these methods because most of these methods require high operational and maintenance costs, produce secondary pollution (toxic sludge, *etc.*), and require complicated procedures involved in the treatment. Out of these techniques, adsorption is a potential method for defluoridation because of its ease of operation, high productivity, and capability of producing high-quality water. The adsorption technique involves contaminated water entry from the contact bed where fluoride is removed by ion exchange or surface chemical reaction with a solid bed matrix. This technique is prominent over another method of defluoridation because it is straightforward, flexible, and accessible to a wide variety of adsorbents. The adsorption technique is highly efficient and suitable over a wide range of pH to a lower leftover concentration than precipitation [9, 10]. There are a variety of materials reported in the literature used as adsorbents to remove fluoride, which include activated alumina, activated carbon, activated alumina coated silica gel, calcite, activated sawdust, activated coconut shell powder, groundnut shell, activated fly ash, coffee husk, rice husk, magnesia, serpentine, tri-calcium phosphate, activated soil sorbent, bone charcoal, defluoron-1, defluoron-2, *etc.* [11 - 19]. Besides that, there are materials like schwertmannite, which can be used as nano-adsorbent, with an adsorption capacity of 17.24 mg/g, for defluoridation of water [20, 21]. Recently, iron oxide hydroxide has shown its potential for fluoride removal from an aqueous medium with a fluoride removal capacity of 11.3 mg/g [22]. In nano-adsorbents, metal oxide nano-particles are quite promising for the adsorption for their large surface area and porous structure in addition to the short diffusion route [23]. Due to larger surface areas, it is evident that nanosized adsorbents with a strong affinity towards fluoride ions can be suitable tools for enhancing the adsorption capacity in drinking water treatment. However, because of their small size, the isolation of nanosized adsorbents from matrices is difficult for practical use. The Nalgonda Technique is found to be effective and economical in that there is sequential addition of lime, bleaching powder (for disinfection), an alum, or aluminum chloride, but it has some problems associated with operation and sludge disposal [24]. The most common materials currently used in the defluoridation process are activated alumina and activated carbon. The fluoride removing efficiency of these materials is influenced by hardness, pH, and surface loading. The present review paper focuses on the potential of activated alumina and alumina-based adsorbents for fluoride removal from water and wastewater. The adsorption capacities of the modified alumina along with adsorption isotherm models, kinetic models with some of the latest important findings, and a source of up-to-date literature are provided and discussed in this review.

An aqueous solution of fluoride ions containing activated alumina may not be clearly soluble and form various aluminum species, including several fluorides and hydroxyl-aluminum complexes. A series of generalized stepwise equilibrium equations for Al–F complex formation can be expressed as follows:

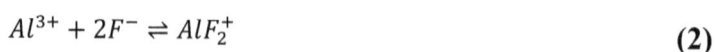

$$Al^{3+} + F^- \rightleftharpoons AlF^{+2} \tag{1}$$

$$Al^{3+} + 2F^- \rightleftharpoons AlF_2^+ \tag{2}$$

$$Al^{3+} + 3F^- \rightleftharpoons AlF_3^+ \qquad\qquad (3)$$

$$Al^{3+} + 4F^- \rightleftharpoons AlF_4^+ \qquad\qquad (4)$$

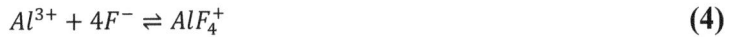

The complex formation of aluminum ions with hydroxyl ions in aqueous solution can be written as:

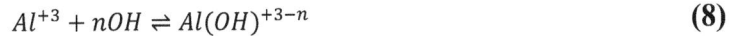

$$Al^{+3} + OH \rightleftharpoons Al(OH)^{+2} \qquad\qquad (5)$$

$$Al^{+3} + 2OH \rightleftharpoons Al(OH)_2^+ \qquad\qquad (6)$$

$$Al^{+3} + 3OH \rightleftharpoons Al(OH)_3^+ \qquad\qquad (7)$$

$$Al^{+3} + nOH \rightleftharpoons Al(OH)^{+3-n} \qquad\qquad (8)$$

## ADSORPTION TECHNIQUES AND THEORY OF ISOTHERM MODELS

The adsorption method for fluoride adsorption has gained considerable importance due to its greater accessibility and lower cost. This method is capable of removing more than 90% of fluoride ions from water [25, 26]. Several adsorbents can be used for the treatment of wastewater and water, like alumina, activated alumina, carbon, ion exchange resins, carbonates, metal oxides, hydroxides, and clay. The adsorption capacity of any adsorbent is sensitive to temperature, pH, presence of co-existing ions, ionic strength, properties of the adsorbents, initial concentration of the adsorbates, contact time, *etc.* [27]. Out of these parameters, the pH of the solution is a major controlling parameter as it affects the surface charge on the adsorbent and the interaction between ions and adsorption sites [28, 29]. At higher pH, the adsorption of anions decreases due to an increase in OH⁻ ions, whereas it increases due to protonated surfaces. The adsorption process is also governed by the zero point charge of the adsorbent in an aqueous medium. The net neutrality of the adsorbent surface can be determined by pHzpc (pH at zero point charge), where acidic or basic groups do not contribute to the pH of the solution. When pH increases above zero point charge, the surface acquires a net negative charge, and the surface will either attract cations or participate in cation exchange reactions. At a pH value below the zero point charge, the surface acquires a net positive charge and it will attract anions. The nature, metal chemistry in the solution, and ionic state of the functional group will decide the dependence of adsorption upon pH. The adsorption capacity of the adsorbents is also dependent upon surface area, total pore volume, and other chemical characteristics of the adsorbent. The performance of the adsorbent can be controlled by physical and chemical oxygen functional groups. The presence of co-existing ions possesses different adsorption properties and competes with each other, which reduces adsorption. These ions may include anions like sulfate, nitrate, carbonate, chloride, bicarbonate, and phosphate, which affect fluoride adsorption by adjustment of the electrostatic charge at the solid surface because of the same negative ions [30]. The adsorption process for practical adsorbents requires consideration of adsorption capacity in dilute solutions, time for fluoride removal, pH, stability of adsorbent, regeneration, loading capacity in the presence of other ions, consideration of leaching of any metal in treated water, and finally, the overall cost for fluoride removal. There are very few research papers that report on all these factors and their interdependence.

There are four widely used theoretical isotherms that are used to examine adsorption equilibrium data and set up a standard model, the theoretical Langmuir, empirical Freundlich, Temkin, Redlich–Peterson, and Dubinin–Radushkevich isotherms. In order to express the mechanism of fluoride adsorption onto the surface of the adsorbent, the kinetic model's like pseudo-first-order, pseudo-second-order, intraparticle diffusion, and Elovich models are used to analyze the present adsorption data to determine various kinetic parameters.

*Langmuir adsorption isotherm* [20] is the most widely used isotherms amongst others and can be applied in solid/liquid systems for describing saturated monolayer adsorption. This adsorption can be expressed as:

$$q_e = \frac{q_m K_a C_e}{1 + K_a C_e}$$

(9)

Where $C_e$- the equilibrium concentration (mg/L);

$q_e$- the amount of ion adsorbed (mg/g);

$q_m$-$q_e$for a complete monolayer (mg/g);

$K$a- adsorption equilibrium constant (L/mg).

The adsorption capacity of a given adsorbate can be calculated by considering the following linear form of equation (9):

$$\frac{C_e}{q_e} = \frac{1}{q_m} C_e + \frac{1}{K_a q_m} \tag{10}$$

The values of the constants $q_m K_a$ can be calculated by determining the slope of the linear plot of $C_e/q_e$ *versus* $C_e$.

*Freundlich adsorption isotherm* [21] is generally based on the adsorption of material on heterogeneous surfaces. The following equation describes the relationship of the adsorption equilibrium:

$$q_e = K_F C_e^{1/n} \tag{11}$$

Where $q$e - the amount of ion adsorbed (mg/g);

$C$e - the equilibrium concentration (mg/L);

$KF$ and $1/n$ are the constants that indicate the adsorption capacity and adsorption intensity, respectively.

The linear form of equation (11) can be represented as follows-

$$\log q_e = \log K_F + \frac{1}{n} \log C_e \tag{12}$$

When we plot the graph of log $q$e *versus* log $C$e of Eq. (12), it results in a straight line. From the slope and intercept of the plot, the values for $n$ and $KF$ can be obtained.

The *Temkin model* is generally based on the assumption that the interaction of an adsorbent with the surface decreases the heat of adsorption with an increase in coverage. This adsorption is expressed by a uniform distribution of binding energy. The Temkin equation represents the equilibrium binding constant AT corresponding to the maximum binding energy and another constant BT associated with the heat of adsorption. Dubinin–Radushkevich's (D.R.) isotherm model is also used to determine the physical and chemical nature of the adsorbent. This model is represented by a linear plot of $\ln(Qe)$ *vs.* $\varepsilon2$. The mean free energy for adsorption of fluoride per mole of adsorbate is given by the following equation-

$$E_S = -1/\sqrt{2K} \tag{13}$$

## ACTIVATED ALUMINA

The activated alumina is an excellent adsorbent with highly porous structure and offers a large surface area. The activated alumina is prepared by the heat treatment of some form of hydrated alumina (*i.e.*, a crystalline hydroxide, oxide-hydroxide, or hydrous alumina gel). It has been known for many years that certain forms of activated alumina can also be used as powerful desiccants for the adsorption of various vapors. It was observed that the adsorbent activity of an adsorbent depends on the conditions of heat treatment. The activated alumina is a commercial porous material with a specific surface area of 200-300 m2 g$^{-1}$ and can be used for fluoride, phosphate, and arsenic removal from groundwater. The activated alumina adsorbent is specific for these anions and can be regenerated with dilute NaOH and sulfuric acid [31].

Defluoridation of drinking water by activated alumina was first suggested by Boruff and further studied by Swope and Hess, Maier, Savinelli and Black, Harmon and Halechman, Zebbon and Jewett, Wu and Nityaand Balusu, and Nawlakhe demonstrated the effectiveness of activated alumina for defluoridation of water. However, the adsorption capacity of activated alumina seems to depend upon water quality parameters like pH, alkalinity, TDS, *etc.*, ionic environment *i.e.*, presence of other anions and cations, the physicochemical properties of activated alumina, and regeneration procedures [32].

In order to study the adsorption kinetics of fluoride, we consider the activated alumina surface on which the adsorption of fluoride takes place to be composed of a number of discrete sites that undergo ionization in the aqueous medium. During the process, the surface mobile H$^+$ ions enter the solution, and the surface gets charged. The fluoride ions contained in the solution form complexes with these positively charged sites. There exists an electrostatic field between the surface and the aqueous environment, which leads to the ionization of surface sites and formation of the complexes between these sites and other ionic species. The adsorption of the fluoride ions incorporates three distinct phenomena: ionization of the surface, complexation between the ionized sites and fluoride ions, and establishment of the electrical double layer in the aqueous environment immediately adjacent to the surface. The 2pK model represents the adsorption and absorption of protons on the surface of alumina, which leads to a charge on the surface due to the amphoteric nature of aluminium oxide [33]. The amount of charge developed on the surface is determined by the acidic or basic nature of the solution, *i.e.*, the pH of the solution. According to this model, the ions H$^+$, the cations $C^+$, and the anions $A^-$ of the basic solution form the following surface complexes: $SOH^0$, $SOH_2^+$, $SO^-C^+$, and $SOH_2^+A^-$, where S represents a surface metal atom. The surface concentrations of these complexes are denoted by $[SOH^0]$, $[SOH_2^+]$, $[SO^-C^+]$, and $[SOH_2^+A^-]$, respectively, whereas $[SO^-]$ is the surface concentration of the free sites (unoccupied surface oxygen). The following reaction occurs at the oxide surface. The equilibrium constants for these reactions, given by the law of mass action, are as follows:

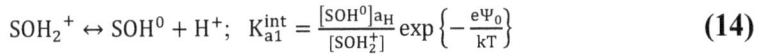

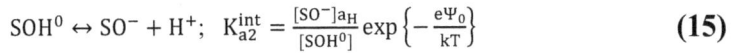

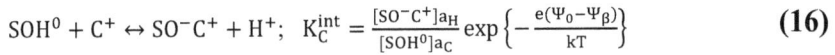

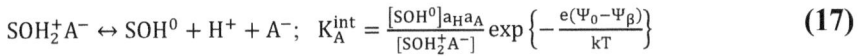

$$SOH_2^+ \leftrightarrow SOH^0 + H^+; \quad K_{a1}^{int} = \frac{[SOH^0]a_H}{[SOH_2^+]}\exp\left\{-\frac{e\Psi_0}{kT}\right\} \tag{14}$$

$$SOH^0 \leftrightarrow SO^- + H^+; \quad K_{a2}^{int} = \frac{[SO^-]a_H}{[SOH^0]}\exp\left\{-\frac{e\Psi_0}{kT}\right\} \tag{15}$$

$$SOH^0 + C^+ \leftrightarrow SO^-C^+ + H^+; \quad K_C^{int} = \frac{[SO^-C^+]a_H}{[SOH^0]a_C}\exp\left\{-\frac{e(\Psi_0-\Psi_\beta)}{kT}\right\} \tag{16}$$

$$SOH_2^+A^- \leftrightarrow SOH^0 + H^+ + A^-; \quad K_A^{int} = \frac{[SOH^0]a_H a_A}{[SOH_2^+A^-]}\exp\left\{-\frac{e(\Psi_0-\Psi_\beta)}{kT}\right\} \tag{17}$$

In the above equations, aH is the proton activity in the equilibrium bulk solution, whereas aA and aC are the activities of the anion and cation in the bulk phase. The positively charged surface of the activated alumina serves as a site for negatively charged fluoride ions, which removes fluoride from the aqueous medium. The binding of these adsorbed fluoride ions onto the alumina surface can be described by the above complexation reactions. In addition to this, the surface charge density and surface potential on the alumina surface are related by the Gouy-Chapman theory of diffuse layer. There is the possibility to dissolve aluminium from the alumina matrix into the solution, which results in the formation of complexes with fluoride ions present in the solution. The amount of aluminium dissolved in the solution is determined by the formation of the dissolved free aluminium ions, various aluminium fluoride complexes, aluminium hydroxides, and aluminium hydroxyl fluoride complexes. Depending upon the pH of the solution, the aluminum ions separated from alumina can form their hydroxyl complexes like $AlOH^{2+}$, $Al(OH)^{2+}$, $Al(OH)_3^0$, $Al(OH)^4$ and $Al_2(OH)_2^{2+}$, *etc.* The fluoride ions in the solution under low pH conditions form complexes with aluminum such as $AlF_2^+$, $AlF^{2+}$, $AlF_4^-$, $AlF_5^{2-}$ and $AlF_6^{3-}$. Besides that, some aluminum-fluoride hydrates also formed in a dissolved form such as $AlOHF^+$ and $Al(OH)_2F_2$. The solubility product determines the solubility of the aluminum in the solution. The reaction equations for the formation of the complexes are represented by their thermodynamic equilibrium constants $K_{a1}{}^{int}$, $K_{a2}{}^{int}$, $K_C{}^{int}$ and

$K_A^{int}$ with respect to their reaction equations shown in equations (14) to (17). The simulation, with the help of a suitable model, provides information about the amount of all ionic aluminium fluoride and hydroxyl fluoride complexes formed during the defluoridation process with respect to different fluoride concentrations, alumina dosage, and pH [34].

During the defluoridation process, aluminium fluoride and aluminium hydroxyl fluoride complexes dissolve residual aluminium into the treated water. In neutral pH, the solubility of aluminum from the alumina surface is very low (Solubility product = 10-32). However, the presence of high fluoride concentration and pH variation results in an increase in aluminum concentration in treated water. Moreover, during the regeneration process of exhausted activated alumina, corrosive acids and bases are used, which leads to the release of aluminium fluoride complexes and solid alumina waste that are toxic in nature [35]. These toxic aluminum complexes pollute the environment on exposure and consumption by living beings through various drinking water sources and may lead to various disorders such as amyotrophic lateral sclerosis, Alzheimer's, and autism spectrum disorders [36].

## MODIFIED ACTIVATED ALUMINA

### Calcium and Magnesium Impregnated Activated Alumina

As we discussed in the previous section, the use of activated alumina leaves toxic aluminum-fluoride complexes due to surface leaching during defluoridation. The impregnation of any metal with activated alumina creates a chance of retaining that metal in the treated water, which may be harmful to the living body. The traces of calcium and magnesium in drinking water can be quite beneficial to people on a modern processed food diet, which is low in calcium and magnesium. Therefore, the minerals like calcium and magnesium show potential for amending on the alumina surfaces because of their bioavailability and the permissible limits in drinking water. The alumina modified with magnesium oxide has a higher point of zero charges in comparison to activated alumina, resulting in a wider pH range of application for defluoridation [37].

Adsorption kinetics: The fluoride adsorption capacity and related mechanism of modified activated alumina can be understood by adsorption kinetic study. The adsorption mechanism and fluoride uptake kinetics are determined by pseudo-first-order model, pseudo-second-order reaction model, Elovich equations, and an intraparticle diffusion-based model. Because the activated alumina is porous, the adsorbate molecule may also adsorb on the inner area of the porous adsorbent *via* intraparticle diffusion or film diffusion process. Prakash Kumar Singh *et al.* [38] have successfully prepared activated alumina coated with calcium and magnesium in the form of calcium magnesium aluminium oxide and sintered it at 650°C and 850°C. They found that activated alumina and calcium-magnesium-coated activated alumina (sintered at 650°C and 850°C) followed Langmuir and Temkin adsorption isotherms with maximum fluoride removal capacities of 2.48, 2.61, and 2.74 mg/g respectively, for treating a 10 mg/L of fluoride solution at neutral pH. The experimental data were fitted with various rate models, and the reaction rate associated with each model was studied and shown in Table **1**. The pseudo-first-order model is used to study the reversible reaction between the liquid and the solid phase [39]. The model equation in its linear form is as follows:

**Table 1. Reaction and diffusion kinetic model [38]**

| Isotherms | Parameters | Values | | |
|---|---|---|---|---|
| | | Ca Mg Amended AA (650°C) | Ca Mg Amended AA (850°C) | Activated alumina |
| Langmuir | $Q_e$ | 2.24 | 2.6 | 2.27 |
| | $Q_m$(mg/g) | 2.61 | 2.74 | 2.48 |
| | $K_1$(L/g) | 0.32 | 1.28 | 0.43 |
| | $R^2$ | 0.986 | 0.997 | 0.971 |
| | $\chi^2$ | 0.173 | 0.178 | 0.571 |
| Freundlich | 1/n | 0.439 | 0.340 | 0.314 |
| | N | 2.27 | 2.94 | 3.18 |
| | $K_f$(g/mg/g) | 0.719 | 1.27 | 0.9219 |
| | $R^2$ | 0.912 | 0.940 | 0.921 |
| | $\chi^2$ | 0.1586 | 0.26149 | 0.0922 |
| Temkin | $A_T$(L/g) | 4.25 | 25.70 | 10.47 |
| | $B_T$ | 1.21 | 1.06 | 0.94 |
| | $R^2$ | 0.971 | 0.967 | 0.949 |
| | $\chi^2$ | 0.0817 | 0.077304 | 0.1002 |
| Intra particle diffusion | $Q_d$(mg/g) | 2.618 | 2.008 | 1.89 |
| | $K$(mol$^2$kJ$^{-2}$) | $-2*10^{-8}$ | $-9*10^{-10}$ | $-6*10^{-9}$ |
| | $E_s$(kJmol$^{-1}$) | 5000.00 | 7453.55 | 9128.70 |
| | $R^2$ | 0.933 | 0.751 | 0.739 |
| | $\chi^2$ | 4.3215 | 5.509863 | 3.1804 |

$$\ln(Q_e - Q_t) = \ln(Q_e) - k_1 \bullet t \tag{18}$$

where, $Q_t$- is the amount of fluoride on the surface of the sorbent (mg/g) at time $t$ (min) and $k_1$ is the equilibrium rate constant of pseudo-first-order sorption (min$^{-1}$). A plot of $\ln (Q_e\text{-}Q_t)$ vs $t$ gives a straight line with a slope of $-k_1$ and intercept $\ln(Q_e)$. The pseudo-second order model assumes chemisorption as the rate limiting step. The model equation is as follows:

$$\frac{t}{Q_t} = \frac{1}{k_2 Q_e^2} + \frac{t}{Q_e} \tag{19}$$

where, $k_2$ represents the equilibrium rate constant of pseudo second order sorption (g/mg min$^{-1}$). The plot of $t/Q_t$ vs $t$ is a straight line with a slope of $1/Q_e$ and intercept 1k2Qe2 . Besides that, the intra-particle diffusion model was proposed by Weber and Morris whose equation is given by:

$$Q_t = k_3 t^{0.5} + I \tag{20}$$

where, $k_3$ gives intra-particle rate constant (in mg/g.min$^{-0.5}$). The plot of $Q_t$ vs $t^{0.5}$ produces straight line with a slope $k_3$ and an intercept $I$ (mg/g).

The magnitudes of $Q_e$, $Q_t$, and $k_1$ corresponding to activated alumina and modified activated alumina with calcium-magnesium (sintered at 650°C and 850°C) were determined. The Elovich kinetic model is applicable to the kinetics of chemisorption at the solid-liquid interface [40]. The Elovich model is one of the most useful models for describing adsorption on activated surfaces [41], which is given by the Elovich equation as follows:

$$\frac{dQ_t}{dt} = ae^{-bQ_t} \tag{21}$$

Where a-initial adsorption rate

b-Elovich constant

$Q_t$- amount of adsorption at time

In equation (8), the constant a regarded as the initial rate, because as $Q_t$

approaches 0, dQtdt approaches to a. At $Q_t = Q_t$, at $t = t$, and $Q_t = 0$ at $t = 0$, the integrated form of equation (8) is given by-

$$Q_t = \left(\frac{a}{b}\right)\ln(t + t_0) - \left(\frac{1}{b}\right)\ln t_0 \tag{22}$$

Where $t_0 = 1/ab$. If $t \gg t_0$, we can simplify equation (9) as,

$$Q_t = \left(\frac{1}{b}\right)\ln(ab) + \left(\frac{1}{b}\right)\ln(t) \tag{23}$$

The validity of equation (10) for the assumption can be checked by the linear plot of $Q_t$ *vs* ln(t). For the given adsorption system, let $t_{ref}$ be the longest time in the adsorption process, and $Q_{ref}$ solid phase concentration at time $t = t_{ref}$. We can rewrite equation (10) as follows-

$$Q_{ref} = \left(\frac{1}{b}\right)\ln(ab) + \left(\frac{1}{b}\right)\ln t_{ref} \tag{24}$$

From equations (10) and (11) we can easily obtain-

$$\left(\frac{Q_t}{Q_{ref}}\right) = \left(\frac{1}{Q_{ref}b}\right)\ln\left(\frac{t}{t_{ref}}\right) + 1 \tag{25}$$

This equation is called Elovich equation. Using the Elovich model, sorption rate α (mgg$^{-1}$h$^{-1}$), the desorption constant β (gmg$-1$) of the fluoride ions and $Q_t$ (mgg$^{-1}$) the amount of fluoride ions adsorbed at time t can be determined by fitting the model.

Of the kinetic models discussed so far, the applicability of each model to the kinetic data can be evaluated based on their corresponding linear regression coefficient R2 values. The values of the regression coefficient R2 and squared sums of errors (SSE) determine the best suited kinetic model to represent the fluoride adsorption mechanism. The chi-square test is performed for accuracy of model fit, which is the sum of the square of the difference between experimental data and model fit data divided by the corresponding data obtained from models as:

$$\chi 2 = \sum \frac{(Q_e - Q_c)^2}{Q_c} \tag{26}$$

The experimental adsorption data were tested by four adsorption models: Langmuir, Freundlich, Temkin, and Dubinin-Radushkevich (D-R) isotherms, to determine the best suit model and their parameters are shown in Table **1**. It can be observed from Table **1** that calcium and magnesium amended alumina follow Langmuir adsorption and Temkin adsorption isotherm models indicating monolayer adsorption with chemisorption.

The amalgamation of calcium and magnesium onto the activated alumina forms calcium and magnesium complexes with the aluminium oxide surface, which in turn increases the reactivity of the activated alumina towards fluoride. From the work carried out by Prakash Kumar Singh *et al.*, calcium magnesium amended activated alumina treated at 850°C had a highly crystalline porous leaf-like structure and a higher adsorption capacity (2.74 mg/g) than that treated at 650°C (2.61 mg/g) and that of pure activated alumina (2.48 m/g). The modified activated alumina also had a higher point of zero charges in comparison to unmodified activated alumina, which offers a wider pH range application for fluoride removal. The calcium-magnesium amended activated alumina shows better adsorption capacities over warm water, which finds its application for defluoridation in the tropical region. Besides that, this material does not show any signs of alumina leaching, which were observed in the case of activated alumina Thus, calcium-magnesium amended activated alumina is suitable for field-scale applications with high capacity, no aluminum toxicity in treated water, the bioavailability of the amended product, applicability for wider pH range and health benefits than pure activated alumina.

**Alum Impregnated Activated Alumina**

In order to increase the adsorption efficiency of activated alumina, the surface of alumina can be modified by impregnation with alum, which has a high fluoride adsorption capacity [42]. This modification can be done by mixing the appropriate amount of alum into activated alumina during the sol formation process. Rafique *et al.* prepared immobilized activated alumina (MIAA) by the sol-gel method and studied the removal of fluoride from drinking water [43]. The fluoride removal efficiency of MIAA was found to be improved (1.35 times) as compared to normal immobilized activated alumina. An adsorption study was performed on MIAA as a function of contact time, adsorbent dose, initial fluoride concentration, and stirring rate. They have reported more than 90% fluoride removal capacity for a contact time of 60 minutes. Table **2** compares activated charcoal and MIAA in terms of Freundlich and Langmuir constants, which also indicates that the removal efficiency of MIAA was about 10% more than the activated charcoal. Both the Langmuir and Freundlich adsorption isotherms fitted well for the fluoride adsorption on MIAA and activated charcoal.

**Table 2. Freundlich and Langmuir isotherms constants [43].**

| Adsorbent | Freundlich Constants | | | Langmuir Constants | | |
|---|---|---|---|---|---|---|
| | $K_f$ | $N$ | $R^2$ | $K$ | $V_m$ | $R^2$ |
| Activated Charcoal | 0.26 | 2.81 | 0.97 | 2.63 | 0.49 | 0.99 |
| MIAA | 0.60 | 3.57 | 0.98 | 4.0 | 0.80 | 0.99 |

The fluoride adsorption for MIAA and activated charcoal can be explained by the Langmuir adsorption isotherm, generating a regression coefficient $R2$ of 0.99. The theoretical values of the adsorption coefficient $K$ along with the monolayer capacity $Vm$ are found out from the Langmuir equation and described in Table **2**. The higher values of $K$ and $Vm$ for MIAA as compared to activated charcoal indicate that the adsorption of fluoride ions over the MIAA surface is more favourable than that on activated charcoal. The Langmuir constant $K$ is a measure of the particular adsorption system, whether it is favorable or unfavourable. The dimensionless quantity separation factor $RL$ is generally employed for this purpose.

$$R_L = \frac{1}{1+KC_i}$$

(27)

Where, $R_L$ = dimensionless separation factor which indicates isotherm shape,

$C_i$ = initial fluoride concentration (mg/L), and

$K$ = Langmuir constant.

If the value of the $R_L$ factor is greater than 1, it indicates unfavorable isotherm and if its value is less than 1, it indicates a favorable isotherm.

The maximum fluoride adsorption capacity of activated charcoal and MIAA activated charcoal were 0.47 mg/g and 0.76 mg/g, respectively, at an initial fluoride concentration (Ci) of 12 mg/L. The maximum adsorption at optimum stirring rates was observed for activated charcoal and MIAA at 125 rpm and 150 rpm, respectively. The contact time required to attain equilibrium for MIAA was 60 minutes, whereas activated charcoal attains equilibrium after 90 minutes. The fluoride removal efficiency of 95% and 84% was obtained for MIAA and activated charcoal respectively, at an optimum dose of 10g/L.

In a similar study conducted by Tripathy *et al.* [44], the ability of the alum-impregnated activated alumina (AIAA) to remove of fluoride from groundwater through adsorption was investigated. It was observed from the kinetic study that removal of fluoride was high during the initial period, *i.e.*, most of the fluoride was removed during 10–60 min and reached a maximum of 92% after 3 h. The fluoride removal capacity increases with an increase initially up to pH 6.5 then decreases with the increasing pH. The maximum fluoride removal efficiency from the water was found at an optimum pH of 6.5, which is suitable for drinking purposes. In addition to this, maximum adsorption was obtained for a dose of 8 g/Lwith a contact time of 3 h when 20 mg/l of fluoride is present in 50 ml of water. Tripathy *et al.* correlated isotherm and variation of adsorbent dose data to the Bradley equation. It was noticed that the adsorption capacity decreases with an increase in pH, correlating the isotherm to the Bradley equation. During the regeneration process, fluoride adsorbed AlAA was treated with 0.1 M NaOH at pH 12 and then neutralized with 0.1 M HCl. This adsorbent can be reused for the removal of fluoride prior to the impregnation with alum. Such alum-treated activated alumina is suitable for the removal of fluoride up to 0.2 mg/L from water contaminated with 20 mg/L fluoride.

## Copper Oxide Incorporated Activated Alumina

The fluoride removal capacity of the alumina can also be enhanced by coating copper oxide over the alumina surface. Bansiwal *et al.* [45] have synthesised copper oxide-coated alumina (COCA) by impregnating alumina with copper sulphate solution followed by calcination at 450°C in the presence of air. They have prepared new sorbent material by modifying activated alumina with a copper oxide coating, and their fluoride removal performance from the water was evaluated. They have conducted batch adsorption studies to determine kinetic parameters and various adsorption techniques and to understand the mechanism of adsorption. It was also attempted to study the impact of various ions commonly encountered in drinking water and to compare the fluoride removal efficiency with unmodified alumina. The copper-coated activated alumina is suitable for defluoridation in a wide pH range between 4 and 9. Further, it does not show the leaching of metals like aluminium or copper into the treated water. The Freundlich, Langmuir, and Dubinin–Radushkevich isotherm models were used to obtain various adsorption parameters, and to understand the adsorption mechanisms [46]. In order to determine which model correctly describes the adsorption mechanism, the experimental data was fitted into the linear form of the isotherms. Table **3** represents the various parameters obtained from the Freundlich, Langmuir, and Dubinin–Radushkevich isotherm models.

**Table 3. Comparative adsorption isotherm parameters of copper oxide coated activated alumina and unmodified alumina [47, 45].**

| Model & Parameters | Fluoride Removal | | Arsenic Removal | |
|---|---|---|---|---|
| | Unmodified Alumina | Copper Oxide Coated Activated Alumina | Unmodified Alumina | Copper Oxide Coated Activated Alumina |
| **Langmuir isotherm** | | | | |
| $q_m$(mg/g) | 2.232 | 7.220 | 2.161 | 2.017 |
| $K_L$(mol/l) | 1.258 | 0.763 | 0.141 | 0.315 |
| $R^2$ | 0.9865 | 0.994 | 0.998 | 0.999 |
| **Freundlich isotherm** | | | | |
| K (mg/g) | 2.058 | 2.646 | 0.430 | 0.724 |
| 1/n | 0.5679 | 0.6219 | 0.602 | 0.720 |
| $R^2$ | 0.9437 | 0.9749 | 0.990 | 0.975 |
| **Dubinin–Radushkevich isotherm** | | | | |
| $Q_m$(mg/g) | 1.2636 | 4.4902 | 3.620 | 2.3376 |
| $k_{ads}$ (mol$^2$kJ$^{-2}$) | $0.4 \times 10^{-8}$ | $7 \times 10^{-8}$ | $2.23 \times 10^{-8}$ | $2.82 \times 10^{-8}$ |
| $R^2$ | 0.9623 | 0.9248 | 0.981 | 0.963 |

When experimental data for both fluoride and arsenic removal were fitted into these three isotherms, it was realised that they were well fitted to the Langmuir model, followed by the Freundlich and Dubinin–Radushkevich models indicating the monolayer adsorption of fluoride on a uniform surface. The copper-coated activated alumina generates the fluoride adsorption capacity of 7.220 mg/g obtained from the Langmuir model, which is larger than that of unmodified activated alumina. This increase in adsorption capacity after copper oxide coating might be due to an increase in zeta potential, resulting in enhanced fluoride sorption. The values generated by the Freundlich isotherm model are also capable of describing the adsorption mechanism. The higher value of Freundlich constant indicates the reactive nature of the adsorbent, and the value of n is close to 1, reflecting favourable adsorption.

Pillewan *et al.* [47] have also prepared a copper oxide incorporated mesoporous alumina adsorbent for the removal of arsenic from water. They had prepared this adsorbent by treating mesoporous alumina with copper sulphate solution and calcining it at 450°C in the presence of air. The mechanism of adsorption, kinetic parameters, and adsorption capacity of the prepared adsorbent were investigated using various adsorption isotherms. The incorporation of copper oxide into activated alumina improves the adsorption capacity from 0.92 to 2.16 mg/g for As(III) and from 0.84 to 2.02 mg/g for As(V). Table **3** also indicates that the adsorption follows the Langmuir isotherm and pseudo-second-order kinetic models for arsenic. It also offers a wide pH range (4 to 10) for arsenic removal from water without any leaching problems.

## Manganese Oxide Coated Alumina

Another adsorbing material, manganese oxide-coated alumina (MOCA) granule, was reported by Maliyekkal *et al.*, which is capable of removing fluoride faster than activated alumina [48]. This MOCA can bring fluoride levels below 1.5 mg/L and also shows great fluoride loading capacity compared to activated alumina. Tripathy and Raichur (2008) also found that manganese dioxide-coated activated alumina can bring fluoride levels down to 0.2 mg/L with an initial fluoride concentration of 10 mg/L. They have obtained maximum adsorption at pH 5.5 [49]. The mechanism of fluoride adsorption by hydrous manganese oxide-coated activated alumina (HMOCA) is based on anion exchange between hydroxyl ions and fluoride ions in the acidic pH range. After adsorption, the liberation of $OH^-$ ions results in increased pH. At pH > 6.0, the anion exchange was not favourable hence fluoride was likely to adsorb by van der Waals forces at an alkaline solution. Above pH 6.0, sodium ions are adsorbed, which release protons that contribute to the decrease in final pH. Further, at pH > 6.0, the surface of HMOCA was supposed to be electronegative as the pHPZC of HMOCA was found to be 5.9. This would prevent the movement of fluoride ions towards the surface of HMOCA due to coulombic repulsion. The strong competition between hydroxide ($OH^-$) ions and fluoride ions on active adsorption sites may result in a decrease in adsorption in alkaline solutions.

The reaction mechanism for the adsorption of fluoride on hydrous manganese oxide-coated activated alumina (HMOCA) can be explained by Shao-Xiang Teng *et al.* as below [50]:

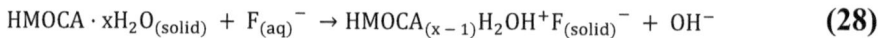

$$HMOCA \cdot xH_2O_{(solid)} + F_{(aq)}^- \rightarrow HMOCA_{(x-1)}H_2OH^+F_{(solid)}^- + OH^- \qquad (28)$$

However, this mechanism of fluoride adsorption is more likely to occur at pH< 6.0. At pH above 6.0 the following reaction would occur-

$$HMOCA \cdot xH2O_{(solid)} + Na_{(aq)}^+ + F_{(aq)}^- \rightarrow HMOCA_{(x-1)}H_2O \cdot OH^-Na^+F_{(solid)}^- + H^+ \qquad (29)$$

From Table **4**, the Langmuir model gives a correlation coefficient of 0.999, which proves a better fit to the experimental data than the Freundlich isotherm. The Langmuir model provides a maximum adsorption capacity of 7.09 mg/g for HMOCA, which is close to the value obtained from the experiments. This result indicates that the adsorption of fluoride on HMOCA offers monolayer coverage of fluoride on the adsorbent surface. The equilibrium parameter, $R_L$, provides an idea about the type of isotherm that has generated values between 0 and 1, indicating a favourable adsorption process [51]. The kinetic constants obtained from the pseudo-first-order and second-order models suggest that the kinetic data fits both types of kinetic models. The values of correlation constant R2 for pseudo-first-order and pseudo-second-order kinetic models were found to be relatively high, and the adsorption capacities obtained by the models were also close to those obtained from experiments.

Table 4. Freundlich and Langmuir isotherms constants [50, 52].

| Adsorbent | Freundlich Constants | | | Langmuir Constants | | |
|---|---|---|---|---|---|---|
| | $K_f$ (mg/g) | N | $R^2$ | $Q_m$(mg/g) | B | $R^2$ |
| HMOCA | 4.360 | 5.6 | 0.706 | 7.09 | 5.3 | 0.999 |
| AOMO | 4.481 | 1.668 | 0.9529 | 18.62 | 0.3834 | 0.9712 |

Alemu *et al.* [52] also demonstrated the capability of aluminium oxide-manganese oxide (AOMO) for fluoride removal. They had observed that fluoride uptake varied with the percentage of manganese oxide in alumina. The effect of factors such as the dose of the adsorbent, initial concentration, and pH on the fluoride uptake capacity of AOMO as an adsorbent was investigated. The AOMO proved a suitable adsorbent for fluoride removal within the pH range of 5–7. The observed experimental data showed good fit to the Freundlich isotherm model and Lagergren second-order kinetic model which explains fluoride adsorption onto AOMO surface with an average rate constant of $3.1 \times 10{-2}$ g/min . AOMO was proved to have better performance as an adsorbent as compared with the other materials reported for fluoride adsorption. The minimum adsorption capacity calculated from the non-linear Freundlich isotherm model was found to be 4.94 mg F$^-$/g whereas the maximum capacity obtained from the Langmuir isotherm was found to be 19.2 mg F$^-$/g. It was also noticed that fluoride removal efficiency was increasing with manganese oxide concentration ranging from 11% to 16% and then found to be decreased at 21% and 26%. The reason for the increase in fluoride removal capacity with manganese oxide concentration might be due to the formation of a porous manganese oxide layer on the outside and inner surfaces of alumina.

## Iron Oxide Impregnated Activated Alumina

Iron oxide-based adsorbent materials are popular for their strong affinity towards arsenic and fluoride removal, whereas alumina is known for its efficient fluoride removal potential [53]. T.C. Prathna *et al.* attempted to compare the feasibility of iron oxide/alumina nano-composites and iron oxide nanoparticles as adsorbents for the removal of fluoride [54]. The beauty of this material as an adsorbent was that it could remove both arsenic and fluoride with remarkable efficiency. It has been reported to have optimum sorption capacity towards arsenic as well as fluoride compared to iron oxide. The adsorption capacity of both iron oxide/alumina nano-composites (NPC) and iron oxide nanoparticles (IONP) increased with an increase in the initial concentration of fluoride. During isotherm studies, the adsorption data were analyzed and found to be well fitted with Freundlich isotherm as shown in Table **5**.

Table 5. Freundlich and Langmuir isotherms constants for iron oxide/alumina nano-composites (NPC) and iron oxide nanoparticles (IONP) [54].

| | Freundlich | | Langmuir | |
|---|---|---|---|---|
| | $k_f$ (mg/g) | $R^2$ | $q_m$ (mg/g) | $R^2$ |
| IONP | 0.78 | 0.64 | 1.47 | 0.48 |
| NPC | 1.19 | 0.92 | 4.82 | 0.86 |

In another study, aluminium hydroxide and iron (III) hydroxide were co-precipitated from a chloride mixture in equal proportion using ammonia. The resulting mixed hydroxide is a more efficient adsorbent for fluoride than either aluminium hydroxide or iron (III) hydroxide [55]. The physical measurements of this compound suggest that these hydroxides were bonded and didn't act independently. It was confirmed from FTIR spectra that there is an Fe–O–Al bond. The SEM image suggested a highly porous and irregular surface morphology with a large surface area. The experimental adsorption data fit best into Langmuir isotherm, indicating monolayer adsorption capacity greater than that of the pure hydroxides. Chubar *et al.* obtained double hydrous oxide ($Fe_2O_3 \bullet Al_2O_3 \bullet xH_2O$) from the sol-gel method from easily available materials and employed it for the adsorption of fluoride and other minerals [56]. In this method, the sols of Al–Fe–$(OH)_5$Cl were formed and dispersed dropwise in undecane, and these small drops were then dropped into an ammonia solution, where they acquiredstrength and formed gel granules. The reaction scheme to form adsorbent granules is as-

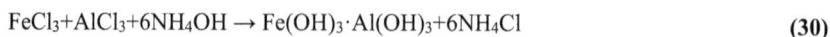

$$FeCl_3 + AlCl_3 + 6NH_4OH \rightarrow Fe(OH)_3 \cdot Al(OH)_3 + 6NH_4Cl \qquad (30)$$

The hydrated form of metal oxides shows cation and anion exchange properties. The cation and anion exchange capacities of the prepared material were found to be 1.38 and 1.8 mEq/g, respectively. The effect of pH on zeta potential was investigated and observed to have a that point of zero charges at pH 3.6, whereas the maximum negative zeta potential was at pH 9. The adsorption capacity at pH range 3-4 (acidic solution) was very high and a sharp decrease was observed beyond pH 4-5. In an acidic medium with pH < 5, the concentration of OH ions is very low which prevents them from competing with fluoride ions for fluoride sorption sites. As the pH of the solution increases, the concentration of OH⁻ ions also increases, which reduces the adsorption capacity of the adsorbent. At pH 4, there is a predominance of anions that results in higher sorption capacity. The adsorption of F⁻ ions on the adsorbent surface can be explained as an exchange reaction against the OH⁻ groups on the surface. Hiemstra and colleagues [57] investigated the sorption of F⁻ ions on goethite and confirmed from IR analysis that the main reaction of the adsorption process is singly coordinated. The FeOH groups lead to the formation of $FeF^{-1/2}$ as-

$$FeOH^{-1/2} + H^+(aq) + F^-(aq) \leftrightarrow FeF^{-1/2} + H_2O \qquad \textbf{(31)}$$

They observed 75% fluoride removal with an initial concentration of 200mg/L with buffer background electrolyte. The experimental data fit well with the pseudo-second-order kinetic model. The rate coefficient calculated from the kinetic model was found to be 0.40 $min^{-1}$. The fluoride adsorption isotherm shows that the adsorbent shows a high affinity for fluoride. This adsorbent proves superior adsorbent for fluoride with an ion exchange capacity (4.2mmol/g) which is four times higher than new ion-exchange fibre and 10 times higher than goethite [58]. At lower fluoride concentrations, F⁻ ions were exchanged with OH⁻ ions in the solution in the single charge scheme, whereas at higher concentrations, OH⁻ ions were echanged with F-ions coordinated on a two-point scheme. During the redistribution of charge on the adsorbent surface, lot of fluoride complexes were formed. These fluoride complexes may be donor type (FeOH, $Fe(OH)^2$) or proton acceptor complexes ($FeF_3$). The fluoride ions precipitate into the $FeF_3$ complex at a relatively higher concentration of fluoride. In isotherm studies, the adsorption data fit well with the Langmuir model, showing a correlation coefficient of 0.9987, an affinity constant of 0.52 L/mg, and a maximum sorption capacity of 90 mg/g. The charge distribution on the cluster was obtained by the model cluster of 4Fe-4Al optimized by HyperChem7. It has been predicted that co-ordinated oxygen atoms possess enhanced electron acceptor and electron donor properties. Using the model, it has been shown that the binding of fluorine occurs on the aluminium atoms with favourable interaction.

## Alumina Cement Granules

An alumina-cement granule (ALC) is a novel adsorbent material prepared from commercially available high alumina cement [59]. This contains alumina and calcium which have strong potential for fluoride scavenging. Ayoob S. *et al.* prepared ALC from high alumina cement at a water-cement ratio of ~0.3 in granular form with a size of 0.212mm. The properties of ALC media are presented in Table **6**.

**Table 6. Properties of ALC media [59].**

| Properties | Quantitative Value |
|---|---|
| Geometric mean size (mm) | 0.212 |
| Bulk density (g cm$^{-3}$) | 2.33 |
| Specific gravity | 2.587 |
| Al$_2$O$_3$ (%) | 78.49 |
| CaO (%) | 15.82 |
| SiO$_2$ (%) | 5.39 |
| Fe$_2$O$_3$ (%) | 0.30 |
| pH of the PZC | 11.32 |
| BET surface area (m$^2$/g) | 4.385 |

Ayoob S. *et al.* also studied the fluoride adsorption properties of ALC on natural groundwater and synthetic water with the same concentrations. It shows fast adsorption capacity initially and reaches equilibrium after 3 hours, showing negligible removal. The dose requirement for initial concentrations of 8.65 and 20 mg/L of fluoride was found to be 2 and 4 g/L, respectively, in order to bring the fluoride level within a safe limit. The fluoride removal capacity using synthetic water was 93%, whereas using natural water it was only 40% with the ALC dose of 2 g/L. To increase this capacity up to 95% in natural water, the ALC dose requirement should be 10 g/L. This indicates that the adsorption process depends upon the availability of adsorption binding sites, which dominates surface-bound sorption in the fluoride removal mechanism. When the adsorption dose was small, all adsorption sites were exposed to adsorption, which saturates the surface. However, when the adsorption dose was high, the availability of higher energy sites was decreased, and a larger fraction of lower energy sites may have been occupied. At higher adsorption dosage, there was lesser availability of higher energy sites and lower energy sites got occupied. The decrease in the binding energy of the surface shows the existence of a reversible reaction between fluoride ions attached to low-energy sites and in an aqueous solution. The uptake capacity of the ALC sorbent measured for synthetic water was found to be 3.91 mg/g and for natural water, it was found to be 0.806 mg/g. For synthetic water, the maximum uptake was 4.75 mg/g at a higher initial fluoride concentration of 20 mg/g which is due to high intra-molecular competitiveness to occupy the unsaturated surface sites at higher fluoride concentrations.

In another study conducted by Ayoob *et al.*, the kinetics and mechanism of removal of fluoride from water by the adsorbent alumina cement granules (ALC) were studied [60]. They have observed that the ALC exhibits a biphasic kinetic profile with an initial rapid uptake phase and a later slow and gradual phase. The pseudo-second-order model offers an excellent fit to the kinetic adsorption profile, generating an impressive R2 value of 0.9987. The kinetic data shows good fitting to the Elovich and intraparticle diffusion models, indicating that diffusion is significant in sorption. The value of the activation energy of the system, obtained from the Arrhenius equation (17.67 kJ mol$^{-1}$) indicates that it was diffusion-controlled and physical adsorption processes. From the poor desorption curve, it was concluded that the process is irreversible, which indicates further that the ion exchange is not the only sorption mechanism involved in the reaction. The rate-limiting step in the adsorption process was determined by analyzing the pH response of the system, the concentration of inert electrolyte, and the desorption pattern of the ALC, instead of assigning it to a single kinetic model. The chemisorptive ligand exchange reaction mechanism in fluoride removal seems dominant and involves the formation of an inner-sphere complexation between fluoride and ALC. The surface reactions include scavenging reactions or mixed surface complex reactions. Zhang and Stanforth [61] suggested that the slow sorption phase, indicative of the rate-limiting step, occurs due to either diffusion or surface reactions. As discussed earlier, the diffusion can be either intraparticle or interparticle, and surface reactions may include surface precipitation and surface site bonding energy heterogeneity. Very few studies have been conducted on the effects of ionic strength on the fluoride removal capacity of an adsorbent. The ionic strength is useful to distinguish whether outer-sphere or inner-sphere surface complexes are formed during the adsorption process.

**Alumina Supported on Carbon Nanotube**

Carbon nanotubes (CNTs) have consistently attracted researchers' interest since their discovery [62]. CNT has remarkable applications in various fields due to its small size, large surface area, electrical conductivity, and high mechanical strength. Carbon nanotube has the capability to adsorb hydrogen and may be a suitable material for hydrogen storage, as suggested by Dillon *et al.* [63]. Yan-Hui Li *et al.* suggested a new material, amorphous alumina supported on carbon nanotube, for the adsorption of fluoride from water [64]. They have studied the fluoride removal efficiency of amorphous alumina supported on carbon nanotubes, and experimental data shows that the fluoride adsorption isotherms matched with the Freundlich isotherm very well, with the correlative coefficients nearly approaching 1. Table **7** gives the various parameter values for the Freundlich isotherm, along with the deviations and correlative coefficients.

**Table 7. Freundlich isotherm parameters for fluoride adsorption onto Al$_2$O$_3$/CNTs [64].**

| pH | K | N | Average Percent Deviation (%) | Correlative Coefficient R$^2$ |
|---|---|---|---|---|
| 3.0 | 1.21 | 1.36 | 9.78 | 0.97 |
| 5.0 | 8.64 | 3.82 | 3.60 | 0.98 |
| 6.0 | 9.68 | 3.96 | 2.32 | 0.98 |
| 7.0 | 7.68 | 3.21 | 4.54 | 0.98 |
| 9.0 | 6.38 | 3.23 | 8.88 | 0.96 |
| 11.0 | 1.84 | 1.59 | 6.36 | 0.98 |

The selection of the appropriate phase of alumina is important as a crystalline form of the alumina influences the adsorption capacity of the alumina adsorbent. Hence, $\gamma$-Al$_2$O$_3$ is preferred over $\alpha$-Al$_2$O$_3$ in order to prepare Al$_2$O$_3$/CNT adsorbent. Smitha K. *et al.* have also evaluated the performance of nano alumina-carbon nanotubes for the removal of fluoride from drinking water [65]. The effect of experimental parameters such as solution temperature, pH, adsorbent dosage, agitation time, and interfering ions on fluoride removal was also determined. The nano alumina-CNTs blend (NCB) shows higher fluoride removal for the pH range of 6-7. At a lower pH value, there was the formation of hydrofluoric acid, which reduced the availability of free fluoride ions and hence fluoride removal was less. Whereas at a higher pH range, competition between fluoride and OH$^-$ ions occurs for the active sites over the adsorbent surface, which reduces fluoride removal capacity. It was also observed that it increased with agitation time and attained equilibrium after 100 min. This might be due to a reduction in the driving force after a long period of operation. The effect of adsorption dosage on removal capacity was also studied, and percent fluoride removal was found to be increased with an increase in adsorption dosage. The increase in the amount of adsorbent increases the sorption surface area for a given initial concentration of fluoride, and more ions can be accommodated on the adsorbent surface. It was also observed that the presence of other ions like sulphates and chlorides has a negligible effect on fluoride removal capacity using nano alumina-carbon nanotubes blend (NCB). The interfering ions increase the competition, and the order of their influence wasdetermined as-

$$HCO_3^- > SO_4^{2-} > Cl^-$$

The Freundlich and Langmuir isotherm model were used to investigate the adsorption mechanism, and data obtained are given in Table **8**. In the case of the Freundlich isotherm model, as the value of $K_f$ increases, the adsorption capacity of the adsorbent also increases.

**Table 8. Freundlich and Langmuir isotherm constants [65].**

| Adsorbent | Freundlich Constants | | | Langmuir Constants | | |
|---|---|---|---|---|---|---|
| | $K_f$ *(mg/g)* | *1/n* | $R^2$ | $Q_m$*(mg/g)* | *B* | $R^2$ |
| NCB | 13.2 | 0.35 | 0.99 | 11 | 0.35 | 0.95 |

Of these isotherm models, the Freundlich isotherm model has described the data fairly well in accordance with experimental values, generating an R2 coefficient value of 0.99. The obtained value of the Freundlich coefficient indicates the favourable adsorption of fluoride onto the adsorbent surface. In contrast, the Langmuir model, which is an ideal localized monolayer model, was developed to represent chemisorption on a set of well-defined, localized identical adsorption sites with the same adsorption energy, independent of surface coverage, without interaction with the adsorbed molecules. In the adsorption with NCB adsorption, the dominant mechanism, was chemisorption which influenced fluoride adsorption.

## Chitosan Coated Activated Alumina

Chitosan is a low-cost natural biopolymer known for its excellent adsorption capacity, used in metal ion removal from aqueous solutions. Chitosan contains an amine group which, when protonated in acidic media, results in the adsorption of metal anions through ion exchange [66]. Due to the presence of OH groups and NHF groups, chitosan proves a suitable material for the removal of fluoride. However, the mechanical strength of the chitosan is very low, hence coating chitosan with alumina boosts the mechanical and physical properties of chitosan. Tang X. L. *et al.* prepared chitosan-coated alumina (CAL) beds by adding a mixture of chitosan and alumina into a NaOH solution [67]. The batch experiment was carried out on CAL beads and the effects of pH, contact time, and adsorption dosage were investigated. It was noticed that the adsorption rate was high initially, up to 1 hour, then decreased and reached equilibrium after 48 hours. The adsorption capacity was 0.67mg/g which was higher than raw chitosan (0.052 mg/g). The removal capacity also increases with adsorption dosage up to 16 g/L then slows down. It almost remains independent within the pH range of 7 to 10. However, the adsorption capacity is reduced from 0.75 mg/g to 0.64 mg/g when solution pH increases from 4 to 7. Thus, CAL adsorbent has certain advantages-low cost and abundance, relatively higher adsorption capacity than raw chitosan, and a spherical shape, which makes it a highly effective adsorbent material for eliminating fluoride ions.

Haifeng Hu *et al.* have prepared a low-cost adsorbent chitosan/Al(OH)$_3$ (CS/Al(OH)$_3$) bead by treating CS/AlCl$_3$·6H$_2$O mixture solution into alkaline solution [68]. To prepare these beads, chitosan and aluminium chloride with a mass ratio of 2:1 were taken as a precursor and *in situ* generations of aluminium hydroxide sorbents in an alkaline sodium hydroxide solution. The beads were dried after washing with distilled water to a neutral pH. The structure of the chitosan/Al(OH)$_3$ adsorbent was determined, and the fluoride adsorption capacity was studied. The defluoridation capacity of the adsorbent was obtained in the acidic medium, and a maximum adsorption capacity of 17.68 mg/g was obtained at pH 4. The reason for this is that in an acidic medium the –NH$_2$ group protonized to –NH3 $^+$, and a portion of Al(OH)$_3$ dissolves into Al$^{3+}$. The adsorption capacity of CS/Al(OH)$_3$ beads could be increased due to the presence of –NH$_3^+$ and Al$^{3+}$ [69 - 71]. The effect of co-existing ions such as HCO$_3^-$, CO$_3^{2-}$, NO$_3^-$, Cl$^-$ and SO$_4^{2-}$ on the defluoridation capacity was also studied. It was discovered that all co-existing anions had a negative effect on the fluoride removal capacity of CS/Al(OH)$_3$ beads, in the following order: HCO$_3^-$ > CO$_3^{2-}$ > SO$_4^{2-}$ > NO$_3^-$ > Cl$^-$. It is evident that CO$_3^{2-}$ has strong adsorption efficiency for H$^+$ in the solution, hence HCO$_3^{2-}$ has shown adverse effects. The presence of HCO$_3^{2-}$ ions also increases the pH of the solution, which reduces the active site for adsorption. The fluoride removal data was modelled using different equations and the results are shown in Table **9**. It was noticed that the adsorption data fitted well with the Langmuir model and the maximum fluoride adsorption capacity was 23.06 mg/g obtained at pH 4 at a dosage concentration of 0.10 g/L.

**Table 9. Parameters of isotherms models for F-adsorption [68].**

| Adsorbent | Temperature | Freundlich Constants | | | Langmuir Constants | | |
|---|---|---|---|---|---|---|---|
| | | $K_f$ | $N$ | $R^2$ | $Q_L$ | $K_L$ | $R^2$ |
| CS/Al(OH)$_3$ | 293 | 0.0909 | 0.6412 | 0.9357 | 23.06 | 0.0266 | 0.9760 |
| | 303 | 0.4295 | 1.5595 | 0.7375 | 24.56 | 0.0504 | 0.9725 |
| | 313 | 1.15179 | 0.6412 | 0.9710 | 27.62 | 0.0595 | 0.9736 |

The whole adsorption process was completed in 2 hours and reached equilibrium. The fluoride adsorption data were fitted into the pseudo-second-order kinetic model, and the parameters associated with it were determined from the slopes and intercepts of the curves, and mentioned in Table **10**. It shows the fluoride adsorption process was spontaneous and endothermic in nature. The fluoride removal capacity of $CS/Al(OH)_3$ reached maximum value after 90 minutes, and it was discovered that the first stage of the reaction was within the contact time, of 60 minutes in which more than 80% fluoride, was removed whereas the second stage was lasted up to 90 minutes.

**Table 10. Parameters of kinetics study for F-adsorption [68].**

| Initial Concentration (mg/L) | Pseudo first-order equation | | | Pseudo second-order equation | | |
|---|---|---|---|---|---|---|
| | $q_e$(mg/g) | $K_1$ (L/min) | $R^2$ | $q_e$(mg/g) | $K_2$ (L/min) | $R^2$ |
| 20 | 14.74 | 0.0353 | 0.9942 | 15.35 | 0.0078 | 0.9991 |
| 40 | 18.28 | 0.0382 | 0.9990 | 18.87 | 0.0077 | 0.9993 |
| 60 | 23.93 | 0.0351 | 0.9972 | 25.17 | 0.0038 | 0.9995 |

## Lanthanum Impregnated Activated Alumina

In most of the adsorption studies, lanthanum (La) impregnation attracted most of the researchers because of its specific affinity towards fluoride [ 72]. Wasay *et al.* were the first to impregnate activated alumina by mixing activated alumina particles into lanthanum solution at pH 7.5. They have observed an increase in fluoride adsorption capacity from 3.31 mg/g to 6.23 mg/g [73]. The La impregnation on activated alumina was 3.3%, which could be increased to improve fluoride adsorption. Later in another study, Shi Q. *et al.* prepared lanthanum activated alumina by impregnating commercially available granulated activated alumina with lanthanum oxide and explained fluoride adsorption at the molecular level [74]. The lanthanum impregnation followed by calcination at 573 K improved the lanthanum content to 19.1% and the fluoride removal from 18.1% by pristine activated alumina to 92.4% by lanthanum impregnated alumina. The experimental techniques revealed that the 5–20 nm thin flakes of LaOOH on lanthanum impregnated alumina were formed in an amorphous form, with 7.6 oxygen atoms around each La. This material was capable of removing fluoride at a rate of about 70.5–77.2% F in the natural pH range of groundwater, which was four times higher than activated alumina. Moreover, the leaching of aluminium into treated water was reduced due to LaOOH formation. During the adsorption process, the XPS peak intensities of La-OH on lanthanum impregnated alumina decreased, indicating the reduction in hydroxyl group concentrations. However, the peak intensity of Al–OH on lanthanum impregnated alumina decreased slightly, showing La–OH played a dominant role in F adsorption as compared to Al–OH. Raman analysis observed a peak at 1162 cm-1 in the lanthanum alumina, indicating the formation of the La-F bond. There was ligand exchange that occurred between the hydroxyl group on La-OH and F which was also supported by a few researchers [75]. Vardhan *et al.* prepared lanthanum impregnated bauxite (LIB) by the thermal impregnation method to remove fluoride from water [76]. The material (LIB) was capable of removing fluoride from 20 mg/L to 0.7 mg/L. When impregnated with lanthanum, this adsorbent (LIB) has removed 99% of fluoride at a dose of 2 g/L compared with bauxite, which could remove only 94% of fluoride at a dose of 6 g/L from the initial fluoride concentration of 20 mg/L. The experimental data follows a pseudo-second-order model for the adsorption of fluoride on the LIB surface. The various values of constants of the kinetic model are represented in Table **11**. It is clear from the value of the R2 constant that the pseudo-second-order model matches well with both LIB and bauxite.

Table 11. Comparison of isothermal constants for adsorption of fluoride onto LIB and bauxite [76].

| Adsorbent | Freundlich Constants | | | Langmuir Constants | | |
|---|---|---|---|---|---|---|
| | $K_f$ *(mg/g)* | *n* | $R^2$ | $Q_m$*(mg/g)* | *B* | $R^2$ |
| Bauxite | 1.902 | 2.085 | 0.861 | 7.722 | 0.379 | 0.992 |
| LIB | 5.794 | 2.753 | 0.805 | 18.18 | 0.541 | 0.997 |

The experimental data shows that the fluoride adsorption isotherms match the Langmuir isotherm model very well, with a maximum sorption capacity of 18.18 mg/g, which is close to observed experimental values. The optimum pH range to operate LIB was 6.5-8.5, which was the pH of naturally occurring water. The addition of coexisting ions like $NO_3^-$, $Cl^-$, $SO_4^{2-}$, $PO_4^{3-}$ and $HCO_3^-$ considerably reduces the fluoride removal, probably due to competition of ions.

Cheng *et al.* have prepared a $La^{3+}$-modified activated alumina adsorbent for the effective removal of fluoride from water [77]. From characterization, the formation of a lanthanum hydroxide coating on activated alumina and strong bonding interaction between Al and $La^{3+}$ atoms was confirmed. The zeta potential value for lanthanum activated alumina particles was higher than pure activated alumina at the same pH. The defluoridation capacity of pure activated alumina and lanthanum activated alumina was governed by the pseudo-second-order kinetics, whereas the adsorption isotherms were described by the Langmuir equation. The adsorption capacity of lanthanum activated alumina was determined using the Langmuir equation and found to be 6.70 mg/g, which was higher than pure activated alumina (2.74 mg/g) at neutral pH. The addition of lanthanum significantly improved fluoride adsorption capacity of activated alumina suitable with an optimum pH range of 4 to 8. The effect of pH on the adsorption of $F^-$ ions was similar to that reported in the literature [78]. The zero point charge of activated alumina and lanthanum activated alumina were at pH 8.94 and 9.57, respectively.

## Hydroxyapatite-modified Activated Alumina

Hydroxyapatite has potential applications for the removal of fluoride at a considerably lower cost [79]. It can be made from bone char or prepared in the laboratory as synthetic nano-hydroxyapatite $[Ca_5(PO_4)_3OH]$ particles. The hydroxyapatite particles cannot be directly used in column studies due to their poor durability and strength. To enhance the strength of hydroxyapatite particles, G. Tomar *et al.* have encapsulated hydroxyapatite particles into activated alumina (AA) [80]. They have prepared and characterized a hybrid adsorbent, hydroxyapatite-modified activated alumina (HMAA), by dispersing nanoparticles of hydroxyapatite inside activated alumina granules. This adsorbent showed remarkably increased fluoride adsorption capacity, even more than virgin-activated alumina. This might be due to the fact that the incorporation of hydroxyapatite nanoparticles into the pores of activated alumina increased the number of active sites for fluoride adsorption. This hybrid adsorbent also allows the passage of water through laboratory columns that run under gravity. To determine equilibrium adsorption, batch isotherm studies were conducted on activated alumina and hydroxyapatite modified activated alumina. It was observed from the kinetic study that the equilibrium state was attained after 8 hours of contact time. The plot between 1/Ce and 1/qe shows its linear nature, which indicates the applicability of the Langmuir isotherm for the adsorption process. Table **12** shows the values of various adsorption parameters corresponding to Freundlich and Langmuir adsorption.

**Table 12. Freundlich and Langmuir adsorption parameters [80].**

| Adsorbent | Freundlich Constants | | | Langmuir Constants | | |
|---|---|---|---|---|---|---|
| | $K_f$ *(mg/g)* | *n* | $R^2$ | $Q_m$*(mg/g)* | *B* | $R^2$ |
| Activated alumina | 1.26 | 3.1 | 0.98 | 3.1 | 0.47 | 0.98 |
| Hydroxyapatite-modified activated alumina | 2.79 | 1.67 | 0.98 | 14.4 | 0.21 | 0.94 |

From isotherm data, both Langmuir and Freundlich's isotherm models explain the adsorption behaviour of activated alumina and HMAA, indicating surface heterogeneity as well as monolayer adsorption onto the surface. In the case of the Freundlich model, we obtain a value of n greater than 1, which shows favourable adsorption. This hybrid adsorbent was consistent in the removal of fluoride for three consecutive cycles of use. The HMAA adsorbent exists in two different phases: activated alumina and dispersed particles of hydroxyapatite within the pores of activated alumina; both phases can adsorb fluoride. The adsorption mechanism on the surface of the activated alumina phase can be explained by the following equations [81, 9]-

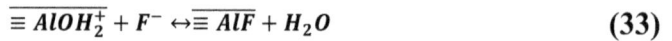

$$\equiv \overline{AlOH} + F^- \leftrightarrow \equiv \overline{AlF} + OH^- \tag{32}$$

$$\equiv \overline{AlOH_2^+} + F^- \leftrightarrow \equiv \overline{AlF} + H_2O \tag{33}$$

The following reactions would take place in hydroxyapatite phase [82]-

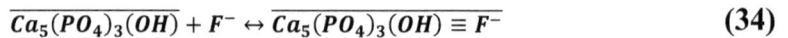

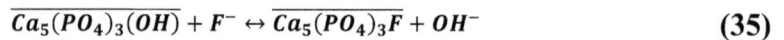

$$\overline{Ca_5(PO_4)_3(OH)} + F^- \leftrightarrow \overline{Ca_5(PO_4)_3(OH)} \equiv F^- \tag{34}$$

$$\overline{Ca_5(PO_4)_3(OH)} + F^- \leftrightarrow \overline{Ca_5(PO_4)_3F} + OH^- \tag{35}$$

$$\overline{Ca_5(PO_4)_3OH_2^+} + F^- \leftrightarrow \overline{Ca_5(PO_4)_3OH_2^+F^-} \tag{36}$$

In another study, Tomar G. *et al.* also prepared a hybrid adsorbent by dispersing hydroxyapatite within the micropores of activated alumina [82]. This adsorbent had shown a maximum adsorption capacity (14.4 mg/g)which was five times higher than activated alumina. With an initial fluoride concentration of 3 mg/L, the column ran for 450-bed volumes before the breakthrough of 1.5 mg/L was observed. The modified adsorbent also shows better regeneration capacity within a six-bed volume using easily available chemicals. Activated alumina acts as a support to the hydroxyapatite nanoparticles and also acts as an adsorbent for fluoride ions. When pH <pHpzc, the activated alumina attains a net non-diffusible positive charge on its surface, which results in a higher concentration of fluoride ions inside the adsorbent phase as compared to the bulk aqueous phase. It was also proposed that this laboratory-based study could be extended to a number of pilot-scale studies in the field to investigate the performance of the adsorbent under different conditions. There is also a need for different optimizing and controlling parameters for the practical use of adsorbents in the field.

## Metallurgical Grade Alumina

The metallurgical grade alumina is a poor quality material composed of $\gamma$-$Al_2O_3$ and $\alpha$-$Al_2O_3$ content, extracted directly from bauxite by the Bayer process. Pietrelli [83] studied the adsorption of fluoride onto metallurgical grade alumina

(MGA) and the effect of various parameters such as pH, contact time, and adsorption concentrations was studied. It was observed that the adsorption capacity was highly affected by pH, and maximum adsorption was obtained at pH 5-6. Alkalinity competes with fluoride ions for the exchange sites. Alumina forms an aluminum complex depending upon solution pH, such as in acidic media $(Al_2O_3)nAl(OH)_3$ and alkaline media $(Al_2O_3)n2H_3AlO_3$ [84]. The effect of carbonate and bicarbonate on adsorption capacity was obtained as these ions could be used for pH adjustment. The fluoride adsorption capacity was decreased by 10-13% with the addition of these ions, which may be due to the completion with fluoride ions to form strong fluoride complexes [85]. One can notice that special consideration has to be given to designing the adsorption unit if there is more hardness in the water. To determine the adsorption capacity of MGA, the adsorption data were analyzed by Langmuir and the Freundlich adsorption isotherms. The data reported in Table **13** shows the values of constants obtained in Langmuir and the Freundlich isotherms. The adsorption data seems to be fitted by both the isotherms and confirms surface heterogeneity and favorable adsorption. The Langmuir model gives a maximum uptake capacity of 12.57 mg/g.

**Table 13. Freundlich and Langmuir adsorption parameters [83].**

| Adsorbent | Freundlich Constants | | | Langmuir Constants | | |
|---|---|---|---|---|---|---|
| | $K_f$ (mg/g) | $1/n$ | $R^2$ | $Q_m$(mg/g) | $B$ | $R^2$ |
| Activated alumina | 6.19 | 0.019 | 0.8227 | 7.09 | 0.048 | 0.993 |
| MGA | 0.63 | 0.197 | 0.982 | 12.57 | 0.023 | 0.991 |

It was concluded that the fluoride removal from wastewater by metallurgical-grade alumina (MGA) has increased by up to 25%, with a capacity of 12.21 mg/g. The breakthrough curve shows the shallow nature due to the high flow rate, which does not allow equilibrium. The increase in alumina bed length or decrease in flow rate results in a mass transfer zone at 5% of breakthrough occupying 70.6% of the total column length. It was also confirmed from the breakthrough curve that at pH = 5.7, 220 BV was required to reach the breakthrough at 2 ppm, with an increase in adsorption capacity of 25%.

## Aluminum Hydroxide Supported on Zeolites

The zeolites consist of three-dimensional frameworks of $SiO_4$ and $AlO_4$ tetrahedra linked together by an oxygen bridge. The $Al^{3+}$ replaces $Si^{4+}$ resulting in a net negative charge on the crystal lattice, which could be balanced by cations like $Na^+$, $K^+$ or $Ca^{2+}$. These cations are exchangeable with other cations, which make zeolite ion exchangers [86 - 88]. The adsorption capacity of alumina adsorbents can be enhanced by modifying their surfaces with zeolite material. The work conducted by Dessalegne *et al.* [89] used natural STI and commercial zeolite (ZY) as support for aluminium hydroxide (AO) and fluoride removal ability was determined. The zeolite-AO composite was synthesised with zeolite to AO ratios of 2:1 and 6:1. The synthesised adsorbent was tested, and the effects of adsorption dose, initial fluoride concentration, and contact time were studied. The adsorbent was capable of removing fluoride content from the initial concentration of 10 mg/L to an acceptable level as suggested by WHO. The isotherm study was carried out for the adsorbent and the adsorption isotherm experiments were plotted for linearized Langmuir, Freundlich, and Dubinin–Radushkevich isotherms as shown in Table **14**. It was noticed that the adsorption isotherm fitted well with the Langmuir isotherm and a maximum fluoride adsorption capacity of 12.12 mg/g and 7.26 mg/g was observed for (2:1) and (6:1) proportions respectively.

**Table 14. Summary of Linear Adsorption isotherm parameters [89].**

| Model & Parameters | Fluoride Removal | |
|---|---|---|
| | Ratio 2:1 | Ratio 6:1 |
| **Langmuir isotherm** | | |
| $q_m$(mg/g) | 12.12 | 7.260 |
| B (L/mg) | 0.182 | 0.251 |
| $R^2$ | 0.987 | 0.982 |
| **Freundlich isotherm** | | |
| K (mg/g) | 4.025 | 2.196 |
| N | 2.614 | 2.105 |
| $R^2$ | 0.969 | 0.955 |
| **Dubinin–Radushkevich isotherm** | | |
| $Q_m$(mg/g) | 0.002 | 0.001 |
| E (kJ/mol) | 11.79 | 10.78 |
| $R^2$ | 0.981 | 0.974 |

The results shown in Table **14** indicate that the AO-zeolite adsorbent shows 7 to 12 times more fluoride adsorption capacity than the parent zeolite adsorbent. The adsorption shows maximum fluoride removal in the pH range of 5-8. Amongst all the isotherms described in Table **14**, the Langmuir isotherm proved the best-suited model for adsorption with a higher correlation coefficient value for both the concentrations. The treated water shows an aluminium content of 0.08 mg/L, which is much less than the guideline value. The fluoride adsorption mechanism can be explained by the following equation:

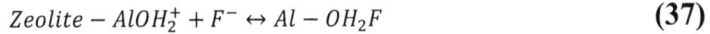

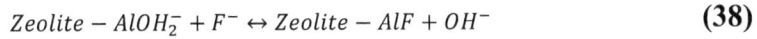

$$Zeolite - AlOH_2^+ + F^- \leftrightarrow Al - OH_2F \qquad (37)$$

$$Zeolite - AlOH_2^- + F^- \leftrightarrow Zeolite - AlF + OH^- \qquad (38)$$

L. I. Yafeng *et al.* also prepared a similar material, aluminium modified zeolite, by acids, bases, and their salts [90]. To prepare the adsorbent, the zeolite was first dipped in 3mol/L of HCl for 10 hours, and then it was rinsed and dried. This dried zeolite was then immersed in 0.2 mol/L of aluminium sulphate solution for 7 hours, washed and dried to obtain the Al-modified zeolite. Due to acid treatment, the acid dissolved impurities hence the surface area of zeolite was enhanced remarkably, which improves the adsorption capacity of the adsorbent. This adsorbent was used to treat water with an initial fluoride concentration of 1.63 mg/L. The fluoride ions were found to be reduced to less than 1 mg/L. This material has proven that it is capable of handling not only low concentrations but also higher concentrations of fluoride. In the fluoride removing process, aluminium ions and fluoride ions can be combined, and their ion-exchange reaction can be stated as:

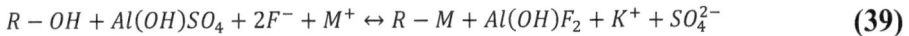

$$R - OH + Al(OH)SO_4 + 2F^- + M^+ \leftrightarrow R - M + Al(OH)F_2 + K^+ + SO_4^{2-} \qquad (39)$$

## Acid Treated Activated Alumina

It is reported in much of the literature that acid treatment of alumina improves adsorption as well as catalytic capacity [91, 92]. Recently, Usha Kumari *et al.* prepared nitric acid-activated alumina adsorbent (HNAA) by a simple and convenient process [93]. The hydroxyl and nitrate groups in the adsorbent play a vital role in the adsorption process [94]. These functional groups facilitate ion exchange with fluoride ions as follows:

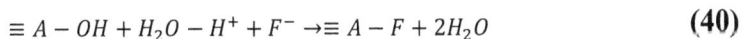

$$\equiv A - OH + H_2O - H^+ + F^- \rightarrow \equiv A - F + 2H_2O \qquad (40)$$

$$\equiv A - OH + 2F^- \rightarrow \equiv A - F_2^- + H_2O \qquad (41)$$

$$\equiv A - NO_3^- + F^- \rightarrow \equiv A - F^- + NO_3^-$$ (42)

Where, $\equiv A$ represents active adsorption site on HNAA.

The HNAA adsorbent possesses superior surface properties to freshly activated alumina. The HNAA had shown a point of zero charges at 5.825 which indicates the electropositive nature of the adsorbent at an optimum pH of 3.5. At low pH, wastewater favours existence of the electropositive complexes like $AlF^{2+}$, $AlF^+$, and weak acid HF, which reduces the adsorption of fluoride ions at low pH $<pH_{zpc}$. The Freundlich isotherm model represents a better fit to the adsorption on HNAA adsorbent than the Langmuir isotherm model. The maximum adsorption capacity of HNAA calculated from the isotherm model was 45.75 mg/g. The various adsorption parameters of HNAA corresponding to the Freundlich isotherm and Langmuir isotherm models are shown in Table **15**. It can be concluded that the dominant adsorption mechanism of fluoride removal using an adsorbent is heterogeneous adsorption, as predicted by other researchers [95]. The pseudo-second-order describes the kinetic of adsorption better than the pseudo-first order with a correlation coefficient value of 0.999. In thermodynamic studies, the values of Gibbs free energy change, enthalpy change, and separation factor suggested spontaneous, endothermic, and favourable adsorption, respectively.

**Table 15. Summary of Linear Adsorption isotherm parameters [93].**

| Adsorbent | Temperature (K) | Freundlich Constants | | | Langmuir Constants | | |
|---|---|---|---|---|---|---|---|
| | | $K_f$ | $n$ | $R^2$ | $Q_L$ | $K_L$ | $R^2$ |
| HNAA | 298 | 0.5876 | 1.7986 | 0.966 | 39.609 | 0.00150 | 0.924 |
| | 308 | 0.8389 | 1.9121 | 0.963 | 42.777 | 0.00287 | 0.915 |
| | 318 | 1.2279 | 2.0679 | 0.957 | 45.749 | 0.00424 | 0.904 |

The nitric acid treatment on activated alumina improved the defluoridation efficiency and adsorption capacity of alumina from 74.18% to 97.43% and 23.42 mg/g to 45.75 mg/g, respectively. The HNAA adsorbent has treated industrial wastewater with an initial fluoride concentration of 17.5 mg/L, which resulted in less than 1.5 mg/L. This shows that HNAA can be a potential adsorbent for fluoride removal from industrial wastewater. Moreover, the HNAA adsorbent shows remarkable regeneration capacity (> 99%).

H Wu, *et al.* prepared HCl-treated alumina by treating alumina particles with acid and base [96]. The alumina particles were washed with deionized water and dipped in 0.1 M HCl solution for 6 hours. Then the particles were washed and dried in a vacuum oven at 80°C for 12 hours. It was observed that acid treatment made alumina particles more amorphous and had significantly improved adsorption capacity. The adsorption capacity was increased by twice when treated with acid, whereas the capacity of base treated alumina was only about half of that of alumina at the solution pH = 7. At pH = 7, the adsorption capacity of acid-treated alumina was 0.89 mg/g with an initial fluoride concentration of 48.40 mg/L. In the case of acid-treated alumina, the $HNO_3$-treated alumina shows better results than the HCl treated alumina.

## $Al_2O_3/ZrO_2$

It is evident that aluminium and zirconium are both known for their strong affinity towards fluoride ions [97]. Patel *et al.* reported the defluoridation of drinking water by preparing $Al_2O_3$-$ZrO_2$ composite xerogel [98]. The adsorbent material was prepared using an environment-friendly sol-gel technique using respective metal salts instead of metal alkoxides as precursors. The prepared adsorbent shows 98% fluoride removal capacity at a normal temperature (27°C) and pH 6-7. At lower pH, the adsorbent surface acquires a positive charge; hence, the protonated surface hydroxyl group of the adsorbent promotes fluoride adsorption through the formation of weak van der Waals bonding [99]. Whereas at higher pH, the repulsion between fluoride ions and the material surface increases, which reduces fluoride uptake capacity in alkaline medium. The following reactions would occur at lower and higher pH:

At lower pH:

$$Metal(Al, Zr)(neutral) - OH + H^+ \leftrightarrow Metal(Al, Zr)\, OH_2^+(protonated) \qquad \textbf{(43)}$$

At higher pH:

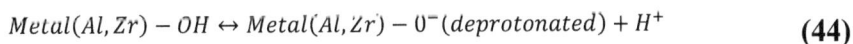

$$Metal(Al, Zr) - OH \leftrightarrow Metal(Al, Zr) - O^-(deprotonated) + H^+ \qquad \textbf{(44)}$$

Thus, the metal ions acquire positive and negative charge at lower and higher pH values, whereas, in the case of neutral pH, the material exhibits a ligand exchange mechanism as follows:

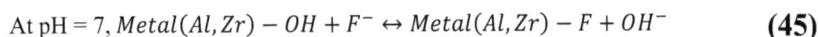

$$At\ pH = 7, Metal(Al, Zr) - OH + F^- \leftrightarrow Metal(Al, Zr) - F + OH^- \qquad \textbf{(45)}$$

It can also be stated that the zirconium ion generates tetra-nuclear ions as well as octanuclear species along with hydroxyl ions and water molecules that can participate in ligand exchange reactions with fluoride ions [100]. In the kinetic study, the adsorbent follows both pseudo-second-order kinetics and an intra-particle diffusion mechanism, showing the effect of material pore structure. The kinetic parameters of the $Al_2O_3$-$ZrO_2$ composite and $ZrO_2$ are shown in Table **16**.

**Table 16. Kinetic parameters for fluoride adsorption at pH 7.0; adsorbent dose: 0.2 g/L, temp: 298 K, initial fluoride conc. 10 mg/L [98].**

| Adsorbent | Pseudo-First Order | | | | Pseudo-Second Order | | | | Intra-Particle Diffusion | | |
|---|---|---|---|---|---|---|---|---|---|---|---|
| | $q_{exp}$ | $K_f$ | $Q_{e,cal}$ | $R^2$ | $K_s$ | H | $Q_{e,cal}$ | $R^2$ | $K_i$ | C | $R^2$ |
| $Al_2O_3$-$ZrO_2$ | 4.91 | 0.16 | 2.695 | 0.9854 | 0.1007 | 2.7578 | 5.23 | 0.99 | 0.5246 | 2.3514 | 0.8274 |
| $ZrO_2$ | 4.85 | 0.0835 | 6.788 | 0.9583 | 0.0715 | 1.9876 | 5.27 | 0.99 | 0.5864 | 1.9142 | 0.9112 |

Patel *et al*. also concluded that the sorption of fluoride increased with an increase in solution temperature, which indicates favourable material-anion interaction. Co-ions such as chloride, nitrate, and sulfate had no effect on fluoride adsorption; however, at higher concentrations of anions, the presence of ions like phosphate and bicarbonate reduces adsorption of fluoride. Zhu J *et al*. [101] prepared pectin/$Al_2O_3$-$ZrO_2$ core/shell bead sorbent from coaxial electronic injection method for defluoridation. The adsorbent was characterised by SEM, EDX, and XPS technique, and its fluoride removal property was investigated by batch experiments. The optimum pH for fluoride removal was found to be 4.0 with a contact time of 8 hours. The adsorption data well fitted with the Langmuir isotherm model with a maximum adsorption capacity of 98.077 mg/g. The isotherm parameters corresponding to the Freundlich and Langmuir isotherms of pectin/$Al_2O_3$-$ZrO_2$ are represented in Table **17**. The pseudo-second-order kinetic model represented sorption kinetics. According to thermodynamic data, $\Delta H_0$, $\Delta S_0$, and $\Delta G_0$, the sorption process was favorable, spontaneous, and exothermic. The adsorption mechanism of the fluoride adsorption was a combined ligand-exchange and ion-exchange mechanism of $Al_2O_3$-$ZrO_2$ and aluminium pectin. The pectin/$Al_2O_3$-$ZrO_2$ absorbent could be a potential material to be used in fluoride removal.

**Table 17. Langmuir and Freundlich model (pH: 4, temperature: 15 °C to 35 °C) [101].**

| Adsorbent | Temperature (K) | Langmuir Constants | | | Freundlich Constants | | |
|---|---|---|---|---|---|---|---|
| | | $Q_L$ | $K_L$ | $R^2$ | $K_f$ | $1/n$ | $R^2$ |
| Pectin/Al$_2$O$_3$-ZrO$_2$ | 288.15 | 59.342 | 0.033 | 0.963 | *6.639* | *0.433* | *0.981* |
| | 293.15 | 67.891 | 0.028 | 0.936 | 5.969 | 0.479 | 0.974 |
| | 298.15 | 70.774 | 0.028 | 0.946 | 6.046 | 0.485 | 0.987 |
| | 303.15 | 93.631 | 0.018 | 0.980 | 4.376 | 0.588 | 0.991 |
| | 308.15 | 98.077 | 0.015 | 0.990 | 3.894 | 0.611 | 0.990 |

The hard-soft acid-base theory states that $Al_3^+$ in the aluminium pectin of the sorbent was served as a hard acid, which prefers to bind with a hard base like $F^-$, which resulted in the sorption of fluoride on the core/shell bead sorbent [102].

## DISCUSSION

The fluoride adsorption capacities, favourable pH range, and type of isotherm followed under various experimental conditions have been represented in Table **18**. It has been studied that virgin activated alumina has a narrow favourable pH range, a tendency to form toxic aluminium fluoride complexes, and aluminium metal leaching problems. These problems could be rectified when activated alumina is modified with some other materials as discussed so far. The modified activated alumina has shown considerable adsorption capacity compared to unmodified activated alumina. In most of the studies conducted with various adsorbents, a pseudo-second-order kinetic model was found to fit well with the experimental data. In isotherm studies, Langmuir, Freundlich, and, in some cases, D-R isotherms seem to fit well for adsorptions of these anions. However, the choice of the particular isotherm to represent the adsorption is still unclear as the extent of adsorption depends on many factors like surface area, nature of the adsorbate and adsorbent, activation of the adsorbent, and experimental conditions under which it has been carried out. With the same fluoride concentration, the isotherm may differ with different adsorbents. Camacho *et al.* [103] have reported activated alumina modified with calcium oxide fitted with the Freundlich isotherm, whereas activated alumina modified with magnesium oxide fitted with the Langmuir isotherm [104]. The fluoride removal capacity highly depends upon pH, and acidic pH is favourable for fluoride adsorption as it offers a protonated surface. As the pH of the solution increases, the concentration of protonated sites decreases, which reduces anion adsorption. Moreover, with an increase in pH and fluoride ions have to compete with $OH^-$ ions, which further reduces adsorption [105]. However, the adsorption significantly varies at high and low pH values due to pH at point zero charge (pHzpc). Fluoride-contaminated natural water always exists with other co-existing ions like nitrate, sulfate, carbonate, chloride, *etc.* The presence of other ions either enhances coulombic repulsion or competes with fluoride ions present in the solution [43]. The chloride and nitrate ions form outer-sphere complexes, whereas phosphate and sulphate ions form inner-sphere complexes with the adsorbent surfaces. In some cases, it has been reported that the sulphate ion partially forms inner-sphere complexes or outer-sphere complexes [106]. The coulombic repulsion due to sulphate ions lessens fluoride interaction with active sites, which reduces fluoride adsorption. The fluoride removal capacity increases in the presence of chloride and nitrate ions due to an increase in the ionic strength of the solution or weakening of lateral repulsion between adsorbed fluoride ions [107, 108]. The carbonate and phosphate ions bring completion with fluoride ions at the same active sites. Of course, these effects strongly depend upon experimental conditions, pH, the nature of the adsorbent, and the concentration of co-existing ions. Ionic strength also plays a vital role in fluoride removal. It gives the idea of whether outer spheres or inner-sphere complexes are formed during the adsorption. The formation of inner-sphere complexes leads to less competition for fluoride ions. The surface area of the adsorbent also plays a key role in fluoride removal as it offers active adsorption sites. It was also observed that high surface area doesn't always guarantee higher adsorption capacity, but it may influence the chemical properties of the adsorbents. The activated alumina modified with different materials proved its effectiveness in fluoride removal because modification of the adsorbents alters physical and chemical properties.

**Table 18. Adsorption capacities and other parameters of various adsorbents for the removal of fluoride.**

| Sr. No. | Name of Adsorbent | Type of Isotherm model | Adsorption Capacity | pH | Contact Time | Reference |
|---|---|---|---|---|---|---|
| 1 | Activated Alumina | Langmuir and Freundlich | 0.86 mmol/g | 7.0 | 16-24 hours | 111 |
| 2 | Calcium and Magnesium impregnated activated alumina (CMAA650) | Langmuir | 2.61 mg/g | 5-7 | - | 38 |
| 3 | Alum Impregnated activated alumina | - | 19.80 mg/g | 9.0 | 3 hours | 44 |
| 4 | Copper oxide incorporated activated alumina | Langmuir | 7.220mg/g | 4-9 | 24 hours | 45 |
| 5 | Manganese oxide coated alumina | Langmuir | 18.62mg/g | 7 | 4 hours | 52 |
| 6 | Iron oxide impregnated activated alumina | Langmuir | 90 mg/g | 4 | - | 56 |

| Sr. No. | Name of Adsorbent | Type of Isotherm model | Adsorption Capacity | pH | Contact Time | Reference |
|---|---|---|---|---|---|---|
| 7 | Alumina Cement granules | Langmuir | 4.25 mg/g | 3-11.5 | - | 59 |
| 8 | Alumina supported on carbon nanotube | Freundlich | 14.9 mg/g | 5-9 | - | 64 |
| 9 | Chitosan coated activated alumina | Langmuir | 23.06 mg/g | 4 | 2 hours | 68 |
| 10 | Lanthanum impregnated activated alumina | Langmuir | 18.18 mg/g | 6.5-8.5 | - | 76 |
| 11 | Hydroxyapatite-modified activated alumina | Freundlich and Langmuir | 3.1 mg/g | 7.2 | 8 hours | 80 |
| 12 | Metallurgical grade alumina | Langmuir | 13mg/g | 5 | >1hour | 83 |
| 13 | Aluminum hydroxide supported on zeolites | Langmuir | 12.12 mg/g | 5-8 | 180 minutes | 89 |
| 14 | Acid treated activated alumina | Freundlich | 45.75mg/g | 3.5 | 3 hours | 93 |
| 15 | $Al_2O_3/ZrO_2$ | Dubinin-Radushkevich (D-R) | 98% | 6-7 | 20 minutes | 98 |

## CONCLUSIONS AND FUTURE SCOPE

After a rigorous qualitative and quantitative literature review of the effectiveness of activated alumina and a wide range of modified activated alumina for fluoride removal, we can put forth the following conclusions:

- Currently, both poor and developing countries are facing drinking water toxicity issues due to fluoride contamination, which is affecting the food chain by creating hazardous health crises at the level of mass poisoning.
- Adsorption is the most appropriate and widely used technique for fluoride removal because it is flexible, cost-effective, environmentally friendly, simple in design, relatively easy to operate, and capable of producing high water quality. It is a technique in which fluoride is adsorbed onto a membrane or fixed bed packed with resin or other materials.
- Activated alumina is a potential adsorbent used in fluoride removal as classified by WHO and USEPA [109]. The performance of the activated alumina is strongly affected by operating pH, contact time, ionic strength, and the presence of co-existing ions in water like phosphates, sulfates, chlorides, silicates, carbonates, *etc.* Many studies are being carried out to modify the defluoridation performance of alumina by impregnating it with other metal oxides. The leaching of aluminium into treated water could be a great challenge to the researchers as it is a neurotoxin and could be dangerous to human health.

Therefore, it is essential to modify the alumina surface in order to prevent such issues.

- The calcium and magnesium amended activated alumina have no aluminium toxicity in treated water, is applicable over a wider pH range, contributes to daily Ca and Mg intake, and the bioavailability of the amended product makes it cost-effective. The alum impregnated activated alumina showed better efficiency towards the removal of fluoride at an optimum pH of 6.5. The modification of alumina with manganese oxide, iron oxide, zirconium oxide, and copper oxide significantly improves fluoride uptake capacity, and such material is suitable for fluoride removal in high concentration areas as an alternative to existing technologies. In alumina cement granules, the presence of natural organic matter has significantly reduced fluoride uptake capacity. Rare earth oxide-based materials (like lanthanum) show better fluoride removal efficiency in batch mode, but in some cases, these materials have been found expensive and cannot be used for large-scale treatment. The carbon nanotube-based alumina possesses a higher adsorption capacity of 13.5 times higher than AIC-300 carbon, 4 times higher than $\gamma$-alumina, and even higher than IRA-410 polymeric resin [110]. The hydroxyapatite-modified activated alumina is a promising material for practical use in the remediation of fluoride. There is future scope to extend laboratory-based studies into pilot-scale studies in the fields to evaluate the performance of the adsorbent under different operating conditions. The chitosan-coated alumina is also a highly effective adsorbent due to its spherical shape, low cost, and abundance, and relatively higher adsorption capacity than virgin chitosan. Zeolite-AO composites also have very good equilibrium adsorption capacity as compared with most of the reported materials. The nitric acid-activated alumina enhances the defluoridation efficiency from 74.18% to 97.43% and the adsorption capacity from 23.42 mg/g to 45.75 mg/g.

- It was also concluded that adsorbent capacity of the specific adsorbent and adsorbate depends upon many factors like surface area, pore-volume, surface chemistry of the material, and experimental conditions like pH, contact time, co-existing ions, and so on; thus when selecting the adsorbent; one must consider the combination of these factors.

- Amongst various isotherm models, the Langmuir isotherm model explains the adsorption of many adsorbents, indicating saturated monolayer adsorption between solid/liquid systems. In most of the cases, the fluoride adsorption data were fitted into a pseudo-second-order kinetic model and various parameters associated with it, determined from the slopes and intercepts of the curves. In a few cases, the values of thermodynamic parameters indicated endothermic, spontaneous, and favourable adsorption.

## DISCLOSURE

The authors did not receive any specific grant to carry out this work from any funding agencies in the public or commercial or nonprofit organizations.

## CONSENT FOR PUBLICATION

Not applicable.

## CONFLICT OF INTEREST

The author declares no conflict of interest, financial or otherwise.

## ACKNOWLEDGEMENT

Declared none.

## REFERENCES

[1]     Sengupta Priyanka, Saha Suparna, Banerjee Suchetana, Dey Ayan, Sarkar Priyabrata. Removal of fluoride ion from drinking water by a new Fe(OH)$_3$/nanoCaO impregnated chitosan composite adsorbent. Polymer-Plastics Technology and Materials 2020; 1-13.

[2]     Manjunatha CR, Nagabhushana BM, Narayana A, Pratibha S, Raghu MS. Effective and fast adsorptive removal of fluoride on CaAl2O4:Ba nanoparticles: Isotherm, kinetics and reusability studies. Mater Res Express 2019; 6(11): 115089.
        [http://dx.doi.org/10.1088/2053-1591/ab4b23]

[3]     Smith Frank A, Hodge Harold C, Dinman BD. Airborne fluorides and man: Part I. CRC Crit Rev Environ Control 2009; 8(1–4): 293-371.

[4]     World Health Organisation Guidelines. Guidelines for drinking water quality. 4th ed. Geneva 2011; pp. 370-2.

[5]     Saxena KL, Sewak R. Fluoride consumption in endemic villages of India and its remedial measures. International Journal of Engineering Science Inventions 2015; 4: 58-73.

[6]     Näsman P, Granath F, Ekstrand J, Ekbom A, Sandborgh-Englund G, Fored CM. Natural fluoride in drinking water and myocardial infarction: A cohort study in Sweden. Sci Total Environ 2016; 562: 305-11.
        [http://dx.doi.org/10.1016/j.scitotenv.2016.03.161] [PMID: 27100011]

[7]     Bundschuh J, Maity JP, Mushtaq S, *et al.* Medical geology in the framework of the sustainable development goals. Sci Total Environ 2017; 581-582: 87-104.
        [http://dx.doi.org/10.1016/j.scitotenv.2016.11.208] [PMID: 28062106]

[8]     Choubisa SL, Sompura K. Dental fluorosis in tribal villages of Dungarpur district (Rajasthan). Pollut Res 1996; 15: 45-7.

[9]     Ghorai S, Pant KK. Equilibrium, kinetics and breakthrough studies for adsorption of fluoride on activated alumina. Separ Purif Tech 2005; 42(3): 265-71.
        [http://dx.doi.org/10.1016/j.seppur.2004.09.001]

[10]    Das N, Pattanaik P, Das R. Defluoridation of drinking water using activated titanium rich bauxite. J Colloid Interface Sci 2005; 292(1): 1-10.
        [http://dx.doi.org/10.1016/j.jcis.2005.06.045] [PMID: 16126217]

[11]    Kennedy AM, Arias-Paic M. Fixed-Bed adsorption comparisons of bone char and activated alumina for the removal of fluoride from drinking water. J Environ Eng 2020; 146(1): 04019099.
[http://dx.doi.org/10.1061/(ASCE)EE.1943-7870.0001625]

[12]    Bose S, Harish Y, Puranik M. Novel materials for defluoridation in India: A systematic review. Journal of Dental Research and Review 2019; 6(1): 3.
[http://dx.doi.org/10.4103/jdrr.jdrr_55_18]

[13]    Gedekar KA, Wankhede SP, Moharil SV, Belekar RM. $Ce_3^+$ and $Eu_2^+$ luminescence in calcium and strontium aluminates. J Mater Sci Mater Electron 2018; 29(6): 4466-77.
[http://dx.doi.org/10.1007/s10854-017-8394-0]

[14]    Wani MA, Dhoble SJ, Belekar RM. Synthesis, characterization and spectroscopic properties of some rare earth activated $LiAlO_2$ phosphor. Optik (Stuttg) 2021; 226(1): 165938.
[http://dx.doi.org/10.1016/j.ijleo.2020.165938]

[15]    Belekar RM. Suppression of coke formation during reverse water-gas shift reaction for $CO_2$ conversion using highly active $Ni/Al_2O_3$-$CeO_2$ catalyst material. Phys Lett A 2021; 395: 127206.
[http://dx.doi.org/10.1016/j.physleta.2021.127206]

[16]    Wang Y, Reardon EJ. Activation and regeneration of a soil sorbent for defluoridation of drinking water. Appl Geochem 2011; 16(5): 531-9.
[http://dx.doi.org/10.1016/S0883-2927(00)00050-0]

[17]    Nava CD, Rios MS, Olguin MT. Sorption of fluoride ions from aqueous solutions and well drinking water bythermally heated hydrocalcite. Separ Purif Tech 2003; 38(1): 31-147.

[18]    Padmavathy S, Amali J, Raja RE, Prabavathi N, Kavitha B. A study of fluoride level in potable water of Salem district and an attempt for defluoridation with lignite. Indian Journal of Environmental Protection 2003; 23(11): 1244-7.

[19]    Thergaonkar VP, Nawalakhe WG. Activated magnesia for fluoride removal. Indian J Environ Health 1971; 16: 241-3.

[20]    Goswami A, Purkait MK. Removal of fluoride from drinking water using nano magnetite aggregated schwertmannite. J Water Process Eng 2014; 1: 91-100.
[http://dx.doi.org/10.1016/j.jwpe.2014.03.009]

[21]    Dey A, Singh R, Purkait MK. Cobalt ferrite nanoparticles aggregated schwertmannite: A novel adsorbent for the efficient removal of arsenic. J Water Process Eng 2014; 3: 1-9.
[http://dx.doi.org/10.1016/j.jwpe.2014.07.002]

[22]    Goswami A, Purkait MK. Defluoridation of water by schwertmannite. World Acad Sci Eng Technol 2013; 73: 1156-61.

[23]    Zhao X, Shi YL, Cai YQ, Mou SF. Cetyltrimethylammonium bromide-coated magnetic nanoparticles for the pre-concentration of phenolic compounds from environmental water samples. J Environ Sci Technol 2008; 1139: 178-84.

[24]    Srimurali M, Pragathi A, Karthikeyan J. A study on removal of fluorides from drinking water by adsorption onto low-cost materials. Environ Pollut 1998; 99(2): 285-9.
[http://dx.doi.org/10.1016/S0269-7491(97)00129-2] [PMID: 15093323]

[25]    Mohapatra M, Anand S, Mishra BK, Giles DE, Singh P. Review of fluoride removal from drinking water. J Environ Manage 2009; 91(1): 67-77.
[http://dx.doi.org/10.1016/j.jenvman.2009.08.015] [PMID: 19775804]

[26]    Mohan D, Pittman C U Jr. Arsenic removal from water/wastewater using adsorbents—a critical review. Journal of Hazardous Materials 2007; 142(1-2): 1-53.

[27]    Bhatnagar A, Kumar E, Sillanpa M. Fluoride removal ̈ from water by adsorption—a review. Chem Eng J 2011; 171(3): 811-40.
[http://dx.doi.org/10.1016/j.cej.2011.05.028]

[28]    Vardhan VMCKJ. Removal of fluoride from water using low cost materials Proceedings of the Fifteenth International Water Technology Conference.

[29]    Di Natale F, Erto A, Lancia A, Musmarra D. Experimental and modelling analysis of As(V) ions adsorption on granular activated carbon. Water Research 2008; 42(8-9): 2007-16.

[30]    John Y, David VE, Mmereki D. A comparative study on removal of hazardous anions from water by adsorption: A review. Int J Chem Eng 2018; 2018: 1-21.
[http://dx.doi.org/10.1155/2018/3975948]

[31]    Belekar RM, Dhoble SJ. Activated alumina granules with nanoscale porosity for water defluoridation. Nano-Structures & Nano-Objects 2018; 16: 322-8.
[http://dx.doi.org/10.1016/j.nanoso.2018.09.007]

[32]    Srimurali M, Karthikeyan J. Activated alumina: Defluoridation of water and household application – a study Twelfth International Water Technology Conference, IWTC12. Alexandria, Egypt. 2008.

[33]    Władysław Rudzinski, Robert Charmas. 1-pK and 2-pK protonation models in the theoretical description of simple ion adsorption at the oxide/electrolyte interface: A comparative study of the behavior of the surface charge, the individual isotherms of ions, and the accompanying electro-kinetic effects. J Phys Chem B 2001; 105: 9755-71.
[http://dx.doi.org/10.1021/jp011299q]

[34]    Suja George AB, PoonamMondal Gupta. Overview of activated alumina defluoridation process. National Conference for Water Quality Management. https://www.researchgate.net/publication/282441116

[35]    Subhashini Ghorai, Pant KK. Investigations on the column performance of fluoride adsorption by activated alumina in a fixed-bed. Chem Eng J 2004; 98(1): 165-73.

[36]    Shaw CA, Tomljenovic L. Aluminum in the central nervous system (CNS): Toxicity in humans and animals, vaccine adjuvants, and autoimmunity. Immunol Res 2013; 56(2-3): 304-16.
[http://dx.doi.org/10.1007/s12026-013-8403-1] [PMID: 23609067]

[37]    Rashmi Devi R, Iohborlang M, Umlong Prasanta K, Raul B Das, Saumen Banerjee, Lokendra Singh. Defluoridation of water using nano-magnesium oxide. J Exp Nanosci 2014; 9(5): 512-24.

[38]    Singh PK, Saharan VK, George S. Studies on performance characteristics of calcium and magnesium amended alumina for defluoridation of drinking water. J Environ Chem Eng 2018; 6(1): 1364-77.
[http://dx.doi.org/10.1016/j.jece.2018.01.053]

[39]    Ho Y, Ng J, McKay G. Kinetics of pollutant sorption by biosorbents. Separ Purif Methods 2000; 29(2): 189-232.
[http://dx.doi.org/10.1081/SPM-100100009]

[40]    Mehta D, Mondal P, Saharan VK, George S. Synthesis of hydroxyapatite nanorods for application in water defluoridation and optimization of process variables: Advantage of ultrasonication with precipitation method over conventional method. Ultrason Sonochem 2017; 37: 56-70.
[http://dx.doi.org/10.1016/j.ultsonch.2016.12.035] [PMID: 28427668]

[41]    Ali A, Hassan K, Dehghani Anahita, Karimzadeh Sima. Preparation and characterization and application of activated alumina (AA) from alum sludge for the adsorption of fluoride from aqueous solutions: New approach to alum sludge recycling. Water Sci Technol Water Supply 2018; 18: ws2018006.

[42]    Alagumuthu G, Veeraputhiran V, Venkataraman R. Fluoride sorption using cynodondactylon—based activated carbon. Hem Ind 2011; 65(1): 23-35.
[http://dx.doi.org/10.2298/HEMIND100712052A]

[43]    Aneeza Rafique, Ali Awan M, Ayesha Wasti, Ishtiaq A, Qazi Muhammad Arshad. Removal of fluoride from drinking water using modified immobilized activated alumina, journal of chemistry. Article ID 386476 2013; 1-7.

[44]  Tripathy SS, Bersillon J-L, Gopal K. Removal of fluoride from drinking water by adsorption onto alum-impregnated activated alumina. Separ Purif Tech 2006; 50(3): 310-7.
[http://dx.doi.org/10.1016/j.seppur.2005.11.036]

[45]  Bansiwal A, Pillewan P, Biniwale RB, Rayalu SS. Rayalu, Copper oxide incorporated mesoporous alumina for defluoridation of drinking water. Microporous Mesoporous Mater 2010; 129(1-2): 54-61.
[http://dx.doi.org/10.1016/j.micromeso.2009.08.032]

[46]  Weber WJ. Physico-chemical Process for Water Quality Control. New York: Wiley Interscience Publication 1972.

[47]  Pillewan P, Mukherjee S, Roychowdhury T, Das S, Bansiwal A, Rayalu S. Removal of As(III) and As(V) from water by copper oxide incorporated mesoporous alumina. J Hazard Mater 2011; 186(1): 367-75.
[http://dx.doi.org/10.1016/j.jhazmat.2010.11.008] [PMID: 21186080]

[48]  Maliyekkal SM, Sharma AK, Philip L. Manganese-oxide-coated alumina: A promising sorbent for defluoridation of water. Water Res 2006; 40(19): 3497-506.
[http://dx.doi.org/10.1016/j.watres.2006.08.007] [PMID: 17011020]

[49]  Tripathy SS, Raichur AM. Abatement of fluoride from water using manganese dioxide-coated activated alumina. J Hazard Mater 2008; 153(3): 1043-51.
[http://dx.doi.org/10.1016/j.jhazmat.2007.09.100] [PMID: 17996364]

[50]  Teng S-X, Wang S-G, Gong W-X, Liu X-W, Gao B-Y. Removal of fluoride by hydrous manganese oxide-coated alumina: Performance and mechanism. J Hazard Mater 2009; 168(2-3): 1004-11.
[http://dx.doi.org/10.1016/j.jhazmat.2009.02.133] [PMID: 19329249]

[51]  Nigussie W, Zewge F, Chandravanshi BS. Removal of excess fluoride from water using waste residue from alum manufacturing process. J Hazard Mater 2007; 147(3): 954-63.
[http://dx.doi.org/10.1016/j.jhazmat.2007.01.126] [PMID: 17363157]

[52]  Alemu S, Mulugeta E, Zewge F, Chandravanshi BS. Water defluoridation by aluminium oxide–manganese oxide composite material. Environ Technol 2014; 35(15): 1893-903.
[http://dx.doi.org/10.1080/09593330.2014.885584]

[53]  Qiao Z, Cui Z, Sun Y, Hu Q, Guan X. Simultaneous removal of arsenate and fluoride from water by Al-Fe (hydr) oxides. Front Environ Sci Eng 2014; 8(2): 169-79.
[http://dx.doi.org/10.1007/s11783-013-0533-0]

[54]  Prathna TC, Sharma S, Kennedy M. Arsenic and fluoride removal by iron oxide and iron oxide/alumina nanocomposites: A comparison, conference. The 3rd World Congress on New Technologies.
[http://dx.doi.org/10.11159/icnfa17.118]

[55]  Biswas K, Saha SK, Ghosh UC. Adsorption of fluoride from aqueous solution by a synthetic iron(III)–aluminum(III) mixed oxide. Ind Eng Chem Res 2007; 46(16): 5346-56.
[http://dx.doi.org/10.1021/ie061401b]

[56]  Chubar NI, Samanidou VF, Kouts VS, *et al.* Adsorption of fluoride, chloride, bromide, and bromate ions on a novel ion exchanger. J Colloid Interface Sci 2005; 291(1): 67-74.
[http://dx.doi.org/10.1016/j.jcis.2005.04.086] [PMID: 15964584]

[57]  Hiemstra T, Van Riemsdijk WH. Fluoride adsorption on goethite in relation to different types of surface *sites.* J Colloid Interface Sci 2000; 225(1): 94-104.
[http://dx.doi.org/10.1006/jcis.1999.6697] [PMID: 10767149]

[58]  Ruixia L, Jinlong G, Hongxiao T. Adsorption of fluoride, phosphate, and arsenate ions on a new type of ion exchange fiber. J Colloid Interface Sci 2002; 248(2): 268-74.
[http://dx.doi.org/10.1006/jcis.2002.8260] [PMID: 16290531]

[59]  Ayoob S, Gupta AK. Performance evaluation of alumina cement granules in removing fluoride from

natural and synthetic waters. Chem Eng J 2009; 150(2-3): 485-91.
[http://dx.doi.org/10.1016/j.cej.2009.01.038]

[60]  Ayoob S, Gupta AK, Bhakat PB, Bhat VT. Investigations on the kinetics and mechanisms of sorptive removal of fluoride from water using alumina cement granules. Chem Eng J 2008; 140(1-3): 6-14.
[http://dx.doi.org/10.1016/j.cej.2007.08.029]

[61]  Zhang J, Stanforth R. Slow adsorption reaction between arsenic species and goethite (alpha-FeOOH): Diffusion or heterogeneous surface reaction control. Langmuir 2005; 21(7): 2895-901.
[http://dx.doi.org/10.1021/la047636e] [PMID: 15779964]

[62]  Iijima S. Helical microtubules of graphitic carbon. Nature 1991; 354(6348): 56-8.
[http://dx.doi.org/10.1038/354056a0]

[63]  Dillon A, Jones K, Bekkedahl TCH, Kiang CH, Bethune DS, Heben MJ. Heben,Storage of hydrogen in single-walled carbon nanotubes. Nature 1997; 386(6623): 377-9.
[http://dx.doi.org/10.1038/386377a0]

[64]  Li Y-H, Wang S, Cao A, *et al.* Adsorption of fluoride from water by amorphous alumina supported on carbon nanotubes. Chem Phys Lett 2001; 350(5-6): 412-6.
[http://dx.doi.org/10.1016/S0009-2614(01)01351-3]

[65]  Smitha K, Thampi SG. Experimental investigations on fluoride removal from water using nanoalumina-carbon nanotubes blend. J Water Resource Prot 2017; 9(7): 760-9.
[http://dx.doi.org/10.4236/jwarp.2017.97050]

[66]  Barik B, Nayak PS, Achary LSK, Kumar A, Dash P. Synthesis of alumina-based cross-linked chitosan–HPMC biocomposite film: An efficient and user-friendly adsorbent for multipurpose water purification. New J Chem 2020; 44(2): 322-37.
[http://dx.doi.org/10.1039/C9NJ03945G]

[67]  Tang XL, Zhang HZ, Zhao S, Gong SF. Removal of fluoride on chitosan coated alumina (CAL) from aqueous solution. Adv Mat Res 2012; 518(523): 518-523, 797-800.
[http://dx.doi.org/10.4028/www.scientific.net/AMR.518-523.797]

[68]  Hu H, Yang L, Lin Z, Zhao Y, Jiang X, Hou L. A low-cost and environment friendly chitosan/aluminum hydroxide bead adsorbent for fluoride removal from aqueous solutions. Iran Polym J 2018; 27(4): 253-61.
[http://dx.doi.org/10.1007/s13726-018-0605-x]

[69]  Li W, Cao CY, Wu LY, Ge MF, Song WG. Superb fluoride and arsenic removal performance of highly ordered mesoporousaluminas Journal of Hazardous Materials 2011; 143-50.

[70]  López Valdivieso A, Reyes Bahena JL, Song S, Herrera Urbina R. Temperature efect on the zeta potential and fuoride adsorption at the $\alpha$-Al$_2$O$_3$/aqueous solution interface. Journal of Colloidal Interface Science 2006; 298: 1-5.

[71]  Wang SG, Ma Y, Shi Y, Gong WX. Defuoridation performance and mechanism of nano-scale aluminum oxide hydroxide in aqueous solution. J Chem Technol Biotechnol 2009; 84(7): 1043-50.
[http://dx.doi.org/10.1002/jctb.2131]

[72]  Belekar RM, Athawale SA, Gedekar KA, Dhote AV. Various techniques for water defluoration by alumina: Development, challenges and future prospects. AIP Conf Proc 2019; 2104: 030004.
[http://dx.doi.org/10.1063/1.5100431]

[73]  Wasay SA, Tokunaga S, Park S-W. Removal of hazardous anions from aqueous solutions by La(lll)- and Y(lll)-lmpregnated alumina. Sep Sci Technol 1996; 31(10). 1501-14.
[http://dx.doi.org/10.1080/01496399608001409]

[74]  Shi Q, Huang Y, Jing C. Synthesis, characterization and application of lanthanum-impregnated activated alumina for F removal. J Mater Chem A Mater Energy Sustain 2013; 1(41): 12797.
[http://dx.doi.org/10.1039/c3ta12548c]

[75]    Kasprzyk-Hordern B. Chemistry of alumina, reactions in aqueous solution and its application in water treatment. Adv Colloid Interface Sci 2004; 110(1-2): 19-48.
[http://dx.doi.org/10.1016/j.cis.2004.02.002] [PMID: 15142822]

[76]    Vivek Vardhan CM, Srimurali M. Removal of fluoride from water using a novel sorbent lanthanum-impregnated bauxite. Springerplus 2016; 5(1): 1426.
[http://dx.doi.org/10.1186/s40064-016-3112-6] [PMID: 27625980]

[77]    Cheng J, Meng X, Jing C, Hao J. La$_3^+$-modified activated alumina for fluoride removal from water. J Hazard Mater 2014; 278: 343-9.
[http://dx.doi.org/10.1016/j.jhazmat.2014.06.008] [PMID: 24996152]

[78]    Tang Y, Guan X, Su T, Gao N, Wang J. Fluoride adsorption onto activated alumina: Modeling the effects of pH and some competing ions. Colloids Surf A Physicochem Eng Asp 2009; 337(1-3): 33-8.
[http://dx.doi.org/10.1016/j.colsurfa.2008.11.027]

[79]    Sairam Sundaram C, Viswanathan N, Meenakshi S. Fluoride sorption by nano-hydroxyapatite/chitin composite. J Hazard Mater 2009; 172(1): 147-51.
[http://dx.doi.org/10.1016/j.jhazmat.2009.06.152] [PMID: 19646815]

[80]    Sarkar S, Guibal E, Quignard F, Sen Gupta AK. Polymersupported metals and metal oxide nanoparticles: Synthesis, characterization, and applications. J Nanopart Res 2012; 14(2): 715.
[http://dx.doi.org/10.1007/s11051-011-0715-2]

[81]    MacDonald LH, Pathak G, Singer B, Jaffe PR. An integrated approach to address endemic fluorosis in Jharkhand. J Water Resource Prot 2011; 3(7): 457-72.
[http://dx.doi.org/10.4236/jwarp.2011.37056]

[82]    Tomar G, Thareja A, Sarkar S. Fluoride removal by a hybrid fluoride-selective adsorbent. Water Sci Technol Water Supply 2014; 14(6): 1133-41.
[http://dx.doi.org/10.2166/ws.2014.072]

[83]    Pietrelli L. Fluoride wastewater treatment by adsorption onto metallurgical grade alumina. Ann Chim 2005; 95(5): 303-12.
[http://dx.doi.org/10.1002/adic.200590035] [PMID: 16477938]

[84]    Wu YC, Nitya A. Water defluoridation with activated alumina. J Environ Eng Div 1979; 105(2): 357-67.
[http://dx.doi.org/10.1061/JEEGAV.0000895]

[85]    Kut KMK, Sarswat A, Srivastava A, Pittman CU Jr, Mohan D. A review of fluoride in african groundwater and local remediation methods. Groundw Sustain Dev 2016; 2: 190-212.
[http://dx.doi.org/10.1016/j.gsd.2016.09.001]

[86]    Onyango MS, Kojima Y, Aoyi O, Bernardo EC, Matsuda H. Adsorption equilibrium modeling and solution chemistry dependence of fluoride removal from water by trivalent-cation-exchanged zeolite F-9. J Colloid Interface Sci 2004; 279(2): 341-50.
[http://dx.doi.org/10.1016/j.jcis.2004.06.038] [PMID: 15464797]

[87]    Onyango MS, Kojima Y, Kumar A, Kuchar D, Kubota M, Matsuda H. Uptake of fluoride by Al$_3^+$ pretreated low-silica synthetic zeolites: Adsorption equilibrium and rate studies. Sep Sci Technol 2006; 41(4): 683-704.
[http://dx.doi.org/10.1080/01496390500527019]

[88]    Waghmare S, Arfin T, Rayalu S, Lataye D, Dubey S, Tiwari S. Adsorption behaviour of modified zeolite as novel adsorbents for fluoride removal from drinking water: Surface phenomena, kinetics and thermodynamics studies. IJSETR 2015; 4: 4114-24.

[89]    Dessalegne M, Zewge F, Diaz I. Aluminum hydroxide supported on zeolites for fluoride removal from drinking water. J Chem Technol Biotechnol 2017; 92(3): 605-13.
[http://dx.doi.org/10.1002/jctb.5041]

[90]   Yafeng L, Wenqing L. Preparation of aluminium modified zeolite and experimental study on its treatment of fluorine-containing water. IOP Conf Ser Earth Environ Sci 2018; 199: 032023.
[http://dx.doi.org/10.1088/1755-1315/199/3/032023]

[91]   Dutta M, Mishra S, Kaushik M, Basu JK. Application of various activated carbons in the adsorptive removal of methylene blue from aqueous solution. Res J Environ Sci 2011; 5(9): 741-51.
[http://dx.doi.org/10.3923/rjes.2011.741.751]

[92]   Gong WX, Qua JH, Lia RP, Lan HC. Adsorption on different type of aluminas. Chem Eng J 2012; 189-190: 126-33.
[http://dx.doi.org/10.1016/j.cej.2012.02.041]

[93]   Kumari U, Behera SK, Siddiqi H, Meikap BC. Facile method to synthesize efficient adsorbent from alumina by nitric acid activation: Batch scale defluoridation, kinetics, isotherm studies and implementation on industrial wastewater treatment. J Hazard Mater 2020; 381: 120917.
[http://dx.doi.org/10.1016/j.jhazmat.2019.120917] [PMID: 31376661]

[94]   Wu T, Mao L, Wang H. Adsorption of fluoride from aqueous solution by using hybrid adsorbent fabricated with Mg/Fe composite oxide and alginated *via* a facile method. J Fluor Chem 2017; 200: 8-17.
[http://dx.doi.org/10.1016/j.jfluchem.2017.05.005]

[95]   Lima EC, Adebayo MA, Machado FM. Kinetics and Equilibrium Models of Adsorption, Carbon Nanomaterials As Adsorbent for Environmental and Biological Application. Springer 2015; pp. 33-69.

[96]   Wu H, Chen L, Guo Gao, Yan Zhang, Tingjie Wang, Shouwu Guo. Image processing for the CCD based lateral flow strip detector. Nano Biomed Eng 2010; 2(4): 231-5.
[http://dx.doi.org/10.5101/nbe.v2i4.p231-235]

[97]   Deng H, Yu X. Fluoride sorption by metal ion-loaded fibrous protein. Ind Eng Chem Res 2012; 51(5): 2419-27.
[http://dx.doi.org/10.1021/ie201873z]

[98]   SimpiBhawna Patel, Amulya Prasad Panda, TanushreePatnaik , *et al.* Development of aluminum and zirconium based xerogel for defluoridation of drinking water: Study of material properties, solution kinetics and thermodynamics. Journal of Environmental Chemical Engineering. Elsevier 2018; 6: pp. (5)6231-42.

[99]   Swain SK, Padhi T, Patnaik T, Patel RK, Jha U, Dey RK. Kinetics and thermodynamics of fluoride removal using cerium-impregnated chitosan. Desalination Water Treat 2010; 13(1-3): 369-81.
[http://dx.doi.org/10.5004/dwt.2010.995]

[100]  Zhang G, He Z, Xu W. A low-cost and high efficient zirconium-modified-Na-attapulgite adsorbent for fluoride removal from aqueous solutions. Chem Eng J 2012; 183: 315-24.
[http://dx.doi.org/10.1016/j.cej.2011.12.085]

[101]  Zhu J, Lin X, Wu P, Luo X. Pectin/Al2O3-ZrO2 core/shell bead sorbent for fluoride removal from aqueous solution. RSC Advances 2016; 6(33): 27738-49.
[http://dx.doi.org/10.1039/C5RA26404A]

[102]  Zhou QS, Lin XY, Li B, Xuegang L. Fluoride adsorption from aqueous solution by aluminum alginate particles prepared *via* electrostatic spinning device. Chem Eng J 2014; 256: 306-15.
[http://dx.doi.org/10.1016/j.cej.2014.06.101]

[103]  Camacho LM, Torres A, Saha D, Deng S. Adsorption equilibrium and kinetics of fluoride on sol-gel -derived activated alumina adsorbents. J Colloid Interface Sci 2010; 349(1): 307-13.
[http://dx.doi.org/10.1016/j.jcis.2010.05.066] [PMID: 20566204]

[104]  Goswami A, Purkait MK. The defluoridation of water by acidic alumina. Chem Eng Res Des 2012; 90(12): 2316-24.
[http://dx.doi.org/10.1016/j.cherd.2012.05.002]

[105]   Goswami PK, Raul PK, Purkait MK. Arsenic adsorption using copper (II) oxide nanoparticles. Chem Eng Res Des 2012; 90(9): 1387-96.
[http://dx.doi.org/10.1016/j.cherd.2011.12.006]

[106]   Eskandarpour A, Onyango MS, Ochieng A, Asai S. Removal of fluoride ions from aqueous solution at low pH using schwertmannite. J Hazard Mater 2008; 152(2): 571-9.
[http://dx.doi.org/10.1016/j.jhazmat.2007.07.020] [PMID: 17719175]

[107]   John Yasinta, David Victor Emery Jr, Mmereki Daniel. A comparative study on removal of hazardous anions from water by adsorption: A review International Journal of Chemical Engineering 2018; Article ID 3975948: 1-21.
[http://dx.doi.org/10.1155/2018/3975948]

[108]   Chen N, Zhang Z, Feng C, Sugiura N, Li M, Chen R. Fluoride removal from water by granular ceramic adsorption. J Colloid Interface Sci 2010; 348(2): 579-84.
[http://dx.doi.org/10.1016/j.jcis.2010.04.048] [PMID: 20510421]

[109]   Tressaud A, Ed. Advances in Fluorine Science, Fluorine and the Environment, Agrochemicals, Archaeology, Green Chemistry & Water. Elsevier 2006; 2.

[110]   Li Y-H, Wang S, Cao A, *et al.* Adsorption of fluoride from water by amorphous alumina supported on carbon nanotubes. Chem Phys Lett 2001; 350(5-6): 5-6, 412-416.
[http://dx.doi.org/10.1016/S0009-2614(01)01351-3]

[111]   Ku Y, Chiou H-M. The adsorption of fluoride ion from aqueous solution by activated alumina. Water Air Soil Pollut 2002; 133(1/4): 349-61.
[http://dx.doi.org/10.1023/A:1012929900113]

*Water Pollution Sources and Purification*, 2022, 75-89

# CHAPTER 3

# Degradation of Substituted Benzoic Acids Related to Structural Reactivity

**Pratibha S. Agrawal**[1,*]**, M. K. N. Yenkie**[1]**, M.G. Bhotmange**[1]**, B.D. Deshpande**[1] **and S.J. Dhoble**[2]

[1] *Department of Applied Chemistry, Laxminarayan Institute of Technology, RTM Nagpur University, Nagpur, Maharashtra, India-440010*

[2] *Department of Physics, R.T.M. Nagpur University, Nagpur, India-440033*

**Abstract:** The existence of organic acids in aqueous waste continues to be an important environmental concern because of the odor and toxicity they impart to water. The photochemical degradation of benzoic acids (BA) and some of the substituted benzoic acids (SBA), which act as environmental pollutants, are studied in the present investigation using the Advanced Oxidation Processes (AOPs) and combinations of different oxidants and UV irradiation (UV/$H_2O_2$, UV/$TiO_2$, UV/ZnO, and Fe(III)-oxalate complex). The photo-oxidative degradation of these pollutants was followed by studying their concentration decay over the period of exposure to the UV-oxidant combination. The degradation kinetics of substituted benzoic acids (SBA) is observed to be dependent on the directory nature of the substituent groups, analyzed by the Hammett constant ($\sigma$), where electron-withdrawing groups (EWGs) show positive values and electron-donating groups (EDGs) account for its negative values. These observations figured out the processes that can be efficiently used for the system. Thus, this paper aims to examine parameters that affect the photocatalytic degradation of substituted benzoic acids.

**Keywords:** Advanced oxidation process, Degradation kinematics, Hydroxyl radicals, Organic acid waste, Photochemical degradation, Photo-oxidative degradation, Substituted benzoic acids, UV/$H_2O_2$.

## INTRODUCTION

Water remains a vital source for the sustenance of life. It is a medium for carrying out all metabolic and physical activities. Though large water bodies are surrounding our planet, less than 0.3% of freshwater resources in the form of

---

* **Corresponding author Pratibha S. Agrawal:** Department of Applied Chemistry, Laxminarayan Institute of Technology, RTM Nagpur University, Nagpur, Maharashtra, India-440010; Tel: +91-9763844660; E-mail: pratibha3674@gmail.com

**R. M. Belekar, Renu Nayar, Pratibha Agrawal and S. J. Dhoble (Eds)**
**All rights reserved-© 2022 Bentham Science Publishers**

lakes, rivers, and springs are consumable [1]. Biological bodies and manufactured products contain very little amount of freshwater.

The water cycle plays an important role in regulating water above and below the ground [2, 3].

Lately, due to the population explosion, the demand for water resources has increased, largely disturbing the ratio between the supply and demand. Hence, humans have put tremendous pressure on water resources worldwide. This stress is due to the combined effect of hydrological unpredictability (climatic changes due to global warming, deforestation, floods, and droughts) and extension of human activities (damming, agriculture, diversion, unjudicial use, recreational use, and pollution) [4].

Water can be assumed as an important pillar of the world economy. Approximately 70% of fresh resources are utilized for agricultural activities. Sea, river, and lake routes are common and effective means for the export and import of goods in local, national, and international markets. Water bodies provide a source of seafood as well as act as a means of transport. Industries use gallons of water and its phases as coolants and for heating purposes. Water, an excellent solvent, is used in industries and for domestic work. A number of recreational activities are associated with water, for example, yachting, sailing, rowing, boating, rafting, surfing, *etc*. There are several water parks worldwide that facilitate tourism and help in revenue generation [5].

Science and technology have given a wide outlook to the world. However, for meeting the demands of a large population and in the haste of creating plenty in less time, are not we loosing on something? An increase in industrial activity has led to the unjudicial use of fresh water resources, causing its scarcity. Moreover, industrialization has also led to the pollution of water bodies. Pollution means undesirable and unwanted bodies that change the physical and chemical composition of water air, soil, *etc*. Nature has its purification process, which removes the biodegradable waste by biological process. The broken contaminants settle and are absorbed by the soil. Though the self-purification process cannot make the water drinkable, it can be used for cleaning, washing and gardening purposes. However, large dumping of industrial, domestic and sewage wastes into fresh water resources make them unfit for consumption, and self-purification fails after a certain degree of pollution in water. Industrial waste (chemicals such organics and inorganics, resins, oils), agricultural runoff (pesticides, herbicides, weedicides), and untreated sewage (microbes, pathogens, virus,) contain non-biodegradable compounds that affect the entire ecosystem. Surface and ground water degradation due to these chemicals even in low concentrations can cause serious problems for humans and wildlife. Clean drinking water is the right of every individual, after knowing the source of water contamination to be industrial effluent, certain legal standards were laid in 1972, CWA (Clean Water Act) having the goals as Zero discharge, Fishable and Swimmable water and No Toxicity in safe drinking water [6, 7].

To limit water pollution we must know the source and ways to overcome it and be the part of the solution. Pollutants in water bodies lead to bacterial contamination and eutrophication. Industries effluents containing toxic compounds enter the food chain of aquatic organisms that ultimately affect higher hierarchy. Sudden temperature change due to industrial water may affect the oxygen content leading to high fish mortality. The runoff from construction sites and farms makes the water unclear, hence, affecting photosynthesis by obstructing the sunlight to reach aquatic plants. As compared to ground water that moves miles unseen, surface water can be easily cleaned. Ground water is self-purified when it passes through the porous and fine-grain aquifer. It is more susceptible to contamination where population density is more [8]. Groundwater chemicals can enter through the Point source (oil spills, leakage in pipe, treatment plants effluent, factories, refineries) and Nonpoint source (fertilizers and pesticides runoff by rain, contamination through soil/ground water system by improper disposal) or atmosphere (rain, gaseous emissions from automobiles, factories, and even bakeries) [9].

Pollution is imperative to industrialization as it is to population explosion. Environmentalists, conservationists, NGOs, and government and non-government organizations are imposing stringent rules and regulations on the wastewater management and conservation and treatment of effluent before disposing of it in the large water bodies. The degree of treatment required for a wastewater depends mainly on discharged requirements of effluents and the nature of water pollution it contains. The processes used for the treatment of water are primary, secondary and ternary [10].

Primary methods are used for removing suspended solids and floating materials. This can be done using Screening, Sedimentation, Flotation, Equalization, Neutralization, Stripping and Coagulation. Primary methods help to reduce biological oxygen demand ($BOD_5$), total suspended solid (SS) and oil and grease up to 25-50%, 50 to 70%, and 65% respectively. Organic nitrogen and phosphorous are removed also considerably [11 - 13].

Secondary methods further treat the effluent to remove chemical oxygen demand COD by 90%, $BOD_5$ by 85% and SS to 70%. Some processes used to treat primary wastewater are extended aeration, contact stabilization, membrane osmosis, biological treatment and pure oxygen aeration [14 - 17].

Tertiary treatments are used to attain high quality water. Tertiary processes used to treat industrial and municipal wastewater are different for suspended solids: micro screening ultrafiltration, chemical coagulation and clarification, and for organic matters: adsorption (carbon, charcoal, $TiO_{2)}$, ion exchange, electrodialysis, sonozone process- oxidation (Chlorination, Ozonation, and UV light) [18 - 26].

In developing industrializations, industrial activity has encouraged interest in finding more efficient technologies for wastewater treatment. Aromatic acids such as benzoic acids (BA) and their derivatives are common contaminants in industrial effluents. Many food industries use Benzoic acid (BA) as preservatives, which may be existing in the aqueous wastes. If it exists in a substantial amount, it may be hazardous to human health [27]. Therefore, such compounds need to be degraded from wastewater before it gets discharged. Physicochemical methods for BA removal suffer from major shortcomings due to their high operational costs and still limited effectiveness [28, 29]. Further, the conventional processes only relocate the contaminants from one phase to another causing further problems like sludge disposal or regeneration of used adsorbents. Thus, current work has been aimed to develop other techniques for the removal of organic pollutants from wastewater. Amongst the oxidation, AOPs have gained immense attention in the last decade. Glaze *et al.* [30, 31] defined AOPs as "the generation of hydroxyl radicals in adequate quantity for water purification near ambient temperature and pressure" [32]. Earlier AOPs were used for portable water treatment only, currently, AOPs are gaining prominence for removing a broad spectrum of persistent organic pollutants. In water treatment using AOPs, oxidative degradation is carried out using hydroxyl (•OH) radicals produced by the combination of oxidants like hydrogen peroxide, ozone, ultraviolet light(UV), a photo catalyst, ultra-sound, or Fenton reagent. These processes are user friendly, fast and show potential for applications in pollutant degradation. However, AOPs are not used as a disinfectant as the hydroxyl radicals are short lived, but when subjected to industrial wastewater, they act as strong oxidizing agents and are expected to destruct the pollution to less toxic compounds. AOP offers flexibility in producing OH radicals in different possible ways, allowing its wider usage, hence, is one of the popular techniques [33]. The hydroxyl radical (˙OH) is a non-selective powerful chemical oxidant, which acts very rapidly with most organic compounds. Once generated, the hydroxyl radicals attack virtually all organic compounds unselectively [34]. Two types of hydroxyl radical attacks are possible, abstraction of hydrogen atom from water as in alkanes or alcohols, or addition to olefins or aromatic compounds [35]. The selection of mechanism depends upon the nature of the organic species. The rate constants of different organic compounds with molecular ozone are given in Table **1**. These reaction rate constants vary in quite a wide range from 0.01 to $10^4$ $M^{-1}$ $s^{-1}$.

**Table 1. Reaction rate constants ($k$, $M^{-1}$ $s^{-1}$) of ozone *vs.* hydroxyl radical [60].**

| Compound | $O_3$ | OH |
|---|---|---|
| Chlorinated alkenes | $10^3$–$10^4$ | $10^9$–$10^{11}$ |
| Phenols | $10^3$ | $10^9$–$10^1$ |
| N-containing organics | $10$–$10^2$ | $10^8$–$10^{10}$ |
| Aromatics | $1$–$10^2$ | $10^8$–$10^{10}$ |
| Ketones | $1$ | $10^9$–$10^{10}$ |
| Alcohols | $10^{-2}$–$1$ | $10^8$–$10^9$ |

Photodecomposition by UV irradiation (excitation and degradation), oxidation by photocatalysis (with $Fe^{3+}$ or $TiO_2$ or ZnO), and oxidation by the direct action of $O_3$ and $H_2O_2$ are the photochemical processes for destroying organic pollutants. Organic compounds absorb UV light [36 - 39], get excited to higher energy level and they either become more reactive with strong chemical oxidants or also decompose due to direct photolysis. Aliphatic and aromatic hydrocarbons, halocarbons, phenols, ethers, ketones, *etc.* have been treated successfully by AOP. The radiation output of an effective lamp to carry out degradation should be 254 nm [40]. In recent years, new excimer lamps producing OH and H· radicals by the direct photolysis of water, at the emission wavelengths of 172 and 222 nm have been proven to be effective in UV-oxidation processes. Various AOPs proposed in the literature include (oxidant/catalyst, when present/light): $H_2O_2$/UV [41, 42]; $O_3$/UV [34, 43 - 45]; $O_3$-$H_2O_2$/UV [31, 46, 47] [$TiO_2$]/UV [48 - 50]; Fe(III)/[$TiO_2$]/UV-Vis [51]; direct photolysis of water with vacuum UV [52]; Fenton reaction or $H_2O_2$-Fe(II) [53 - 57] and the photo-Fenton reaction or $H_2O_2$ [Fe(II)/Fe(III)]/UV [58] $H_2O_2$/UV/ ZnO [59]; and the photo-Fenton reaction or $H_2O_2$ [Fe(II)/Fe(III)]/UV.

In the Fenton process, iron is cycled within +2 and +3 oxidation states, hydroxyl radical is dependent on the availability of light and $H_2O_2$ concentration, however, at the acidic pH the process yield is relatively low. The quantum yield was accelerated with Fe (III) complexed with carboxylic anions, such as oxalate. The ferrioxalate complex $[Fe(C_2O_4)_3]^{3-}$ is used as a chemical actinometer. The ligand to metal charge transfer can take place using UV and visible light [61, 62].

In the present work, the comparative analysis of the degradation of these pollutants by AOPs (UV/$H_2O_2$, UV/$TiO_2$, Fe (III)-oxalate complexes) is studied for the optimization of the concentration of oxidants, photo catalysts, Fe (III)-ligand ratio, pH and pollutant removal rate using reaction rate kinetics.

## EXPERIMENTAL

### Materials

$FeCl_3 \cdot 6H_2O$, $H_2O_2$, $H_2SO_4$, NaOH, Oxalic acid dehydrate, 2- hydroxyl benzoic acid (salicylic acid-SA), p- nitro benzoic acid (pNB), p- amino benzoic acid (pAB), BA, m- nitro benzoic acid (mNB), m- amino benzoic acid (mAB), m-hydroxyl benzoic acid (mHB), p- hydroxyl benzoic acid (pHB) were all of analytical grade. pH was adjusted for the system using sodium hydroxide and sulphuric acid. All degradations were examined in a 1.0 liter thermostatic photo reactor (diameter 81 mm and height 320 mm) encased in a quartz tube placed axially at its center in batch mode. The reactor was placed over a magnetic stirrer at the constant rpm using a dimmerstat. At atmospheric pressure, constant reactor temperature was maintained by a cryostat bath (Fourtech Systems, Mumbai, India).

### Methodology

For studying the photodegradation of organic acids in batch mode, the photoreactor was charged with 800 ml of $3 \times 10^{-4}$ mol/L solution of the pollutant in deionised water and the required concentrations of $FeCl_3$ as a source of Fe (III) were added along with a 30% aqueous hydrogen peroxide solution. During the experiments, aliquots were withdrawn from the reactor at desired time intervals and directly used for further exploration. The change in pollutant concentration was measured with UV-visible double beam spectrophotometer (Spectrascan UV 2600, Chemito, India). The experimental data of time *versus* [Ct/Co], (where Ct is the concentration of pollutant at time 't' and Co is the initial concentration of pollutant) is used to study the reaction kinetics of degradation of pollutants.

## RESULTS AND DISCUSSIONS

Optimization experiments on degradative oxidation of substituted BA were carried out and analyzed by kinetic model with pseudo first order. The observed degradation rate of the pollutants was affected due to the nature of the substituent present which can be analyzed using the Hammett equation. Hence, to understand the mechanistic difference between aromatic substitutions and kinetics, reaction rate constants of disubstituted benzene were studied, keeping one substituent specifically –COOH.

### Optimization of initial pH

The oxidation potential of •OH [63] is dependent on pH value. Experiments were carried out from acidic to the basic range for optimization of pH. At high pH values, the chances of recombination of OH radical is high [64]. Maximum degradation efficiency was observed at pH 4 (Table **2**).

**Table 2. Optimization Studies of Photo Fenton Degradation Process of SBA[a].**

| Conditions | Variations | K, min$^{-1}$ for Different Pollutants | | | | | | | |
|---|---|---|---|---|---|---|---|---|---|
| | | BA | mAB | pAB | mHB | pHB | SA | mNB | pNB |
| pH | 3.0 | 0.00541 | 0.00371 | 0.00162 | 0.00613 | 0.00267 | 0.00191 | 0.01676 | 0.01889 |
| | 4.0 | 0.00922 | 0.00700 | 0.00299 | 0.01129 | 0.00491 | 0.00375 | 0.03085 | 0.03475 |
| | 5.0 | 0.00845 | 0.00642 | 0.00273 | 0.01031 | 0.00447 | 0.00349 | 0.02816 | 0.03173 |
| | 6.0 | 0.00763 | 0.00579 | 0.00247 | 0.00932 | 0.00405 | 0.00294 | 0.02548 | 0.02871 |
| | 7.0 | 0.00682 | 0.00518 | 0.00221 | 0.00834 | 0.00365 | 0.00279 | 0.02280 | 0.02569 |
| | 9.0 | 0.00466 | 0.00353 | 0.00149 | 0.00564 | 0.00245 | 0.00209 | 0.01542 | 0.01738 |
| UV,[$H_2O_2$]/COD ratio | 2 | 0.00324 | 0.00244 | 0.00104 | 0.00393 | 0.00170 | 0.00154 | 0.01073 | 0.01209 |
| | 3 | 0.00421 | 0.00320 | 0.00136 | 0.00515 | 0.00224 | 0.00195 | 0.01408 | 0.01587 |
| | 4 | 0.00541 | 0.00381 | 0.00165 | 0.00613 | 0.00266 | 0.00201 | 0.01676 | 0.01889 |
| | 5 | 0.00783 | 0.00694 | 0.00253 | 0.00957 | 0.00715 | 0.00309 | 0.02715 | 0.03246 |
| | 6 | 0.00592 | 0.00449 | 0.00192 | 0.00724 | 0.00314 | 0.00247 | 0.01977 | 0.02229 |
| | 7 | 0.00417 | 0.00315 | 0.00133 | 0.00504 | 0.00218 | 0.00187 | 0.01375 | 0.01549 |
| UV,$H_2O_2$,TiO$_2$,mg | 10 | 0.00395 | 0.00297 | 0.00127 | 0.00478 | 0.00218 | 0.00195 | 0.01308 | 0.01473 |
| | 30 | 0.00896 | 0.00678 | 0.00289 | 0.01092 | 0.00474 | 0.00362 | 0.02984 | 0.03362 |
| | 50 | 0.01064 | 0.00959 | 0.00409 | 0.01746 | 0.00650 | 0.00572 | 0.04224 | 0.04760 |
| | 70 | 0.00653 | 0.00495 | 0.00211 | 0.00797 | 0.00346 | 0.00295 | 0.02179 | 0.02455 |
| | 90 | 0.00415 | 0.00312 | 0.00133 | 0.00503 | 0.00218 | 0.00178 | 0.01375 | 0.01549 |
| UV,$H_2O_2$,ZnO,mg | 10 | 0.00384 | 0.00289 | 0.00123 | 0.00466 | 0.00202 | 0.00165 | 0.01274 | 0.01435 |
| | 30 | 0.01052 | 0.00797 | 0.00341 | 0.01288 | 0.00559 | 0.00463 | 0.03523 | 0.03966 |
| | 50 | 0.01443 | 0.00944 | 0.00403 | 0.01121 | 0.00662 | 0.00536 | 0.04157 | 0.04884 |
| | 70 | 0.01232 | 0.00929 | 0.00396 | 0.01497 | 0.00644 | 0.00537 | 0.04090 | 0.04609 |
| | 90 | 0.00555 | 0.00416 | 0.00177 | 0.00675 | 0.00293 | 0.00231 | 0.01844 | 0.02078 |
| Fe(III)/oxalate ratio | 1:1 | 0.28743 | 0.21852 | 0.09321 | 0.35211 | 0.15279 | 0.01281 | 0.96224 | 1.08413 |
| | 1:2 | 0.36487 | 0.21789 | 0.11379 | 0.41369 | 0.17583 | 0.15627 | 1.2646 | 1.4478 |
| | 1:5 | 0.35154 | 0.26724 | 0.11400 | 0.43063 | 0.18686 | 0.14561 | 1.17681 | 1.32589 |

[a] Values of rate constant of different pollutants at varying conditions of pH, UV,[$H_2O_2$]/COD ratio, UV,$H_2O_2$,TiO$_2$,mg, UV,$H_2O_2$,ZnO,mg, Fe(III)/oxalate ratio.

The pH in the range 4-5 was best for the Fenton and the $H_2O_2$/UV processes, but it starts decomposing above pH 5 which can also be responsible for the decrease in the degradation rate of [2-HBA] [65]. The results obtained are in good agreement with those previous obtained [66]. At elevated pH, the ferrous catalyst might be deactivated due to the formation of ferric hydroxyl complexes [67].

BA=benzoic acid; mAB=meta-amino benzoic acid; pAB=para-amino benzoic acid; mHB=meta-hydroxy benzoic acid; pHB=para-hydroxy benzoic acid; SA=salicylic acid; m-NB=meta-nitro benzoic acid; p-NB=para nitro benzoic acid.

## Optimization of Initial $H_2O_2$/COD Ratio

In order to fix the optimum ratio of $H_2O_2$/COD, studies were conducted for the ratio varying from 2.0 to 7.0 (Table **1**). As shown in Table **2**, the degradation efficiency first increased with $H_2O_2$ concentration, and after its optimum concentration, the rate suffers, which may be due to the auto decomposition of $H_2O_2$ and the probability of occurrence of auto-scavenging reactions increases. The excess of $H_2O_2$ will react with OH reducing the efficiency of the degradation. The degradation of pollutants increases with increasing $H_2O_2$/COD ratio because, in this process, hydrogen peroxide produces hydroxyl radicals on photolysis, which were mainly responsible for pollutants elimination [54, 68]. From the study, it is found that for the $H_2O_2$/COD ratio 4.0 - 5.0, all the acids are efficiently degraded [68].

## Optimization Studies of $TiO_2$ Photo Catalyst

Heterogeneous photocatalytic oxidation using $TiO_2$, ZnO light has turned into an encouraging route for the successful degradation of many pollutants [69 - 73]. To optimize the catalyst concentration, photo catalyst dosages varied from 10 to 90 mg and the degradation of SBA was studied. The results as illustrated in Table **2** shows that as the dosage of photo catalyst increases, the efficiency increases due to the number of active sites which can be made available for the photocatalytic reaction. The fragmentation of catalyst can produce a higher surface area which may be helpful in increasing the degradation rate. But a further increase in catalyst dosage, deteriorates the rate because of the accumulation of cluster formation of catalyst, retarding the light penetration, and finally resulting in the increased light scattering. This can be the reason for the decreased degradation rate above an optimum catalyst loading.

## Optimization Studies of Fe (III): Ligand Ratio

The photo ability of Fe(III)–oxalate complexes is used as a chemical actinometer [62, 74, 75]. It is efficient up to 500 nm providing a continuous source of Fenton's reagent. AOP uses this source for the degradation keeping other operating variables unchanged. Table **1** clarifies the effect of ($Fe^{3+}$/oxalate) ratio on the substituted phenol degradation efficiency. The best degradation rate has been observed for $Fe^{3+}$/oxalate 1:2 ratio.

## Effect of Molecular Structure on Reactivity

In aromatic compounds, electronic effects at meta and para positions and the steric hindrances at ortho positions play a vital role in studying the course of the entire reaction. Electrophilic reactions are accelerated by EDG, while EWG will favor reactions concerning low electron density. In the present study, the highest rate was observed for p-NB, and the lowest one was for p-AB.

## Comparison of Different AOP

A comparison between degradation rate of various AOPs is illustrated in Table **2**. It can be observed that the degradation accomplished by UV in combination with peroxide and photocatalytic process is almost very little which may be due to the photochemical reaction of the compound after the absorption of the light. During this, electronically excited states of SBA may undergo different electron distributions, resulting in decomposition which doesn't contribute to compound removal. Moreover, the degradation time is still long. Earlier workers [76] also observed that the hydroxyl radical attack is less encouraging in the electron rich atmosphere in the aromatic ring. The rate constant increased from UV/ $H_2O_2$ to the powerful oxidant Fe (III)-oxalates in degrading SBA. The degradation processes for the pollutants followed the order:

$$Fe(III)\text{- Oxalate} > UV+H_2O_2+TiO_2 > UV+H_2O_2+ZnO$$

Among all the degradation processes studied, AOP involving Fe (III)-oxalates is comparatively more efficient in degrading SBA (90%), as evidenced by the reaction constant values. It can be observed that as compared to uncomplexed Fe (II), Fe (II)-oxalate complex reacts much faster with $H_2O_2$ than yielding -OH radicals. Therefore, the addition of Fe (III)-complexes in the water purification techniques is strongly admired.

## Effect of Substituent and the Reactivity

The nature of the substituents atoms marks the degradation rate of the reaction more rather than the number of such substituents [77]. Due to σ-electron withdrawing conductive effect of substituent present in the aromatic ring ($-NO_2$), the electron density is decreased; whereas, the presence of electron donating group ($-NH_2$ and $-OH$) can increase the electron cloud by π-electron donating conjugative effect. These two effects could diminish each other to some extent, but, the steric hindrance effect of the substituent at the ortho position on the aromatic ring (-OH) plays a remarkable role in decreasing the rate [78]. EWG in the aromatic ring at ortho- and para-positions make the favorable conditions for OH$^\bullet$ radicals which are then supposed to attack the aromatic centers. Therefore, the degradation kinetics decreases accordingly following the order:

p-nitro benzoic acid> m- nitro benzoic acid > m- hydroxy benzoic acid > benzoic acid> m- amino benzoic acid > p- hydroxy benzoic acid > o-hydroxy benzoic acid> p- amino benzoic acid.

In the case of –NH$_2$ and -OH, the $\prod$- electron density of the ring increases by resonance due to one of the substituent, while the other one decreases it by induction, resulting in higher Kmeta than K para.

## CONCLUSION

The effectiveness of oxidative degradation of SBA in synthetic wastewater is pH dependent which is higher in an acidic environment. The presence of hydrogen peroxide in optimum ratio shows a positive impact on the degradation rate. The substrate degradation efficiency by UV+ H$_2$O$_2$ was maximum at H$_2$O$_2$/COD ratio of 5. Furthermore, the optimum addition of ferrioxalate complexes (in the ratio 1:2) enhanced the degradation. The advantages of such a step in real applications are easy handling of the method, efficiency in degrading aromatic compounds at NTP also. Thus, such AOP based processes yielding inoffensive byproducts when combined with other treatment technologies can be a boon in wastewater purification methods.

## CONSENT FOR PUBLICATION

Not applicable.

## CONFLICT OF INTEREST

The author declares no conflict of interest, financial or otherwise.

## ACKNOWLEDGEMENTS

The author is very much thankful to the Laxminarayan Institute of Technology; RTM Nagpur University, Nagpur for providing all the necessary facilities. The work is not funded for submission.

## REFERENCES

[1]     Donchyts G, Baart F, Winsemius H, Gorelick N, Kwadijk J, Van De Giesen N. Earth's surface water change over the past 30 years. Nat Clim Chang 2016; 6(9): 810-3.
[http://dx.doi.org/10.1038/nclimate3111]

[2]     Bates B C, Kundzewicz Z W, Wu S, Palutikof J P. *Climate change and water.* 2008.

[3]     Gleick PH, Heberger M. Water and conflict: Events, trends, and analysis (2011-2012) in The World's Water. 2014; 8: pp. 159-71.

[4]     Doungmanee P. ScienceDirect The nexus of agricultural water use and economic development level. Kasetsart J Soc Sci 2016; 37(1): 38-45.
[http://dx.doi.org/10.1016/j.kjss.2016.01.008]

[5]     Gleick PH, *et al.* Improving understanding of the global hydrologic cycle observation and analysis of the climate system : The global. Water Cycle 2013.

[6]     Crinnion WJ. The CDC fourth national report on human exposure to environmental chemicals: What it tells us about our toxic burden and how it assist environmental medicine physicians. Altern Med Rev 2010; 15(2): 101-9.
[PMID: 20806995]

[7]     Tang WZ. Physicochemical treatment of hazardous wastes. 2016. Availble; https://books.google.com/books?id=0pWGmybyyH4C&pgis=1

[8]     Raghav S, Painuli R, Kumar D. Threats to Water : Issues and Challenges Related to Ground Water and Drinking Water. Springer International Publishing 2019.

[9]     Gustavsson A M, Cederberg U, Sonesson R, Otterdijk van. "Global Food Losses and Food Waste, Food and Agric. Rome: Organ. of the U. N. 2011.

[10]    Gupta DA. UNIT-I WASTEWATER TREATMENT , PRIMARY TREATMENT OF WASTEWATER WASTEWATER ENGINEERING OBJECTIVES OF

[11]    Jiang J-Q. The role of coagulation in water treatment. Chem Eng 2015; 8: 36-44.
[http://dx.doi.org/10.1016/j.coche.2015.01.008]

[12]    Ramavandi B. Treatment of water turbidity and bacteria by using a coagulant extracted from Plantago ovata. Water Resour Ind 2014; 6: 36-50.
[http://dx.doi.org/10.1016/j.wri.2014.07.001]

[13]    Prisca DG, Robert Didier. Lanciné, Combination of coagulation-flocculation and heterogeneous photocatalysis for improving the removal of humic substances in real treated water from Agbô River (Ivory-Coast), Catal Today. Part 2017; Vol. 281: pp. 2-13.

[14]    Water W. EPA_WASTE WATER TERTIARY MANUALS.

[15]    Ghangrekar MM. Aerobic Secondary Treatment of Wastewater. Ind.: Water Pollut. Control 2014; pp. 1-26.

[16]    Iit N, Web K. Module 19 : Aerobic Secondary Treatment Of Wastewater Lecture 30 : Aerobic Secondary Treatment Of Wastewater (Contd ).

[17]    Gray and N.f Water Technology An Introduction for Environmental Scientists and Engineers,. 3rdTaylor & Francis 2017.

[18]    Ljunggren M. Micro screening in wastewater treatMent – an overview. Vatten 2006; 62: 171-7.

[19]    Falsanisi D, Liberti L, Notarnicola M. Ultrafiltration (UF) Pilot Plant for Municipal Wastewater Reuse in Agriculture: Impact of the Operation Mode on Process Performance. 2010; pp. 872-85.

[20]    Sillanpää M, Ncibi MC, Matilainen A, Vepsäläinen M. Removal of natural organic matter in drinking water treatment by coagulation: A comprehensive review. Chemosphere 2018; 190: 54-71.
[http://dx.doi.org/10.1016/j.chemosphere.2017.09.113] [PMID: 28985537]

[21]    Rashed M N. Adsorption technique for the removal of organic pollutants from water and wastewater Adsorpt J Int Adsorpt Soc 2013; 28.

[22]    Oller I, Malato S, Sánchez-Pérez JA. Combination of advanced oxidation processes and biological treatments for wastewater decontamination--a review. Sci Total Environ 2011; 409(20): 4141-66.

[http://dx.doi.org/10.1016/j.scitotenv.2010.08.061] [PMID: 20956012]

[23]     Lipnizki J, Adams B, Okazaki M, Sharpe A. Combining reverse osmosis and ion exchange. Filtr Sep 2012; 49(5): 30-3.
[http://dx.doi.org/10.1016/S0015-1882(12)70245-8]

[24]     Malkin S, Guo C. Theory and applications 2008; 3: 367.

[25]     McGraw-Hill. Is ozone the way to treat sewage? business week 1973.

[26]     Segneanu AE, *et al.* Waste Water Treatment Methods. Intech 2013; pp. 53-80.

[27]     Haahtela MHT. Hypersensitivity reactions to food additives wiley Online Libr. 1987; 42: pp. 561-75.

[28]     Husain Q. Peroxidase mediated decolorization and remediation of wastewater containing industrial dyes: a review. Rev. Environ. Sci. Bio/Technology Vol 2010; Vol. 9: pp. 117-40.

[29]     Fernando J.Beltrán J F G, JoséM Encinar. Industrial wastewater advanced oxidation. Part 2. Ozone combined with hydrogen peroxide or UV radiation. Water Res 1997; 31(10): 2415-28.
[http://dx.doi.org/10.1016/S0043-1354(97)00078-X]

[30]     Glaze WH. Reaction products of ozone: A review. Environ Health Perspect 1986; 69: 151-7.
[http://dx.doi.org/10.1289/ehp.8669151] [PMID: 3545802]

[31]     Glaze WH, Kang J-W, Chapin DH. The chemistry of water treatment processes involving ozone, hydrogen peroxide and ultraviolet radiation. Ozone Sci Eng 2016; 9: 335-52.
[http://dx.doi.org/10.1080/01919518708552148]

[32]     Ollis DF. Comparative aspects of advanced oxidation processes. Emerg Technol Hazard Waste Manag 1993; III(1): 18-34.

[33]     Buthiyappan A, Raman A, Aziz A, Mohd W, Wan A. Recent advances and prospects of catalytic advanced oxidation process in treating textile effluents 2015.

[34]     Liu K, *et al.* New concepts on UV/H$_2$O$_2$ oxidation. J Hazard Mater 2013; 127(June): 402-8.

[35]     Trapido M, Veressinina Y, Munter R. Ozonation and advanced oxidation processes of polycyclic aromatic hydrocarbons in aqueous solutions - A kinetic study. Environ Technol 1995; 16(8): 729-40.
[http://dx.doi.org/10.1080/09593331608616312]

[36]     Legrini O, Oliveros E, Braun AM. Photochemical processes for water treatment. Chem Rev 1993; 93(2): 671-98.
[http://dx.doi.org/10.1021/cr00018a003]

[37]     von Sonntag C, Schuchmann H-P. The Elucidation of peroxyl radical reactions in aqueous solution with the help of radiation-chemical methods. Angew Chem Int Ed Engl 1991; 30(10): 1229-53.
[http://dx.doi.org/10.1002/anie.199112291]

[38]     Ambashta RD, Sillanpää M. Water purification using magnetic assistance: A review. 2010; 180(1-3): 38-49.
[http://dx.doi.org/10.1016/j.jhazmat.2010.04.105] [PMID: 20488616]

[39]     Fischbacher A, von Sonntag C, Schmidt TC. Hydroxyl radical yields in the Fenton process under various pH, ligand concentrations and hydrogen peroxide/Fe(II) ratios. Chemosphere 2017; 182: 738-44.
[http://dx.doi.org/10.1016/j.chemosphere.2017.05.039] [PMID: 28531840]

[40]     Sharma S, Ruparelia J, Patel M. A general review on advanced oxidation processes for waste water treatment Int Conf Curr. 8-10.

[41]     Miodrag Belosevic MGEDZSJRB. Degradation of alizarin yellow R using UV/H$_2$O$_2$ advanced oxidation process. Environ Sci Technol 2014; 33(2): 482-9.

[42]     Ried A, Wieland A, Mielcke J, Rohring D. Advanced oxidation processes (AOP) — Comparison of different treatment scenarios based on processes combining Ozone, UV and hydrogen peroxide IWA

World Water Congr Exhib. Montréal. 2010; pp. 2010; 1-9.

[43]    Mokrini A, Oussi D, Esplugas S. Oxidation of aromatic compounds with UV radiation/ozone/ hydrogen peroxide. Water Sci Technol 1997; 35(4): 95-102.
[http://dx.doi.org/10.2166/wst.1997.0095]

[44]    Esplugas S, Giménez J, Contreras S, Pascual E, Rodríguez M. Comparison of different advanced oxidation processes for phenol degradation. Water Res 2002; 36(4): 1034-42.
[http://dx.doi.org/10.1016/S0043-1354(01)00301-3] [PMID: 11848342]

[45]    Liu P, Zhang H, Feng Y, Yang F, Zhang J. Removal of trace antibiotics from wastewater: A systematic study of nanofiltration combined with ozone-based advanced oxidation processes. Chem Eng J 2014; 240: 211-20.
[http://dx.doi.org/10.1016/j.cej.2013.11.057]

[46]    Wert EC, Rosario-Ortiz FL, Drury DD, Snyder SA. Formation of oxidation byproducts from ozonation of wastewater. Water Res 2007; 41(7): 1481-90.
[http://dx.doi.org/10.1016/j.watres.2007.01.020] [PMID: 17335867]

[47]    Pera-Titus M, García-Molina V, Baños MA, Giménez J, Esplugas S. Degradation of chlorophenols by means of advanced oxidation processes: A general review. Appl Catal B 2004; 47(4): 219-56.
[http://dx.doi.org/10.1016/j.apcatb.2003.09.010]

[48]    gaya Ibrahim U, Halim A. Journal of photochemistry and photobiology C : Photochemistry reviews heterogeneous photocatalytic degradation of organic contaminants over titanium dioxide : A review of fundamentals , progress and problems. 2008; 9: 1-12.

[49]    Zhao BX, Li XZ, Wang P. 2,4-dichlorophenol degradation by an integrated process: Photoelectrocatalytic oxidation and E-Fenton oxidation. Photochem Photobiol 2007; 83(3): 642-6.
[http://dx.doi.org/10.1562/2006-09-05-RA-1030] [PMID: 17132072]

[50]    Matthews RW. Photo-oxidation of organic material in aqueous suspensions of titanium dioxide. Water Res 1986; 20(5): 569-78.
[http://dx.doi.org/10.1016/0043-1354(86)90020-5]

[51]    Machulek A, Quina F H, Gozzi F. Fundamental mechanistic studies of the photo-fenton reaction for the degradation of organic pollutants 2010.

[52]    Gonzalez MG, Oliveros E, Wörner M, Braun AM. Vacuum-ultraviolet photolysis of aqueous reaction systems. J Photochem Photobiol Photochem Rev 2004; 5(3): 225-46.
[http://dx.doi.org/10.1016/j.jphotochemrev.2004.10.002]

[53]    De Laat J, Dao YH, El Najjar NH, Daou C. Effect of some parameters on the rate of the catalysed decomposition of hydrogen peroxide by iron(III)-nitrilotriacetate in water. Water Res 2011; 45(17): 5654-64.
[http://dx.doi.org/10.1016/j.watres.2011.08.028] [PMID: 21920579]

[54]    Pignatello JJ, Oliveros E, MacKay A. Advanced oxidation processes for organic contaminant destruction based on the fenton reaction and related chemistry. Crit Rev Environ Sci Technol 2006; 36(1): 1-84.
[http://dx.doi.org/10.1080/10643380500326564]

[55]    Kwon BG, Lee DS, Kang N, Yoon J. Characteristics of p-chlorophenol oxidation by Fenton's reagent. Water Res 1999; 33(9): 2110-8.
[http://dx.doi.org/10.1016/S0043-1354(98)00428-X]

[56]    Horváth E, Szabó-Bárdos H. Czili, and Attila, Photocatalytic oxidation of oxalic acid enhanced by silver deposition on a TiO$_2$surface. J Photochem Photobiol Chem 2003; 154(2–3): 195-201.

[57]    Pontes RFF, Moraes JEF, Machulek A Jr, Pinto JM. A mechanistic kinetic model for phenol degradation by the Fenton process. J Hazard Mater 2010; 176(1-3): 402-13.
[http://dx.doi.org/10.1016/j.jhazmat.2009.11.044] [PMID: 20005036]

[58]   Bokare AD, Choi W. Review of iron-free Fenton-like systems for activating $H_2O_2$ in advanced oxidation processes. J Hazard Mater 2014; 275: 121-35.
[http://dx.doi.org/10.1016/j.jhazmat.2014.04.054] [PMID: 24857896]

[59]   Kamani H, Bazrafshan E, Ghozikali M G, Askari M. Photocatalyst decolorization of C . I . Sulphur red 14 from aqueous solution by uv irradiation in the presence of zno nanopowder photocatalyst decolorization of C . I . Sulphur Red 14 From Aqueous Solution by UV Irradiation in the Presence of ZnO Nanopowder 2015.

[60]   Gogate PR, Pandit AB. A review of imperative technologies for wastewater treatment I: Oxidation technologies at ambient conditions. Adv Environ Res 2004; 8(3–4): 501-51.
[http://dx.doi.org/10.1016/S1093-0191(03)00032-7]

[61]   Clarizia L, Russo D, Di Somma I, Marotta R, Andreozzi R. Homogeneous photo-Fenton processes at near neutral pH: A review. Appl Catal B 2017; 209: 358-71.
[http://dx.doi.org/10.1016/j.apcatb.2017.03.011]

[62]   Hislop KA, Bolton JR. The photochemical generation of hydroxyl radicals in the UV-vis/ferrioxalate/$H_2O_2$ system. Environ Sci Technol 1999; 33(18): 3119-26.
[http://dx.doi.org/10.1021/es9810134]

[63]   Kim SM, Geissen SU, Vogelpohl A. Landfill leachate treatment by a photoassisted Fenton reaction Water Sci Technol 1997; 35: 239-48.
[http://dx.doi.org/10.2166/wst.1997.0128]

[64]   Chiou C-H, Wu C-Y, Juang R-S. Influence of operating parameters on photocatalytic degradation of phenol in UV/$TiO_2$ process. Chem Eng J 2008; 139(2): 322-9.
[http://dx.doi.org/10.1016/j.cej.2007.08.002]

[65]   Bigda RJ. Consider fenton's chemistry for wastewater treatment. Chem Eng Prog 1995; 89: 62-6.

[66]   Villegas LGC, Mashhadi N, Chen M, Mukherjee D, Taylor KE, Biswas N. A short review of techniques for phenol removal from wastewater. Curr Pollut Rep 2016; 157-67.
[http://dx.doi.org/10.1007/s40726-016-0035-3]

[67]   Chen SS, Nguyen NC, Chen YM, Li CW. Coagulation enhancement of nonylphenol ethoxylate by partial oxidation using zero-valent iron/hydrogen peroxide. Desalin. Water Treat 2013.
[http://dx.doi.org/10.1080/19443994.2012.699232]

[68]   Zanjanchi MA, Ebrahimian A, Arvand M. Sulphonated cobalt phthalocyanine-MCM-41: An active photocatalyst for degradation of 2,4-dichlorophenol. J Hazard Mater 2010; 175(1-3): 992-1000.
[http://dx.doi.org/10.1016/j.jhazmat.2009.10.108] [PMID: 19939562]

[69]   Hoffmann MR, Martin ST, Choi W, Bahnemannt DW. Environmental Applications of Semiconductor Photocatalysis. 1995; pp. 69-96.

[70]   Gaya UI, Abdullah AH. Heterogeneous photocatalytic degradation of organic contaminants over titanium dioxide: A review of fundamentals, progress and problems. J Photochem Photobiol Photochem Rev 2008; 9(1): 1-12.
[http://dx.doi.org/10.1016/j.jphotochemrev.2007.12.003]

[71]   Pichat P. Photocatalysis and water purification: From fundamentals to recent applications. 2013.
[http://dx.doi.org/10.1002/9783527645404]

[72]   Fox MA, Dulay MT. Heterogeneous photocatalysis Chem Rev 1993; 93: 341-57.

[73]   Linsebigler AL, Linsebigler AL, Yates JT Jr, Lu G, Lu G, Yates JT. Photocatalysis on $TiO_2$ Surfaces: Principles, mechanisms, and Selected results. Chem Rev 1995; 95(3): 735-58.
[http://dx.doi.org/10.1021/cr00035a013]

[74]   Safarzadeh-Amiri A, Bolton JR, Cater SR. Ferrioxalate-mediated solar degradation of organic contaminants in water. Sol Energy 1996; 56(5): 439-43.
[http://dx.doi.org/10.1016/0038-092X(96)00002-3]

[75]   Safarzadeh-Amiri A, Bolton JR, Cater SR. Ferrioxalate-mediated photodegradation of organic pollutants in contaminated water. Water Res 1997; 31(4): 787-98.
[http://dx.doi.org/10.1016/S0043-1354(96)00373-9]

[76]   Tang WZ, Huang CP. The effect of chlorine position of chlorinated phenols on their dechlorination kinetics by Fenton's reagent. Waste Manag 1995; 15(8): 615-22.
[http://dx.doi.org/10.1016/0956-053X(96)00022-0]

[77]   Czaplicka M. Photo-degradation of chlorophenols in the aqueous solution. J Hazard Mater 2006; 134(1-3): 45-59.
[http://dx.doi.org/10.1016/j.jhazmat.2005.10.039] [PMID: 16325999]

[78]   Song-hu Y, Xiao-hua L. Comparison treatment of various chlorophenols by electro-Fenton method: Relationship between chlorine content and degradation. J Hazard Mater 2005; 118(1-3): 85-92.
[http://dx.doi.org/10.1016/j.jhazmat.2004.08.025] [PMID: 15721532]

CHAPTER 4

# Analysis of Seasonal and Spatial Variations of Water Quality of Dulhara and Ved Ponds in Ratanpur, Chhattisgarh, India

**Renu Nayar[1,*], G.D. Sharma[2], Ritesh Kohale[3] and S.J. Dhoble[4]**

[1] *Department of Chemistry, D.P. Vipra College, Bilaspur, C.G., India-495001*

[2] *Atal Bihari Vajpayee University, Bilaspur, India-495 001*

[3] *Department of Physics, Sant Gadge Maharaj Mahavidyalaya, Hingna, Nagpur, India, 441110*

[4] *Department of Physics, R.T.M. Nagpur University, Nagpur, India-440033*

**Abstract:** Pollution load, rising population, and scarcity of water have drawn special attention for the management of water resources such as pond water. The present investigation was carried out at Dulahra and Ved ponds in Ratanpur, Bilaspur District. The seasonal deviations in water such as transparency, temperature, pH, dissolved oxygen, nitrate, phosphate, biological oxygen demand (BOD), total dissolved solids (TDS), and total suspended solids (TSS) were evaluated. In the contemporary study, the BOD standards were considerably higher than World Health Organization (WHO) standards (5 mg/l). The water samples were collected from each site at outer (about 100-150 meters) and internal (10 m from the shoreline) localities. The highest mean value of BOD, *i.e.*, 32±4.6, was found at the north peripheral S-1 in the summertime. In the summer season, maximum mean BOD 39±2.1 was found at S-2 (West peripheral) in Ved Pond. It indicates the biological pollution load on the water body in the site of North peripheral in Dulhara pond and West peripheral in Ved Pond in the summer season. Low Secchi depth readings such as 20±1.0 at S-2 North peripheral site during summer seasons are indicative of reduced water clarity that is habitually related to the existence of suspended particles and algal tinges. We also found the maximum value of total suspended solids on the north side of the pond, where the transparency of water was also very low. A transparency value of 37.0±0.40 was noted at S-1 in the East marginal sites and 30±0.22 at the East inner sites in the rainy period at the Ved pond. The transparency of the water physique is exaggerated by the elements like planktonic growth, rainfall, the sun's location in the sky, the angle of incidence of rays, cloudiness, electiveness, and turbidity due to deferred inert particulate material. Our outcomes suggest that the lowermost water transparency value was 16.0 ± 0.41 at S-1 in the North marginal sites for the duration of summer. The concentration of Calcium ions was much above the WHO recommended value of 75 mg/l at almost all the sites and both ponds during the study period. Numerous indicators and catalogues have been

---

\* **Corresponding author Renu Nayar:** Department of Chemistry, D.P. Vipra College, Bilaspur, C.G., India-495001; Tel: +91-9977413760; E-mail: nayar.renu@yahoo.co.in

**R. M. Belekar, Renu Nayar, Pratibha Agrawal and S. J. Dhoble (Eds)**
**All rights reserved-© 2022 Bentham Science Publishers**

established in this particular study to evaluate water quality in intermediate water bodies. For transparency studies, low-cost Secchi disk was used.

**Keywords:** Dulahra tank, East marginal sites, Monthly variation, Physico-chemical parameters, Pond water, Seasonal variation, Secchi disk, Water bodies, Water quality parameters.

## INTRODUCTION

Domestic inlets, such as aquatic ecological units where entire water assembles with fresh water from significant sources, endure a resemblance to superficial ponds as they regularly proceed as shallow water bodies disjointed from the enormous water pools by an obstruction. Their physicochemical appearances and irregularity of the entering mutability differ significantly between annual progressions. During the earlier epochs, the formation of human societies and metropolitan growth on the earth's surface had straightforward influences on these ecological units. The modification of ecological systems, as well as the endless accumulative nutrient releases from agronomic and industrialized undertakings, have commanded the ecological regions in the surrounding. Water superiority may be influenced by the resident geology and environment, as well as human usages like sewage dispersal, industrialized contamination, and use of water bodies as a cleansing, for cloths, animals and motor vehicles cleansing, *etc*. Water is not only essential to life but is also essential for industrial processes. In industries, it is used in steam generation. Water is used in chemical plants for cooling purposes. A considerable amount of water is also used for solution, and dilution purposes in chemical plants [1]. The majority of people in developing countries do not have access to clean water or any other form of sanitation. Therefore, millions of people are suffering from diseases related to water, sanitation, and hygiene, such as diarrhea, skin diseases, and trachoma [2]. Man uses water for the production of steel, rayon, papers, and textiles, but also for air conditioning and irrigation purposes [3]. Water is therefore important for everything in our sphere to cultivate and grow. Water that is free of disease-generating micro-organisms and biochemical constituents poisonous to well-being is called potable water. Water polluted with either internal or industrialized garbage is called non-potable or polluted water [4]. As we all know, natural life was initially originated in water. The worldwide water shortage is increasing, and as such, it is essential to yield new water bases that we may previously have thought of as unapproachable. Water is one of the most significant elements of the existing creatures in the world [5]. Water is a widespread and outstanding solvent as it liquefies in a wide variety of materials than other solvents [6]. So, no other composite can be associated with water as a solvent. Water has numerous exclusive and worldwide physiognomies that make it appropriate for living beings to endure dissimilar circumstances for life. Water excellence eminence is a solitary term that means lending a hand to the assortment of suitable management procedures to encounter varied problems [7]. The hydrosphere covers more than 75% of the earth's surface as freshwater. Seawater, rivers, oceans, lakes, and pond water form a hydrosphere that encompasses more than 35 ppt of liquefied solids. Normal water always contains liquefied and suspended substances of biological and inorganic mineral. Most pulverized water surrounds dissolved materials while superficial water is rich in the adjourned matter [8].

The rapid growth of urban areas directly or indirectly affected the existence of the ponds, such as over-exploitation of resources and improper waste disposal practices. The objective of primary concern is to provide potable water. Although humans tend to protect themselves from harmful micro-organisms and undesirable or harmful chemicals, they pollute our rivers, lakes, and oceans [22]. Regardless of advancements in drilling, irrigation, and decontamination, the position, excellence, magnitude, possession, and mechanism of drinkable water residues remains a major anthropoid concerns (Cunningham, W. P.). Ratanpur is one of the famous holy places that is situated around 25 km from Bilaspur –Korba main road, Chhattisgarh, India, as indicated in Fig. (**1a**). It is a well-known fact that Ratanpur is in Bilaspur district in the Indian state of Chhattisgarh, situated at 22.30N and 82.170E, with an average rainfall varying from 135mm to 445mm and a humidity of 34 percent. The town is popular as a religious center, and many Hindu devotees come here to offer their prayers. Perhaps due to various temples of different eras/periods and other religious purposes, so many small and large ponds/ tanks are available. That is why Ratanpur is popularly known as 'Temple City' as well as 'City of Ponds'. Retrospectively, temples correlate with more than a hundred small and large ponds. As quoted in the history, there were more than 150 ponds during the early days. Of these ponds, Dulhara, Bikma, Ved, Ratneshwar, and Krishnajuni are large ponds.

## Objectives of the Present Work

The physico-chemical investigation of pond water in regions of India where widespread human activities are taking place is extremely essential [12]. The pollution of pond water by inorganic chemicals, organics, and micro-organisms evidently takes place since inappropriate drainage schemes, septic tanks, and solid waste disposal are still being utilized. A literature review reveals that no organized study has been statistically analyzed concerning the excellence of surface water in Ratanpur and its nearby rustic expanses. Hence, contemporary work was carried out to examine the quality of water in Ratanpur and its surrounding areas, where some major ponds are located and used by people in the form of different activities [13]. To uncover the specimen locations throughout the prevailing analysis, the all-inclusive zone of Ratanpur in Bilaspur District has been divided into four regions (A, B, C, and D), in Ratanpur, which comprise major and minor ponds, as shown in Fig. (1b) and mentioned in Tables 1 and 2. Dulahara is one of the largest ponds that is situated in Zone A. Zone-B entails two huge ponds, termed Ved and Ratneshwer. Nevertheless, for the current exploration, we have selected the major pond Ved in this zone. The dominant anthropological actions take place at the boundaries of the Ved pond. Due to incorrect drainage systems, septic tanks, and solid waste disposal, pollution of groundwater by organic chemicals and micro-organisms is expected to transpire. Likewise, pollution due to inorganic chemicals from leakage of overflows from industries is also promising. The pond is anthropogenic, and its water is used for domestic purposes and irrigation. The pond is encircled by semi-urban and semi-agricultural growing areas. Conversely, for the present exploration, we have preferred the major ponds of Dulhara in zone A and Ved pond in zone B.

**Fig. 1(a).** Map showing the location of Ratanpur in Chhattisgarh Map.

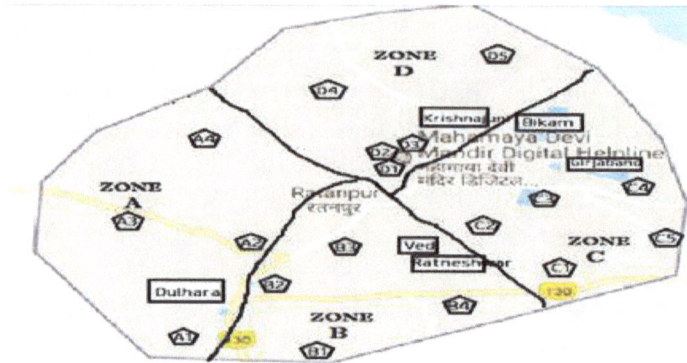

**Fig. 1(b).** Map of Ratanpur showing study zones and Samples sites and water samples collection in four directions.

**Table 1. Dulahra Pond and selected sites.**

| Ponds Name | Directional Area (m²) | Peripheral Sites | Inner Sites |
|---|---|---|---|
| Dulhara | NORTH (600) | S-1,S-2,S-3 | S-A, S-B |
| | EAST(200) | S-1,S-2 | S-A, S-B |
| | WEST(400) | S-1, S-2 | S-A, S-B |
| | SOUTH (1100) | S-1,S-2, S-3, S-4 | S-A, S-B |

**Table 2. Ved Pond and selected sites.**

| Pond Water | Peripheral Length | | Sampling Sites | | |
|---|---|---|---|---|---|
| | Direction | Length (Appr. In meter) | Marginal sites | INSIDE | |
| | | | | Depth (in feet) | Inner site |
| Ved | North | 430 | S-1, S-2 | 05 | S-A, S-B |
| | East | 370 | S-1, S-2 | 06 | S-A, S-B |
| | South | 380 | S-1, S-2 | 05 | S-A, S-B |
| | West | 270 | S-1, S-2 | 05 | S-A, S-B |

## Description of Study Sites

The present work is based on water samples collected seasonally. Study locations were carefully chosen on the basis of various aspects of significance for community advantage, directly or indirectly. Zone-A, the zone of the study area, comprises several ponds, including a very large pond, popularly known as Dulhara. It is one of the largest pond or wetlands (180-acre land area), having vast historical importance linked with a deep sense of the religious sentiment of regional people. Its name also has a special meaning because Sage Mahatma observed two types of waves and advocated for the union of the Ganga and Yamuna (two major Indian rivers). During the ancient period, this pond was significant to the local people for economic, social, and religious reasons. This pond is still important today in almost every aspect of society, providing irrigation water, facilities for domestic and daily work, and economic aid through a large production of fish, lotus stems (dhens), lotus leaves, lotus flowers, motha grass, and so on. This pond provides irrigation water for domestic and daily work, and economic aid through its large production of fishes, lotus stems, lotus leaves, lotus flowers, mothagrass, *etc*. Fig. (**1c**) indicates the effect of deadly water contamination on fish in ponds.

**Fig. (1c).** Dead fishes due to water contamination in ponds.

A number of random sample places (Fig. **1**) were identified everywhere around this pond. Water examples were collected every month for a period of one year [14]. The proposed investigation will enable us to assess the pollution status of the water of the pond and the suitability of water of these ponds for various domestic purposes, including drinking. Consistent observations of these ponds will undoubtedly aid in capturing long-standing safeguards, which will be beneficial in protecting human health in the study area. In the contemporary study, the physicochemical analysis of Dulahra pond water testing was done in the rainy, winter, and summer seasons, and the mean standards of the different samples were matched with the World Health Organization (WHO) Guideline Values (GVs) for intake water eminence [15]. There is considerable accessible work on the physico-chemical excellence of pond water in the countryside. On the other hand, there are still a few zones that should be targeted to regulate surface water superiority where different human activities are captivating. Due to inappropriate drainage systems, septic tanks, and solid waste disposal, contamination of pond water by organic, inorganic chemicals, and micro-organisms is likely to occur. Hence, it is planned to explore the quality of water in Ratanpur in Chhattisgarh and its nearby areas where some major ponds are used by people for different activities such as laundry solutions, washing plates, bathing animals, *etc.* Regular observation of ponds will help in capturing long-term protections that will be accommodating in defensive applications of public health in the study area. In the contemporary study, the physicochemical investigation of Ved pond water examples was done in the rainy, winter, and summertime, and the meaningful standards of the diverse measured examples were matched with the World Health Organization (WHO) Guideline Standards (GVs) for drinking water superiority.

## Method and Material for Experimental Analysis

Eight exterior sites and eight interior locations covering all the directions were selected for the physicochemical examination of the water in this pond. Eight specimen sites were chosen, individually 100 m apart. These places are named as peripheral or marginal sites in all directions. Approximately 100m from the coastline, eight places were preferred in the middle/inner of the Dulhara and Ved pond, as shown in Table **1(a)** and **(b)**.

## Methodology

### Sampling Sites

To collect water samples from all study sites, specific sampling sites have been determined at the margin and inside the ponds. For large ponds (Dulhara pond), two, three, and four sampling sites at a 100-meter distance along four directions in the peripheral/marginal area have been located; simultaneously, 100 meters inside the margin, two sampling sites at the surface and bottom region of water have been located. Sampling sites and a picture of four directions in Dulahra pond and Ved pond are shown in Figs. (**2 - 4a** and **b**).

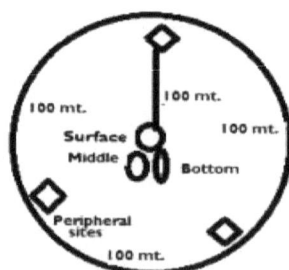

**Fig. (2).** Sampling site at margin and inside the water depth.

**Fig. (3).** Locations in different directions of Dulhara pond.

**Fig. (4). (a)** Ved pond [south] **(b)** Ved pond [north].

## Collection of Water Samples in Different Directions

There are 3 distinct seasons namely winter {November –February}, summer-{March-June}, and rainy {July-September}. In the contemporary effort, we selected 11 sampling pockets at outlying locations in the zone, *i.e.*, 3 sites in the North direction, 4 sites in the South direction, and 2 sites each in the West and East, and 8 sampling sites on the inner sites {center} in the Dulhara pond (Table 1a).

In the Ved pond, we chose, eight sampling sites, each 100 m apart. These sites are termed peripheral/marginal sites in all directions. Nearly 100m from the shore, 08 sites were chosen in the middle and inner part of the pond (Table **1b**). The samples collected were analyzed for 13 physico-chemical parameters to appraise their appropriateness for household solicitation. Numerous constraints touch the water superiority, such as temperature, pH, electrical conductivity, total liquefied solids, dissolved oxygen, organic oxygen requests, clearness, alkalinity, nitrate, phosphate, calcium, and magnesium. Constraints such as pH, temperature, EC, TDS, and DO were examined on the spot by means of the Water Quality Analyzer Kit (Systronic model no. 371). Spectrophotometric work was accomplished using a Carl-Zeiss Spekol-10 spectrophotometer for the examination of nitrate, and phosphate [10, 12]. Determination of other constraints was carried out using standard approaches as recommended by APHA, Manivaskam, and the NEERI guidebook on water and wastewater examination [16, 17]. The chemicals employed in the investigational effort were of analytical grade. Double purified water from all glass distillation will be used for the preparation of all the reagents/solutions. Water was experimented within 1 L polyethylene vessels. Samples of surface water were collected from the outer extents, internal surface, and extremities of the pond just once a month for the period of August 2018 to June 2019. All water samplings were conducted between 10 and 3 o'clock.

For chemical analysis, water samples were moved from the laboratory into the icebox, a shielded vessel, and additional samples were headed within 24 hours. The samples were preserved in a freezer below 5°C temperature that was used for advanced investigation. Specimen sites and the representation of four directions in Ved pond are shown in Figs. (**4a** and **4b**). The approaches implemented for examining the various physicochemical parameters are listed in Table **3**. Temperature, pH, electrical conductivity, and dissolved oxygen were determined at the sampling site itself using Elico's portable soil and water analysis kit (Mode No. PE-136).

**Table 3. Methods used for determination of physic-chemical parameters.**

| S. No. | Parameters | Abbreviations | Units | Method Used |
|---|---|---|---|---|
| 1 | Temperature | Temp. | °C | Fish finder with temp and depth and water analyzer kit modelno 371 |
| 2 | pH | pH | | Water analyser kit (Systronic) Model No. 371 |
| 3 | Electrical Conductivity | EC | mho/cm | Water analyser kit (Systronic) Model No. 371 |
| 4 | Total dissolved solid | TDS | mg/l | Water analyser kit (Systronic) Model No. 371 |
| 5 | Dissolved oxygen | DO | mg/l | Water analyser kit (Systronic) Model No. 371 |
| 6 | Biological oxygen demand | BOD | mg/l | 5days incubation at 20°C and titration of initial and final DO |
| 7 | Transparency | Trans. | cm | By Secchi disc |
| 8 | Alkalinity | Alk. | mg/l | Titrimetric method (with 0.1 N $H_2SO_4$) |
| 9 | Nitrate | NO3- | mg/l | Spectrophotometer |
| 10 | Phosphate | PO4- | mg/l | Spectrophotometer |
| 11 | Calcium ions | $Ca^{++}$ | mg/l | EDTA 0.05 N (Tirimetric method) |

(Table 3) cont.....

| 12 | Magnesium ions | $Mg^{++}$ | mg/l | EDTA 0.05 N (Tirimetric method) |
| 13 | Total suspended solid | TSS | mg/l | Gravimetric methods |

## RESULTS AND DISCUSSION

### Appearance of Dulahraand Ved Pond

Water is more transparent in the south direction compared to the other directions. High turbidity appears in the north and west directions. The color of the water is yellow-brown and has the smell of rotten eggs in the north direction. The Dulahra pond is a non–aerated pond that stratifies in the summer and winter, locking the bottom layer of water for a month. With no circulation, the oxygen is quickly used up, resulting in an anaerobic digestion process. Slow-moving anaerobic bacteria on the bottom use enzymes to ferment and digest muck on the bottom. These bacteria produce waste products, including carbon dioxide and hydrogen sulfide. This buildup of gas stays locked away on the bottom. It is when the weather changes that the foul-smelling gases are released. No algae was observed in the water pond. The water appeared clear and less turbid from the surface to the bottom in the south direction.

The water of the Ved pond looks chaotic and greenish in color. The color of the water is yellow-green in all four directions, such as north, east, west, and south, and it has the odour of awful eggs, possible in the north and south directions. A green coloration was floating on the exterior in stationary and shallow expanses, but the coloration was undetectable underneath the depth. Compact resident of fish are known to be present. Sorts such as Rohu, Katla, Mirgal Lali, and Pathri fishes were found to be present in both ponds. Numerous wandering avian kinds along with inhabited species emphasize the assortment of the pond.

### Transparency

The first measurement done was determining the depth of light perception in the body of water using the Secchi disc. The low-cost disc was self-fabricated and aimed (Fig. 5) as its quoted price was extraordinary. The Secchi disc depth gives a bumpy estimate of clarity of the water. A Secchi disc contains a spherical white plate made of any non-corrosive inelastic material and has a diameter of $30 \pm 1$ cm. To decrease the special effects of currents on the angle of view, a mass of $3.0 \pm 0.5$ kg is suspended below the center of the disc on a rigid rod 15 cm long.

**Fig. (5).** Secchi Disk for checking the transparency of water.

The disc is painted with quadrants in flat black and flat white waterproof paint. The disc is normally attached to a non-stretch rope, which has been marked at appropriate intervals of depth with waterproof markings. As the waters to be measured were of variable clarity, the assessment was made as to the scale of measurement to be used. In turbid waters, markings at 10 cm intervals would be appropriate, whereas in clearer waters, markings at 50 cm intervals would be adequate. Generally, where the disc cannot be seen (disappearance of the black and white quadrants) is where effective light penetration is extinguished. A Secchi disc depth is a measure of the limit of vertical visibility in the upper water column and is therefore a direct function of water clarity. High Secchi depth readings of 43±2.0 at the S-3 south peripheral site during the rainy season correspond to high water clarity [20]. Conversely, low Secchi depth readings such as 20±1.0 at S-2 north peripheral site (Fig. **6**) during the summer season are indicative of reduced water clarity that is often associated with the presence of suspended particles and algal blooms. Low Secchi transparency measurements indicates limited light penetration and limited primary production.

The transparency of the Ved pond measurement was done by measuring the depth of light penetration into the body of water using the Secchi disc. A higher transparency value was observed during the summer due to the absence of rain, runoff, and floodwater as well as the gradual settling of suspended particles [20]. The maximum transparency value of 37.0±0.40 was recorded at S-1 in the East marginal sites and 30±0.22 at the East inner sites in the rainy season. The transparency of the water body is affected by the factors like planktonic growth, rainfall, the sun's position in the sky, the angle of incidence of rays, cloudiness, visibility, and turbidity due to suspended inert particulate matter. Our results revealed that the lowest water transparency value was recorded at 16.0±0.41 at S-1 in the North marginal sites during the summer season (Fig. **7**).

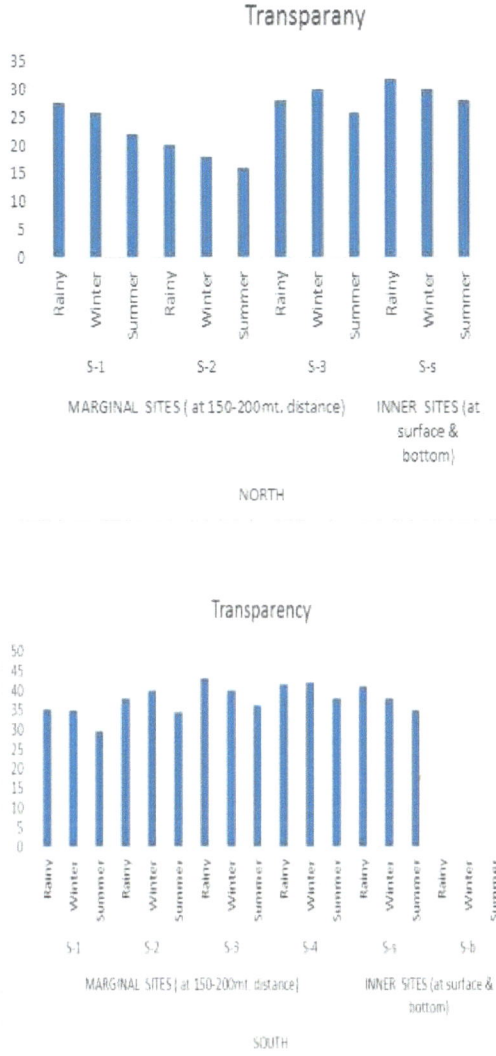

Fig. (6). Transparency of water in north and south peripheral site in Dulharapond.

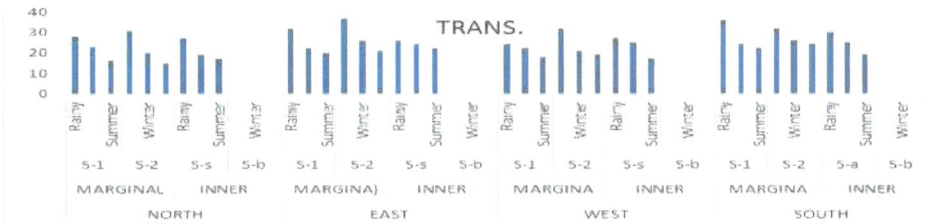

**Fig. (7).** Analysis of transparency in different directions of Vedpond.

The variation in physicochemical parameters of Dulhara and Ved pond water of Ratanpur were shown in Table **4** and **5**.

**Table 4. Physico-chemical condition and water quality of Dulahra pond.**

| Samplin Sites & Samples | | | Season | pH | Temp. | TDS | Cond | ALK | TSS | DO | BOD | Ca²⁺ | Mg²⁺ | NO3⁻ | PO4⁻ | Trans. |
|---|---|---|---|---|---|---|---|---|---|---|---|---|---|---|---|---|
| North | MARGINAL SITES (at 150-200mt. distance) | S-1 | Rainy | 7.6±0.12 | 30.9±2.6 | 107.1±4.0 | 257.0±4.1 | 566.8±5.7 | 29.75±5.4 | 1.4±0.10 | 11±2.6 | 34.1±3.9 | 11.2±2.1 | 34.5±1.6 | 23.8±1.2 | 27.6±1.0 |
| | | | Winter | 7.6±0.13 | 24.7±1.5 | 110.5±2.5 | 199.0±6.7 | 667.7±3.8 | 37.2±3.7 | 2.5±0.22 | 29±6.5 | 46.4±8.4 | 18.7±3.1 | 32.7±6.1 | 22.1±3.0 | 26±3.0 |
| | | | Summer | 7.8±0.18 | 39.6±2.2 | 232.0±6.2 | 266.0±3.8 | 618.2±2.8 | 44.1±4.6 | 2.2±0.44 | 32±4.6 | 66.2±4.2 | 28.4±2.8 | 38.2±3.8 | 24.6±2.2 | 22±1.1 |
| | | S-2 | Rainy | 7.6±0.07 | 30.4+2.8 | 106.5±12.7 | 260.5±5.1 | 760.8+4.6 | 22.8±1.04 | 1.2±0.25 | 11±2.8 | 92.5±4.2 | 12.9±1.3 | 36.8±3.3 | 18.2+1.5 | 20±1.0 |
| | | | Winter | 7.8±0.09 | 25.9±4.5 | 112.5±10.8 | 224.0±8.3 | 751.7±4.3 | 43.5±2.1 | 2.8±1.12 | 28±3.3 | 109.2±9.1 | 17.9±3.5 | 33.6±3.7 | 31.1±1.8 | 18±1.3 |
| | | | Summer | 8.2±0.12 | 40.1±2.6 | 117.1±9.6 | 268.0±6.4 | 766.4±3.8 | 48.2±1.8 | 3.1±0.66 | 31±3.8 | 115.4±4.8 | 26.6±2.8 | 38.4±3.2 | 33.4±1.6 | 16±1.1 |
| | | S-3 | Rainy | 7.4±0.16 | 30.6±2.3 | 104.1±1.6 | 276.0±5.4 | 583.5±2.9 | 20.5±4.4 | 1.4±0.24 | 12±2.4 | 43.8±2.2 | 18.5+1.2 | 33.5±1.9 | 17.6±2.4 | 28±1.0 |
| | | | Winter | 7.9±0.48 | 25.5±3.6 | 90.1±3.0 | 234.0±8.9 | 625.0± 9.9 | 33.2±3.5 | 2.5±0.72 | 24±6.6 | 47.6±2.8 | 34.6±1.4 | 32.4±2.8 | 31.5±2.2 | 30±1.1 |
| | | | Summer | 8.2±0.32 | 37.4±2.1 | 112.6±2.4 | 255.0±6.8 | 592.0±4.8 | 38.2±2.6 | 2.8±0.54 | 22±4.4 | 64.1±2.6 | 28.2±1.2 | 33.6±1.6 | 32.2±2.8 | 26±1.4 |
| | INNER SITES (at surface & bottom) | S-s | Rainy | 7.6±0.10 | 31.8±2.8 | 85.12±3.4 | 189.3±4.5 | 386.8±6.4 | 17.0±3.6 | 1.1±0.19 | 11+1.2 | 45.5±1.8 | 16.0±6.6 | 30.8±5.5 | 14.7±1.2 | 32±2.1 |
| | | | Winter | 7.7±0.19 | 24.5±3.6 | 100.2±4.0 | 139.7±5.4 | 368.5±7.7 | 36.2±3.8 | 2.05±0.89 | 25±6.0 | 59.4±4.2 | 20.4±4.0 | 30.0±3.5 | 17.3±2.1 | 30±2.0 |
| | | | Summer | 7.5±0.22 | 36.5±2.8 | 123.5±5.2 | 143.8±4.3 | 360.2±6.8 | 34.8±2.8 | 1.8±0.42 | 22±3.1 | 66.2±3.6 | 24.8±3.2 | 32.2±3.3 | 18.4±2.4 | 28±1.1 |
| | | S-b | Rainy | 7.7±0.18 | 29.2±1.6 | 86.2±4.2 | 149.5±2.1 | 256.0±3.1 | 23.8+3.7 | 1.1±0.28 | 14+2.2 | 50.4±2.2 | 16.9±1.4 | 37.8±4.4 | 16.8±1.5 | - |
| | | | Winter | 7.8±0.22 | 23.1±3.6 | 96.1±6.4 | 146.5±3.9 | 199.5±2.2 | 33.5±2.8 | 1.9±0.97 | 18±3.6 | 62.05±4.8 | 20.5±3.5 | 36.4±1.7 | 19.4±1.6 | - |
| | | | Summer | 7.6±0.16 | 28.2±2.8 | 117.4±4.8 | 153.9±2.2 | 190.2±2.5 | 31.7±1.8 | 2.3±0.37 | 22±2.8 | 77.8±2.7 | 22.4±3.2 | 36.8±1.5 | 17.6±1.4 | - |
| EAST | MARGINAL SITES ( at 150-200mt. distance) | S-1 | Rainy | 7.6±0.03 | 33.2±1.2 | 85.7±5.7 | 166.4±4.5 | 280.5±6.6 | 15.5±1.6 | 1.2±0.08 | 11±2.6 | 133.2±2.1 | 11.8±1.5 | 37.9±1.8 | 12.8±1.9 | 35.3±1.1 |
| | | | Winter | 7.9±0.24 | 26.3±3.6 | 80.8±3.8 | 168.2±8.7 | 494.2±7.6 | 18.4±1.5 | 2.4±0.54 | 20±4.7 | 171.7±7.7 | 15.9±1.2 | 43.2±2.5 | 9.7±1.1 | 35±1.4 |
| | | | Summer | 8.2±0.33 | 38.4±2.1 | 96.4±2.8 | 209.1±5.2 | 444.2±6.2 | 22.8±1.1 | 2.2±0.33 | 18±3.8 | 177.1±3.2 | 21.2±1.6 | 44.2±2.2 | 10.1±1.4 | 30±1.0 |
| | | S-2 | Rainy | 7.7±0.05 | 33.7±2.3 | 83.75±1.7 | 162.5±9.6 | 287.0±3.24 | 16.8 ± 3.7 | 1.1±0.12 | 13±2.6 | 129±2.7 | 13.2±2.4 | 38.5±3.5 | 12.4±1.6 | 39.6±1.1 |
| | | | Winter | 8.1±0.04 | 25.5±3.9 | 87.6±3.5 | 151.2±4.5 | 426.0±8.4 | 18.2±2.8 | 2.4±0.79 | 20±6.4 | 85.8±9.9 | 15.2±2.1 | 43.9±4.8 | 13.1±2.1 | 40±2.1 |
| | | | Summer | 8.4±0.12 | 36.3±2.5 | 91.6±2.2 | 166.8±3.8 | 402.0±6.2 | 22.4±1.8 | 2.1±0.61 | 26±3.1 | 138.3±6.2 | 16.4±2.4 | 45.5±2.4 | 15.3±1.7 | 33±1.2 |
| | INNER SITES (at surface & bottom) | S-s | Rainy | 7.7±0.04 | 34.7±2.8 | 76.7±2.5 | 169.8±5.1 | 274.8±7.9 | 29.5±2.5 | 1.4±0.10 | 12±2.0 | 126.4±5,2 | 11.8±2.6 | 34.7±3.6 | 14.2±2.9 | 42±2.2 |
| | | | Winter | 7.7±0.16 | 25.8±4.2 | 86.7±6.8 | 172.7±4.2 | 350.5±9.6 | 21.5±3.5 | 2.1±0.56 | 25±5.3 | 83.6± 9.4 | 16.6±2.3 | 39.2±1.2 | 10.3±1.9 | 41±2.0 |
| | | | Summer | 7.9±0.18 | 33.7±3.6 | 118±3.1 | 181.3±2.5 | 341.7±5.8 | 24.4±2.8 | 1.8±0.44 | 27±4.2 | 141.2±7.2 | 20.4±3.1 | 44.1±3.5 | 14.9±1.3 | 40±2.1 |
| | | S-b | Rainy | 7.6±0.04 | 30.4±3.1 | 83.1±1.3 | 188.5±2.9 | 313.0±6.2 | 24.5±2.5 | 1.2±0.20 | 19±.86 | 116.2±5.04 | 14.7±2.6 | 37.5±2.0 | 16.1±1.6 | - |
| | | | Winter | 7.2±0.12 | 23.7±4.9 | 85.9±2.0 | 166.0±5.5 | 349.2±9.4 | 22.5±2.5 | 1.6±0.51 | 23±7.3 | 83.6±9.7 | 18.5±1.9 | 42.3±1.2 | 13.2±1.1 | - |
| | | | Summer | 7.8±0.31 | 31.2±2.2 | 108.1±2.6 | 194.0±4.2 | 333.6±8.2 | 22.8±2.8 | 2.8±0.26 | 24±5.7 | 122.2±3.2 | 18.9±1.2 | 47.2±2.8 | 21.8±2.4 | - |
| WEST | MARGINAL SITES ( at 150-200mt. distance) | S-1 | Rainy | 7.5±0.19 | 23.5±2.6 | 115.6±5.3 | 187.0±6.4 | 276.3±3.7 | 22.3±7.0 | 1.4±0.20 | 12±1.6 | 112.8±3.8 | 14.4±1.6 | 34.5±3.9 | 16.9±1.7 | 35.6±1.1 |
| | | | Winter | 7.7±0.18 | 24.4±3.5 | 110.7±2.3 | 192.5±4.2 | 313.0±9.0 | 14.7±1.9 | 2.2±0.53 | 29±4.5 | 92.2±15.1 | 15.2±2.1 | 39.1±2.0 | 18.6±1.5 | 34±1.4 |
| | | | Summer | 7.4±0.14 | 34.9±2.4 | 122.8±2.2 | 218.6±2.8 | 298.6±4.3 | 22.8±2.2 | 2.4±0.42 | 32±3.8 | 128.3±5.4 | 18.1±1.4 | 42.6±3.2 | 20.9±1.8 | 32±1.2 |
| | | S-2 | Rainy | 7.5±0.12 | 30.2±2.4 | 120.2±1.4 | 185.0±2.2 | 279.0±5.9 | 20.2±5.0 | 1.3±0.08 | 11±1.6 | 117.6±4.4 | 11.2+.91 | 35.8±2.1 | 17.3±1.9 | 35.6±1.1 |
| | | | Winter | 7.9±0.28 | 25.4±3.5 | 116.3±9.1 | 183.0±3.3 | 336.7±6.6 | 14.0±2.5 | 1.9±0.75 | 23±8.3 | 76.8±12.3 | 16.6±3.0 | 41.4±1.6 | 19.5±2.7 | 35±1.2 |
| | | | Summer | 8.4±0.22 | 35.4±2.8 | 143.8±3.3 | 222.4±2.4 | 330.4±8.5 | 21.4±1.8 | 2.2±0.44 | 25±4.4 | 131.8±8.4 | 20.8±3.1 | 48.7±3.2 | 24.3±2.1 | 30±1.4 |
| | INNER SITES (at surface & bottom) | S-s | Rainy | 7.9±0.04 | 30.3±2.1 | 120.5±2.6 | 178.8±3.2 | 311.5±1.6 | 23.5±6.6 | 1.2±0.07 | 11±1.9 | 109.8±3.6 | 12.8±2.1 | 36.8±2.9 | 19.8±1.9 | 43.3±1.6 |
| | | | Winter | 7.8±0.21 | 26.3±4.3 | 114.5±7.5 | 183.7±7.1 | 345.7±8.0 | 19.7±5.2 | 2.0±0.91 | 24±8.8 | 95.8±9.4 | 17.8±1.4 | 39.8±1.9 | 14.9±1.7 | 40±1.4 |
| | | | Summer | 8.1±0.11 | 34.8±3.2 | 166.4±2.3 | 208.4±3.8 | 341.7±6.4 | 23.8±3.6 | 2.1±0.88 | 26±6.2 | 114.1±4.8 | 22.3±2.3 | 49.6±3.6 | 21.6±2.3 | 36±1.4 |
| | | S-b | Rainy | 7.5±0.19 | 29.5±1.4 | 119.5±3.6 | 184±2.44 | 318.3±4.8 | 24.2±4.2 | 1.5±0.18 | 10±1.4 | 112.8±2.3 | 16.7±1.9 | 37.6±4.4 | 18.5±2.5 | - |
| | | | Winter | 7.1±0.21 | 24.6±2.2 | 101.1±3.7 | 191.2±4.8 | 353.2±8.6 | 22.0±3.7 | 1.5±0.66 | 21±8.8 | 107.6±8.5 | 19.1±1.4 | 43.8±2.6 | 17.6±1.5 | - |
| | | | Summer | 7.3±0.21 | 30.6±1.2 | 132.2±3.2 | 221±3.6 | 344.2±4.4 | 25.4±2.6 | 1.8±0.16 | 24±4.2 | 137.5±6.4 | 24.2±2.2 | 48.6±2.7 | 22.4±1.8 | - |

(Table 4) cont.....

| Region | Site group | Site | Season | pH | Temp. | TDS | COND | TSS | TALK | DO | BOD | Ca2+ | Mg2+ | N03- | PO4 | Trans. |
|---|---|---|---|---|---|---|---|---|---|---|---|---|---|---|---|---|
| SOUTH | MARGINAL SITES (at 150-200mt. distance) | S-1 | Rainy | 7.6±0.23 | 30.5±1.8 | 130.8±8.04 | 213.5±9.0 | 258.5±6.4 | 20.5±4.4 | 1.07±0.27 | 11±1.8 | 48.7±3.2 | 10.2±1.0 | 37.5±3.3 | 18.9±1.3 | 35.3±1.2 |
| | | | Winter | 8.2±0.43 | 25.6±2.9 | 115.5±8.9 | 235.5±4.8 | 357.5±7.9 | 37.8±3.6 | 2.3±0.88 | 16±3.3 | 60.2±4.8 | 14.3±4.0 | 36.8±4.7 | 19.7±1.5 | 35±1.0 |
| | | | Summer | 8.0±0.33 | 32.8±2.2 | 138.0±4.2 | 244.1±4.2 | 351.4±3.6 | 40.6±4.2 | 2.2±0.62 | 20±2.8 | 79.2±3.2 | 16.2±2.6 | 41.2±3.8 | 23.8± 1.7 | 29.7±1.4 |
| | | S-2 | Rainy | 7.7±0.26 | 30.8±1.2 | 134.75±3.1 | 224±8.94 | 256.5±7.6 | 17.4±3.6 | 1.2±0.08 | 13±1.6 | 55.05±2.3 | 11.3±1.5 | 35.8±2.1 | 15.3±2.2 | 38±1.2 |
| | | | Winter | 8.4±0.48 | 24.7±2.3 | 135.7±2.9 | 233.5±7.4 | 393.0±8.4 | 27.2±3.5 | 2.4±0.76 | 16.2±1.7 | 73.6±3.5 | 17.8±3.5 | 39.8±4.5 | 17.4±1.7 | 40±1.1 |
| | | | Summer | 8.1±0.33 | 34.1±2.8 | 144.0±2.4 | 256.2±6.2 | 388.2±4.2 | 28.2±2.2 | 2.2±0.22 | 22±2.4 | 88.4±3.1 | 22.1±2.6 | 43.7±3.8 | 18.6±1.4 | 34.5±1.4 |
| | | S-3 | Rainy | 7.7±0.01 | 30.5±1.6 | 142.5±5.6 | 228.5±6.0 | 238.0±7.4 | 23.0±6.7 | 1.8±0.15 | 10±1.9 | 47.7±4.6 | 9.8±1.3 | 36.8±2.9 | 14.6±1.4 | 43±2.0 |
| | | | Winter | 8.0±0.26 | 24.8±2.4 | 122.2±4.6 | 226.7±2.8 | 279.0±6.1 | 31.5±1.6 | 2.2±0.82 | 16±4.6 | 74.7±4.18 | 13.9±3.9 | 38.9±2.2 | 14.7±1.6 | 40±2.1 |
| | | | Summer | 8.3±0.22 | 32.8±2.2 | 155.1±2.8 | 248.4±4.2 | 270.4±6.4 | 33.2±2.4 | 2.6±0.41 | 20±3.3 | 89.4±3.3 | 18.4±1.8 | 40.2±2.6 | 16.3±1.2 | 36.2±2.0 |
| | | S-4 | Rainy | 8.0±0.18 | 30.1±2.4 | 134.5±3.1 | 233.5±3.3 | 248.5±4.3 | 25.4±5.3 | 1.6 ±0.34 | 10±1.4 | 63.1±2.3 | 13.7±1.1 | 34.5±1.6 | 14.2±2.8 | 41.6±2.2 |
| | | | Winter | 8.2±0.30 | 24.6±4.1 | 143.0±3.6 | 237.5±4.9 | 353.0±8.8 | 15.5±1.6 | 2.6±0.82 | 18±4.4 | 76.0±2.8 | 17.5±1.6 | 41.4±2.4 | 16.3±1.0 | 42±2.0 |
| | | | Summer | 8.0±0.22 | 33.8±2.6 | 161.2±2.4 | 248.4±3.6 | 344.8±4.2 | 23.5±2.1 | 2.2±0.16 | 24±2.8 | 93.8±3.2 | 21.1±2.1 | 47.8±3.1 | 16.8±1.4 | 38±2.1 |
| | INNER SITES (at surface & bottom) | S-s | Rainy | 7.8±0.23 | 31.5±3.1 | 155.2±2.1 | 203.8±2.9 | 297.5± 7.5 | 20.4±5.0 | 1.5±0.50 | 12.5±1.6 | 133.1±4.4 | 11.3±2.0 | 37.8±1.9 | 13.3±3.4 | 41±2.6 |
| | | | Winter | 7.9±0.21 | 23.8±4.3 | 153.5±6.3 | 217.7±3.6 | 296.5±5.1 | 23.2±5.3 | 2.5±0.77 | 23±6.12 | 86.8±8.5 | 16.0±2.5 | 39.3±0.9 | 12.5±1.0 | 38±2.1 |
| | | | Summer | 8.2±0.28 | 33.7±3.6 | 158.0±3.2 | 239.8±4.2 | 290.6±6.6 | 23.6±4.6 | 1.9±0.44 | 28±4.8 | 143.2±4.6 | 19.4±1.4 | 41.3±2.4 | 14.1±1.8 | 35±2.2 |
| | | S-b | Rainy | 7.8±0.28 | 29.3±3.8 | 145.2±2.8 | 183.5±4.1 | 259.5±1.65 | 22.5±6.0 | 1.2±0.21 | 10±3.8 | 64.6±2.8 | 17.0±1.5 | 34.5±1.5 | 15.8±3.7 | - |
| | | | Winter | 7.8±0.28 | 22.7±4.6 | 138.0±8.6 | 178.7±4.2 | 285.0± 9.3 | 23.5±2.9 | 2.0±0.67 | 25±8.9 | 66.05±3.3 | 19.5±1.8 | 37.6±4.9 | 22.7±1.4 | - |
| | | | Summer | 7.5±0.22 | 29.8±3.2 | 148.4±4.4 | 218.6±4.4 | 277.6±6.4 | 24.0±3.3 | 0.9±0.47 | 30±6.6 | 71.4±2.8 | 20.4±2.3 | 48.9±4.1 | 26.4±2.7 | - |
| | W H O | | | 7.8-8.5 | - | 1000 mg/l | 300 mhos /cm | 600 mg/l | 50mg/l | 4mg/l | 5mg/l | 75mg/l | 50mg/l | 45mg/l | .01-.5mg/l | - |

## Table 5. Physico-chemical condition and water quality of Ved pond.

| SAMPLING SITES & SAMPLES | | | SEASON | PARAMETERS (MEAN VALUE) | | | | | | | | | | | | |
|---|---|---|---|---|---|---|---|---|---|---|---|---|---|---|---|---|
| | | | | pH | Temp. | TDS | COND | TSS | T ALK | DO | BOD | Ca2+ | Mg2+ | N03- | PO4 | Trans. |
| NORTH | MARGINAL SITES (at 150-200mt. distance) | S-1 | Rainy | 8.5±0.23 | 31.1±1.4 | 453±15.2 | 984±34.2 | 69.6±9.4 | 583.3±64.1 | 1.6±0.23 | 17±2.3 | 62.8±14.1 | 11.6±4.1 | 39.8±13.2 | 17.4±0.13 | 28±0.23 |
| | | | Winter | 8.8±0.22 | 27.4±1.6 | 518±12.8 | 971±29.8 | 72.7±13.9 | 699.7±49.3 | 4.3±0.25 | 20±2.0 | 65.3±10.3 | 18.8±6.3 | 32.7±15.6 | 21.2±0.23 | 23±0.21 |
| | | | Summer | 9.9±0.41 | 41.1±2.3 | 714±21.2 | 1021±38.2 | 88.2±5.7 | 873.7±47.7 | 3.3±0.42 | 34±1.2 | 72.2±24.1 | 21.6±4.9 | 54.2±14.1 | 33.4±0.25 | 16±0.41 |
| | | S-2 | Rainy | 8.1±0.33 | 31.3±2.0 | 458±14.6 | 973±32.6 | 64.1±6.3 | 499.6±41.1 | 2.1±0.37 | 20±1.6 | 56.6±12.6 | 12.4±3.7 | 42.6±11.2 | 18.6±0.20 | 31±0.53 |
| | | | Winter | 8.5±0.40 | 25.2±1.8 | 482±8.8 | 971±42.8 | 78.2±8.4 | 539.3±54.6 | 4.8±0.32 | 23±1.9 | 58.5±16.7 | 17.1±5.5 | 48.3±17.7 | 29.4±0.31 | 20±0.33 |
| | | | Summer | 9.2±0.23 | 40.1±2.4 | 689±18.3 | 1044±44.1 | 127.3±11.4 | 735.8±49.2 | 2.2±0.43 | 33±1.4 | 66.5±22.6 | 27.5±4.5 | 67.5±18.2 | 37.5±0.22 | 17±0.40 |
| | INNER SITES (at surface & bottom) | S-s | Rainy | 8.5±0.43 | 30.9±1.6 | 478±12.6 | 889±28.6 | 62.1±7.1 | 444.3±54.5 | 2.9±0.43 | 22±1.2 | 44.8±14.7 | 13.6±4.1 | 32.3±10.2 | 14.6±0.26 | 27±0.47 |
| | | | Winter | 8.7±0.41 | 26.5±1.4 | 540±16.2 | 897±22.8 | 71.2±7.4 | 461.5±41.4 | 4.1±0.43 | 26±1.1 | 61.3±18.6 | 20.4±6.4 | 30.5±15.3 | 16.7±0.19 | 19±0.32 |
| | | | Summer | 8.9±0.27 | 38.4±1.8 | 591±20.8 | 969±26.2 | 89.1±9.4 | 544.7±49.8 | 1.9±0.43 | 37±2.4 | 76.8±28.1 | 38.6±7.1 | 38.1±12.4 | 22.6±0.21 | 17±0.36 |
| | | S-b | Rainy | 8.2±0.38 | 29.1±1.6 | 448±16.5 | 854±34.3 | 70.3±8.2 | 459.6±51.7 | 3.8±0.43 | 22±2.7 | 48.4±15.4 | 16.7±4.7 | 38.3±14.2 | 13.2±0.16 | - |
| | | | Winter | 8.7±0.37 | 24.2±1.4 | 559±19.5 | 894±32.2 | 78.2±10.6 | 460.2±44.1 | 4.6±0.43 | 26±2.3 | 64.2±19.5 | 20.9±4.4 | 36.4±11.5 | 18.2±0.26 | - |
| | | | Summer | 9.2±0.47 | 37.7±2.6 | 590±12.8 | 914±48.6 | 112.4±14.4 | 473.6±47.6 | 3.6±0.43 | 34±1.8 | 84.3±13.3 | 31.4±4.1 | 49.8±15.1 | 28.4±0.29 | - |
| EAST | MARGINAL SITES (at 150-200mt. distance) | S-1 | Rainy | 7.7±0.42 | 30.7±1.3 | 558±22.8 | 879±42.8 | 63.3±11.8 | 381.3±54.1 | 1.9±0.43 | 15±1.3 | 80.6±18.9 | 12.5±5.3 | 38.6±13.2 | 17.4±0.21 | 32±0.42 |
| | | | Winter | 7.5±0.44 | 26.3±2.1 | 560±10.6 | 875±40.2 | 72.7±12.1 | 357.5±58.7 | 3.9±0.43 | 22±1.4 | 84.1±17.1 | 15.4±3.8 | 44.1±16.8 | 20.8±0.25 | 22±0.37 |
| | | | Summer | 7.9±0.33 | 38.5±2.6 | 659±14.8 | 892±46.8 | 84.2±12.8 | 415.3±55.3 | 3.7±0.43 | 26±2.1 | 121.7±24.3 | 27.1±3.3 | 52.6±14.6 | 31.6±0.28 | 20±0.33 |
| | | S-2 | Rainy | 8.1±0.23 | 30.9±2.2 | 771±16.8 | 861±42.2 | 70.6±7.9 | 586.3±44.6 | 1.6±0.43 | 17±2.3 | 82.5±19.1 | 13.5±4.7 | 40.2±11.7 | 17.3±0.13 | 37±0.40 |
| | | | Winter | 8.6±0.45 | 25.7±2.6 | 592±22.6 | 881±44.8 | 68.5±8.4 | 636.2±47.1 | 4.5±0.43 | 21±2.7 | 127±22.3 | 15.2±4.3 | 51.1±15.8 | 21.5±0.11 | 26±0.41 |
| | | | Summer | 8.8±0.44 | 39.3±2.4 | 698±28.2 | 891±40.1 | 76.4±10.4 | 653.6±45.7 | 3.4±0.43 | 30±2.5 | 133.9±31.0 | 20.5±4.1 | 61.7±17.2 | 33.2±0.15 | 21±0.23 |
| | INNER SITES (at surface & bottom) | S-s | Rainy | 7.1±0.46 | 30.7±1.8 | 569±11.8 | 969±33.9 | 63.3±8.8 | 342.6±43.7 | 2.6±0.43 | 16±2.2 | 95.5±34.8 | 13.6±4.8 | 48.6±16.1 | 17.4±0.24 | 26±0.28 |
| | | | Winter | 7.5±0.36 | 27.5±2.7 | 530±15.4 | 947±32.8 | 73.2±6.7 | 357.5±48.6 | 2.3±0.43 | 21±2.4 | 83.8±24.7 | 16.6±5.2 | 42.7±15.8 | 20.3±0.22 | 24±0.41 |
| | | | Summer | 7.7±0.33 | 36.4±2.2 | 610±24.4 | 984±42.6 | 80.6±13.4 | 382.4±41.5 | 2.7±0.43 | 28±1.9 | 94.6±22.8 | 26.1±6.6 | 58.4±15.0 | 30.9±0.21 | 22±0.46 |
| | | S-b | Rainy | 7.9±0.47 | 29.5±1.6 | 493±21.2 | 983±40.8 | 72±11.7 | 461.3±48.6 | 2.6±0.43 | 23±1.5 | 108.5±19.7 | 12.9±3.1 | 50.5±14.2 | 17.8±0.20 | - |
| | | | Winter | 8.3±0.42 | 25.5±2.1 | 566±18.6 | 955±42.8 | 86.5±17.4 | 480.5±46.5 | 3.7±0.43 | 24±2.4 | 103.3±25.1 | 18.1±3.4 | 47.3±14.9 | 23.7±0.24 | - |
| | | | Summer | 8.5±0.41 | 37.1±2.6 | 627±20.6 | 978±50.4 | 71.5±16.6 | 492.1±49.6 | 2.8±0.43 | 26±2.8 | 141.6±37.2 | 23.3±3.9 | 51.8±16.4 | 34.2±0.26 | - |
| WEST | MARGINAL SITES (at 150-200mt. distance) | S-1 | Rainy | 8.3±0.36 | 30.6±1.8 | 460±17.9 | 962±32.1 | 57.3±9.6 | 688.5±54.1 | 2.4±0.43 | 19±1.7 | 102.6±30.5 | 11.5±4.4 | 47.3±10.2 | 17.2±0.22 | 24±0.22 |
| | | | Winter | 8.5±0.33 | 26.1±2.1 | 489±14.2 | 954±37.5 | 70.7±7.4 | 728.7±58.3 | 3.4±0.43 | 20±1.5 | 97.7±26.5 | 18.6±4.7 | 55.3±13.6 | 20.1±0.26 | 22±0.13 |
| | | | Summer | 8.7±0.23 | 38.9±1.8 | 540±24.6 | 960±33.8 | 76.1±8.2 | 753.2±64.1 | 2.3±0.43 | 32±1.9 | 132.5±32.6 | 31.7±4.1 | 67.2±15.2 | 27.6±0.25 | 18±0.20 |
| | | S-2 | Rainy | 8.9±0.29 | 30.5±2.0 | 445±16.5 | 882±22.8 | 65.6±7.6 | 796.1±34.9 | 3.1±0.43 | 20±2.4 | 121.4±23.6 | 14.3±4.9 | 42.4±14.8 | 18.8±0.21 | 32±0.40 |
| | | | Winter | 8.6±0.27 | 25.7±2.4 | 488±12.3 | 932±29.4 | 72.3±7.4 | 732.5±43.8 | 4.4±0.43 | 25±2.1 | 123.1±30.1 | 15.5±6.1 | 51.3±11.5 | 19.2±0.22 | 21±0.33 |
| | | | Summer | 8.8±0.42 | 40.3±2.9 | 528±18.6 | 984±21.1 | 77.3±6.9 | 746.3±46.5 | 2.8±0.43 | 39±2.1 | 153.6±25.4 | 23.5±6.6 | 58.1±16.2 | 29.6±0.27 | 19±0.20 |
| | INNER SITES (at surface & bottom) | S-s | Rainy | 8.3±0.31 | 31.3±1.7 | 553±21.6 | 974±27.8 | 65.1±9.4 | 669.4±40.6 | 1.5±0.43 | 18±2.3 | 40.9±27.9 | 11.8±6.2 | 37.3±12.8 | 18.1±0.18 | 27±0.23 |
| | | | Winter | 8.9±0.44 | 26.6±1.4 | 549±18.1 | 955±20.5 | 74.5±4.9 | 734.2±43.7 | 2.2±0.43 | 29±2.7 | 50.8±24.3 | 17.8±4.2 | 43.2±15.7 | 23±0.13 | 25±0.25 |
| | | | Summer | 9.1±0.47 | 36.8±2.1 | 596±24.3 | 968±34.8 | 71.5±6.7 | 762.5±44.2 | 1.8±0.43 | 30±3.5 | 73.6±36.6 | 25.1±4.4 | 48.5±13.1 | 33.4±0.17 | 17±0.39 |
| | | S-b | Rainy | 8.1±0.37 | 29.8±1.6 | 583±11.6 | 982±29.2 | 55.6±8.4 | 750.6±40.6 | 2.4±0.43 | 20.6±2.5 | 45.8±44.5 | 17.4±6.5 | 40.7±11.7 | 19.5±0.20 | |

## Temperature

Water temperature directly as well as indirectly influences many abiotic and biotic components of the aquatic ecosystem [14], which play an important role in the metabolic activities of the organisms in the ponds [15]. The variations in the temperature, were influenced by factors such as air temperature humidity, wind, and solar energy. The average temperature ranged between 36.3±2.5 (S-2) (east summer) peripheral and 23.5± 0.49oC (S-1) (west) peripheral during the rainy season, whereas in the winter season it varied between 22.7± 4.6°C (south inner) to 31.2± 2.2°C (east inner) sites in Dulhara pond. The water temperature was found to be suitable for fish growth due to the standing water of those water bodies.

The temperature ranged between 31.7±1.8 at (S-1) (south peripheral) in the rainy season and 24.2± 1.6 (S-2) (north peripheral) in the winter season, whereas for the inner sites it varied between 38.7±2.6(west inner)in the summer season and 23.2±1.4 (north inner) in the winter in Ved pond. The water temperature is generally low, ranging from 5°C -36°C. This is standard for fish and other aquatic organisms. The water temperature range was discovered to be within the standard range (17.2–33.4 oC) and higher during the rainy season due to moisture in the atmosphere and high air temperature. During this study, we observed a difference of 2 to 3°C between the water surface and inner site.

## pH

The mean pH value varies between 7.4±0.16 at the peripheral site S-3 in the north of the pond in the rainy season and 8.4±0.48 at the peripheral site S-2 in the south in the winter season. The inner sites had values ranging from 7.1±0.21 (west) in the winter season to 8.2±0.28 (south) in the summer at Dulahra pond. A higher pH value is normally associated with the high photosynthesis activity in water. The pH of an aquatic system is an important indicator of water quality and the extent of pollution in watershed areas. During the assessment period, the maximum mean pH value in Ved pond was 9.90.41 at S-1 (north peripheral) in the summer season and 9.10.47 at S-s (South inner) surface in the rainy season. The alkaline nature of the pond water values may be due to sewage discharged by surrounding villages and agricultural fields. Sewage and agricultural discharges are generally a complex combination of natural organic and inorganic materials and man-made compounds. It usually contains many fertilizers, metals, sediments, pesticides, nutrients, salt, sodium, calcium, potassium, chlorine, phosphate, bicarbonate, *etc* [21].

## Total Dissolved solids (TDS) and Total Suspended Solids(TSS)

The principal constituents of TDS are usually major cations and anions. The higher values of TDS in the Dulhara pond are an indicator of higher ionic concentrations, probably due to the high anthropogenic activity in the region acquiring a high concentration of dissolved minerals. The highest TDS was recorded at 155.1±2.8 (S-3) south peripheral in the summer season and a minimum value of 80.8±3.8 (S-1) east in the winter season for peripheral sites. In the inner site, the maximum TDS found was 158.0±3.2(S-S) south in the summer season. Total dissolved solids for Dulahra pond were under the standard value of WHO (1000mg/l). In the present study, the maximum average value of electrical conductivity was analyzed at 276.0±5.4 [mhos cm$^{-1}$] north in the rainy season at S-3 for peripheral sites [16, 17]. The present study showed the highest alkalinity was recorded at 766.4±3.8 (S-2) north in the summer season, which was higher than the permissible limit of 600mg/l.The higher values of alkalinity may be due to the presence of bicarbonate, carbonate, and hydroxide compounds of calcium, magnesium, sodium, and potassium [19]. TSS is a significant factor in observing water clarity. The more solids are present in the water, the less clear the water will be. In the summer season, aamximum mean TSS value 48.2±1.8 was observed at S-2, North marginal in Dulhara pond.

In our studies, the total dissolved solids (TDS) were s found to be extremely high in the rainy season. TDS has a positive association with water temperature that is not higher than the typical level. The TDS value of Ved pond is not damaging to marine life. A maximum TDS of 771±16.8 was perceived at S-2 (East Peripheral) in the rainy season. (Fig. **8**). Total deferred solids embrace clay, sand, various silicates, and micro-organisms *etc.* [22]. The deferred TSS matter is resolute by clarifying the sample, parching the remainder, and determining its weight by alteration. Total deferred solids frequently transpire in waste and sewage material. The TSS values were continuously above the WHO allowable limit of 50 mg/l. Total postponed solids consist of the grainier as well as the reparable particulates. Suspended solids containing much organic matter may cause decomposition and, consequently weaken the liquefied oxygen [23]. Mineral matter in a suspended state may cause silting, causing recompense to marine plants and animal life. The existence of adjourned solids also increases the turbidity, making the water unsuitable for domestic use too. In the present investigation, the maximum total suspended solids was 127.3+11.4 at S-2 (North peripheral) in the summer season and 86.5±9.7 (East inner) at S-b in the winter season.

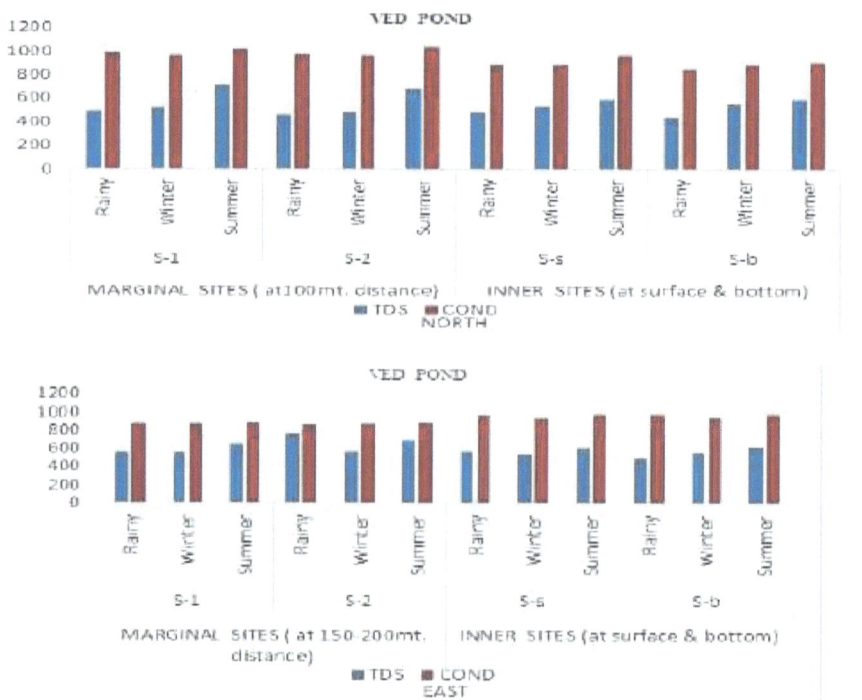

Fig. (8). Water quality characteristics with TDS and conductivity of Vedpond in north and east directions.

## Electrical Conductivity

The conductivity of the Ved pond is a measure of the capability of water to conduct electricity. It is to be determined by the ionic concentration and water temperature. The total capacity of salts in a water body is openly connected to its conductivity. The EC values of the study area had a maximum mean of 1044±44.1 (north peripheral) at S-2 in the summer season and 982±29.2 (west inner) at S-b in the rainy season.

The alkalinity of natural waters is generally due to the occurrence of bicarbonate, carbonate, and hydroxide compounds of calcium, magnesium, sodium, and potassium. Maximum alkalinity 873.7±47.7 was observed at S-1 (North peripheral) in the summer season and 750.66±40.6 at S-b (west inner sites) in the rainy season in Ved Pond. (Fig. 9).

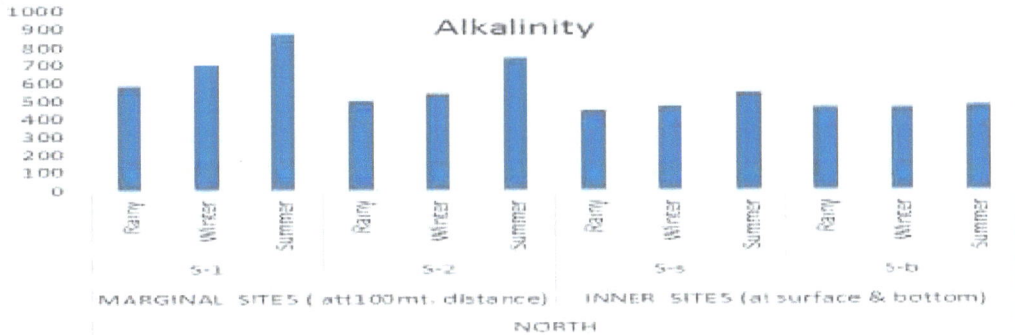

**Fig. (9).** Effect of alkalinity of Vedpond water.

## Dissolved Oxygen and Biological Oxygen Demand

Dissolved oxygen is a characteristic of shallow waters; its values decrease with depth. In groundwater, oxygen comes from the atmosphere and participates in hydro-geochemical processes. The lack of oxygen in the shallow waters proves the pollution of the water [18]. A maximum mean DO value $3.1\pm0.66$ was observed at S-2, north marginal in the summer season. The biochemical oxygen ultimatum is the most commonly used constraint to describe the strength of civic or biological industrial wastewater. BOD is the amount of oxygen used in the respirational procedures of micro-organisms in reacting to the organic matter in the sewage and for the further metabolism (oxidation) of cellular mechanisms manufactured from the wastes. In the present study, the BOD values were significantly higher than the WHO standards (5 mg/l). The maximum mean value of BOD, *i.e.*, $32\pm4.6$ was found at the north peripheral S-1 in the summer season. It indicates the biological pollution load on the water body.

Dissolved oxygen standards were at all times above the WHO permitted limit of 4 mg/l at all specimen locations. It is a physical singularity and is contingent upon the solubility of oxygen, which in turn is prejudiced by water temperature. The maximum value of DO $4.8\pm0.43$ was perceived at S-2 (West Peripheral) in the winter season and $4.6\pm0.43$ at S-b (North Inner) in the winter season in Ved pond. This means DO is reduced with high moisture and an increase in water temperature. Maximum mean BOD value of $39\pm2.1$ was found at S-2 (west peripheral) in the summer season. The significance of the BOD upsurge in the summer season demonstrates an increase in organic load in the water.

## Calcium and Magnesium

The concentration of $Ca^{++}$ ions was much above the WHO recommended value of 75 mg/l at almost all the sites during the study period. The maximum value of

133.1±4.4 was found at S-s in the south inner site in the rainy season in Dulhara pond. The presence of $Ca^{++}$ ion in water is mainly due to their passage through or over deposits of limestone, dolomite, and gypsum. The $Mg^{++}$ ion occurs as a result of chemical weathering and dissolution of dolomite, and other rocks. The $Mg^{++}$ ions are seldom dominant in natural waters, which is also true for the study area. This is due to the fact that $Mg^{++}$ ions have weak biological activity and the highest solubility. The dissolution of Mg-rich minerals is usually a slow process. The concentration of magnesium ions in all the samples was below the WHO standard of 50 mg/l. The maximum mean concentration of $Mg^{++}$ ions was observed at 34.6±1.4 at the north marginal S-3 in the winter season. The maximum value of 133.1±4.4 was found at S-s in the south inner site in the rainy season at Dulahra pond.

During the study, the mean concentration of actions such as $Ca^{2+}$ and $Mg^{2+}$ ions ranged from 27.2±14.8 (S-2, South Marginal in the rainy season) to 133.9±31.0 (S-2, East Marginal in the summer season) and for inner sites, it varied between 9.6±.4.2 (S-1 South marginal in the rainy season) and 31.7±4.1 (S-1 west marginal in the summer season) at Ved pond.

The presence of calcium ions in water is mainly due to their passage through or over deposits of limestone, dolomite, and gypsum. The whole study revealed that there was no significant change in water quality parameters between the stations and seasons in the Dulahra tank, with the exception of biological oxygen demand and phosphate. There existed a positive correlation between electrical conductivity and total dissolved solids, between pH and alkalinity, and between transparency and total suspended solids.

**Nitrates and Phosphates**

Nitrate is one of the several inorganic pollutants along with nitrogenous fertilizers, organic manures, human and animal wastes, and industrial effluents through the biochemical activities of microorganisms. Nitrate enters the human body through the use of groundwater for drinking and causes a number of health disorders, namely methemoglobinemia, gastric cancer, goiter, birth malformations, hypertension, *etc.*, when present in high concentrations in drinking water. A nitrate value of 45.2±.4 was recorded at S-2, east marginal in the summer season, which was below the permissible limit. Phosphate concentration is always higher than the permissible limit at all sampling points and during all seasons. It ranges between 9.7±1.1 at S-1 marginal east in the winter and 33.4±1.6 at S-2 marginal north during the summer. The higher values indicate that the waters are polluted or eutrophic. The control over undesirable blue-green algae can often be obtained only at phosphorus levels below 0.02 mg/L.

In the current study at Ved Pond, the highest mean deliberation of total nitrogen was observed to be 67.518.2 at S-2 north peripheral and a maximum of 58.712.7 at S-2 south inner during the summer season (Fig. **10**). Nitrogen fertilizers are applied in very large quantities to field crops. Since the plants cannot exploit all the nitrogen applied to the fields, some is left in the soil and it can leach into groundwater. In addition, not all the applied nitrogen gets into the soil and some is washed away from the fields in the form of overflow and it flows into surface waters. An increased level of nitrate-nitrogen in water organizations causes nutrient improvement, which results in the huge growth of aquatic filamentous green algae, leading to eutrophication. This huge algae growth creates painkilling circumstances in the water body [24]. Some of the blue-green algae classes are reported to be toxic. Anaerobic circumstances are developed in the water body due to the oxygen uptake during the nighttime, which in turn, is not compensated for photosynthesis. Phosphorus transpires in natural waters and wastewater in the form of numerous phosphates. Surface waters seldom contain high concentrations of phosphates, since they are utilized by plants, while ground waters frequently contain considerable quantities of phosphate [25 - 30].

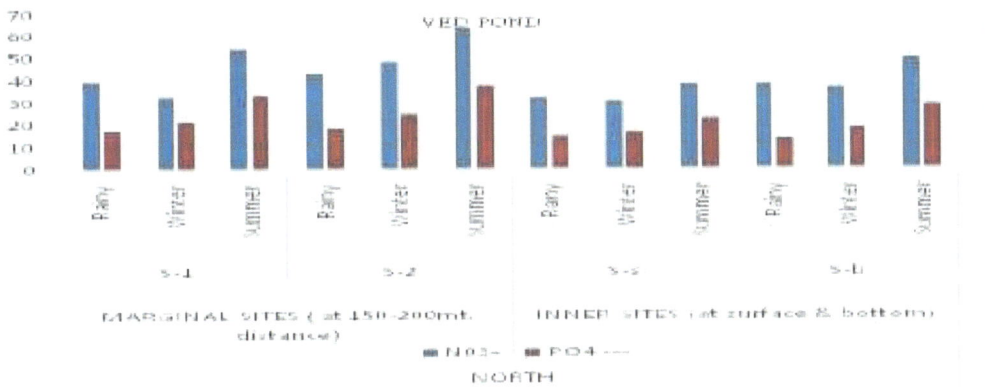

**Fig. (10).** Analysis of nitrate and phosphate in Vedpond.

## CONCLUSION

The entire qualitative and quantitative study reveals that the degree of pollutants in this pond is significantly subordinate in comparison with any other ponds under investigation. This suggests that the water in this pond should be treated so that it can be used for domestic purposes as well as for irrigation. The extreme softness of Dulhara pond water is due to low $Ca^{++}$ and $Mg^{++}$ values. The water when touched is of a considerably soft nature, almost identical to conventional soaps. The population within the pond uses the water for washing clothes and themselves. A perusal of the table shows that a correlation exists between EC and TDS, as electrical conductivity depends directly on the number of salts dissolved in water and the content of two valence ions.

Correlation also exists between pH and alkalinity. Alkalinity may be defined as the acid absorbing property of water. The major acid absorbing constituents that we typically deal with are hydroxide (OH-bicarbonate ($HCO_3^-$) and carbonate ($CO_3^{--}$) ions. As the pH increases between 4.3 and 8.3, the dissolved carbon dioxide starts to convert to bicarbonate ions. This conversion is complete at a pH of about 8.3, where only bicarbonate is present. Increasing the pH beyond 8.3, the bicarbonate ion is converted to a carbonate ion.

Total suspended solids (TSS) are particles that are larger than 2 microns and found in the water column. Anything smaller than 2 microns (average filter size) is considered a dissolved solid. Most suspended solids are made up of inorganic materials, though bacteria and algae can also contribute to the total solid concentration. The turbidity of water is based on the amount of light scattered by particles in the water body. The more particles that are present, the more light will be scattered. As such, turbidity and total suspended solids are related. However, turbidity is not a direct measurement of the total suspended materials in water. Instead, as a measure of relative clarity, turbidity is often used to indicate changes in the total suspended solids concentration in water. The transparency of water is affected by the amount of sunlight available, suspended particles in the water bodies, and dissolved solids such as colored dissolved organic material present in the water. In the present work, we found the maximum value of total suspended solids on the north side of the pond, and the transparency of the water also appeared to be very low.

With workable strategies, water foundations can be sheltered from contamination. For example, the drainage seaways and septic tanks must be assembled far away from the wells. Normally, water ducts and septic pipelines in the town run equivalent to each other in that way, giving probabilities for adulteration of drinking water. Adequate care should be taken by municipal establishments to isolate these lines to inhibit faecal infection. Faecal pollution may take place because of open evacuation practises, and the running of water pipes close to sewage lines. Open-air defecation along the canal bunds should be banned. Good sanitation practices should be initiated near open water sources. Frequent biochemical and microbiological tests should be carried out so as to defend the public from aquatic disease outbreaks. Ozonisation in place of chlorination is suggested. Furthermore, dredging of deposits, construction of new sewerage management plants, and openings of sewerage water should be averted.

The tank water quality parameters are well within the prescribed levels and are suitable for fishing and other domestic human activities. Seasonal changes in the environment were observed, but these were not statistically significant. Regular monitoring of pond water and applying appropriate corrective actions such as immersion of statues and discharging of treated water into the pond should be stopped. This will help in improving the water quality of the pond.

The revision of the present work reveals that the water of the Ved pond is set up to be contaminated as the pond is enclosed on all sides by housing neighborhoods, and sewerage from houses is constantly flowing into the pond. A number of occurrences of skin infections in people who were in interaction with the pond water living nearby have been described. Thus, actions have to be implemented for the management of the pond water.

## CONSENT FOR PUBLICATION

Not applicable.

## CONFLICT OF INTEREST

The author declares no conflict of interest, financial or otherwise.

## ACKNOWLEDGEMENTS

The author acknowledges the support from the State Planning Commission, Raipur (Chhattisgarh) for the financial assistance in the form of major research project and expresses gratitude to the Vice-Chancellor Dr. G.D. Sharma of the Atal Bihari Vajpayee University, Bilaspur, Chhattisgarh for his expert guidance, excellent supervision, and constant encouragement given to him throughout the period of the work. He also expressed his deep gratitude to Dr. D.K. Shrivastav, Professor and Head of the Department of Microbiology, Govt. Science College, Bilaspur for his keen interest shown in the progress of this work and for his constant motivation. He is thankful to NEERI, Nagpur, and Orissa University of Agriculture and Technology, Bhubneshswer for providing library and laboratory facilities.

## REFERENCES

[1]     Sharma BK. Environmental Chemistry. 4th. 1998; pp. 3-11.

[2]     Muhammed AU. The Impact of Drinking Water Quality and Sanitation Behavior on Child Health: Evidence from Rural Ethiopia, ZEF– Discussion Papers on Development Policy. Bonn,Germany: Center for Development Research 2016.

[3]    Venkataeswarlu KS. Water Chemistry. 1st ed. New Age International Publishers 1999; pp. 2-4.

[4]    Vyas , *et al.* Environment impact of idol immersion activity lakes of Bhopal, India. Asian J Exp Sci 2006; 20(2): 289-96.

[5]    Gupta . Assessment of heavy metals in surface water of lower Lake. Poll Res 2009; 24: 805-8.

[6]    Sharma T, *et al.* Int J Adv Res (Indore) 2015; 3(5): 130-9.

[7]    Manjare , *et al.* The monthly changes in physico-chemical parameter in Tamadalge water tanks at Kolhapur district. Maharashtra 2010.

[8]    Toure Al. *et al.*, Comparative study of the physico-chemical quality of water from wells, boreholes and rivers consumed in the commune of pelengana of the region of segou in mali environmental science ;An indian journal, november 2017.

[9]    APHA. Standard methods for examination of water and wastewater. New York: American Public Health Association 2005.

[10]   Manivaskam N. Physico chemical examination of water, sewage and industrial effluents PragatiPrakashan. Merrut 2000.

[11]   Manual on water and waste water analysis. NEERI Publication 1988.

[12]   Trivedy RK, Goel PK. Chemical and biological methods for water pollution studies. Karad, India: Environ. Pub. 1986; pp. 1-23.

[13]   Prasath BB, Nandakumar R, Kumar SD, *et al.* Seasonal variations in physico-chemical characteristics of pond and ground water of Tiruchirapalli, India. J Environ Biol 2013; 34(3): 529-37. [PMID: 24617138]

[14]   Kavita Sahni, Sheela Yadav. Seasonal variations in physico-chemical parameters of bharawas pond, rewari, haryana. Asian J Exp Sci 2012; 26(1): 61-4.

[15]   Dinesh Kumar G, Karthik M, Rajakumar R. Study of seasonal water quality assessment and fish pond conservation in Thanjavur, Tamil Nadu, India. J Entomol Zool Stud 2017; 5(4): 1232-8.

[16]   Dhanalakshmi V, Shanthi K M. Int J Curr Microbiol Appl Sci 2013; 2(12): 219-27.

[17]   Sahni K, Yadav S. Seasonal variations in physicochemical parameters of Bharawas pond, Rewari, Haryana. Asian J Exp Sci 2012; 26(1): 61-4.

[18]   Sunita Verma, Khan JB. Analysis of water quality by physico-chemical parameters in fatehsagartalab in bagar, dist. of jhunjhunu (Raj.). India IOSR, Journal of Pharmacy and Biological Sciences 2015; 10(5): 2278-3008.

[19]   Abhrajyoti Mandal and diptimoyeesahoo, study of physico-chemical parameters of three different urban pond water of nadia district, westbengal, india. Int J Fish Aquat Stud 2017; 5(6): 23-7.

[20]   Pankaj Namaand Dhan Raj. Water quality assessment using physico-chemical parameters of palasani pond. jodhpur district, rajasthan, india 2018; 5(3): 2348-1269. Availble; http://ijrar.com

[21]   Myre ES. The turbidity tube: Simple and accurate measurement of turbidity in the field. in department of civil and environmental engineering. Michigan Technological University Retrieved 2006. Avalble; http://www.cas.umn.edu/assets/pdf/Turbidity%20Tube.pdf

[22]   Monitoring T. Annual average secchi depth. In Tahoe Status & Trend: Monitoring & Evaluation Program 2013. Availble; http://www.tahoemonitoring.org/water/lake-tahoe/360.html

[23]   Monitoring T, Solids T. Tahoe monitoring ,total solids. In Water: Monitoring and Assessment 2012. Availble; http://water.epa.gov/type/rsl/monitoring/vms58.cfm

[24]   Monitoring T. Annual average secchi depth. In Tahoe Status & Trend: Monitoring & Evaluation Program 2013. Availble; http://www.tahoemonitoring.org/water/lake-tahoe/360.html

[25]   Belekar RM, Dhoble SJ. Activated alumina granules with nanoscale porosity for water defluoridation.

Nano-Structures & Nano-Objects 2018; 16: 322-8.
[http://dx.doi.org/10.1016/j.nanoso.2018.09.007]

[26]　Gedekar KA, Wankhede SP, Moharil SV, Belekar RM. Synthesis, crystal structure and luminescence in Ca 3 Al 2 O 6. J Mater Sci Mater Electron 2018; 29(8): 6260-5.
[http://dx.doi.org/10.1007/s10854-018-8603-5]

[27]　Gedekar KA, Wankhede SP, Moharil SV, Belekar RM. $Ce_3^+$ and $Eu_2^+$ luminescence in calcium and strontium aluminates. J Mater Sci Mater Electron 2018; 29(6): 4466-77.
[http://dx.doi.org/10.1007/s10854-017-8394-0]

[28]　Wani MA, Dhoble SJ, Belekar RM. Synthesis, characterization and spectroscopic properties of some rare earth activated $LiAlO_2$ phosphor. Optik (Stuttg) 2021; 226(1)165938
[http://dx.doi.org/10.1016/j.ijleo.2020.165938]

[29]　Belekar RM, Athawale SA, Gedekar KA, Dhote AV. Various techniques for water defluoridation by alumina: Development, challenges and future prospects. AIP Conf Proc 2019; 2104(1): 03004.
[http://dx.doi.org/10.1063/1.5100431]

[30]　Belekar RM. Suppression of coke formation during reverse water-gas shift reaction for CO2 conversion using highly active $Ni/Al_2O_3$-$CeO_2$ catalyst material. Phys Lett A 2021; 395127206
[http://dx.doi.org/10.1016/j.physleta.2021.127206]

CHAPTER 5

# Nanoparticle-aided AOP for Treatment of Benzoic Acid

**Bhavna D. Deshpande[1,*], Pratibha S. Agrawal[1], M.K.N. Yenkie[1] and S.J. Dhoble[2]**

[1] *Department of Applied Chemistry, Laxminarayan Institute of Technology, R.T.M. Nagpur University, Nagpur, India–440010.*

[2] *Department of Physics, R.T.M. Nagpur University, Nagpur, India-440033*

**Abstract:** Advanced oxidation process (AOP) degrades a number of non-degradable organic compounds in low concentrations, saving time and energy. Benzoic acid and its derivatives are readily used in pharmacy, textile, and dyes industries. Through these applications, benzoin acid enters the ecosystem, which leads to its accumulation and various health hazards. In the present study, the degradation of Benzoic acid was studied using Iron nanoparticles as heterogeneous photocatalyst and hydrogen peroxide as an oxidizing agent. This paper also discusses the synthesis of Fe nanoparticles *via* hydrothermal process at ordinary temperature and elevated temperature. The powder samples were characterized by X-ray diffraction (XRD), Scanning Electron Microscope (SEM) and Energy dispersive analysis of X-rays (EDAX). The percentage degradation of benzoic acid using goethite ($\alpha$-FeOOH) and hematite ($\alpha$-Fe$_2$O$_3$) was 49.02% and 90.90% with the nano concentrations of 0.07g and 0.05g, respectively using visible light, in addition, the hydrothermal model of nanoparticle synthesis proved affordable, efficient and eco-friendly.

**Keywords:** Advanced oxidation process (AOP), Hydrothermal process, Nanoparticles, Non-degradable, Photocatalyst, Heterogeneous, Goethite, Hematite, Iron nanoparticles.

## INTRODUCTION

In view of the ever-increasing world population, freshwater scarcity has become the world's most critical environmental issue that plagues the earth today. Although Earth is surrounded by 71% water, 97.5% being salty, only 2.5% of freshwater remains usable [1]. Due to rural migration, industrialization, and the growing needs of an exponentially increasing population, the usage of freshwater

* **Corresponding author Bhavna D. Deshpande:** Department of Applied Chemistry, Laxminarayan Institute of Technology, R.T.M. Nagpur University, Nagpur–440010, India; Tel: +91-9881370448; E-mail: dnd.bhavna@gmail.com

**R. M. Belekar, Renu Nayar, Pratibha Agrawal and S. J. Dhoble (Eds)**
**All rights reserved-© 2022 Bentham Science Publishers**

has increased over the years, which has led to the depletion of freshwater resources that have made it an expensive commodity [2]. Climatic changes due to global warming remain a major factor in freshwater depletion [3]. Freshwater resources are inevitable for the sustainability of society [4]. Various researches suggest that by 2025, the world would be under acute water stress [5, 6]. Regardless of this fact, in most developing countries, industrial effluents are discharged into the water bodies without any proper treatment neglecting the guidelines laid down by governments and international bodies [7]. Wastewater contaminants include inorganic and organic pollutants, heavy metals [8], refractory organic pollutants [9] and many other complex compounds. Aromatic compounds (benzoic acids, phenols and their respective derivatives) are widely used as precursors in pharmaceutical, textile, food and cosmetic industries [10]. The industrial usage of natural resources can be curtailed by reusing and recycling wastewater produced in industries by various methods such as physio-chemical, chemical, biological, electrochemical, oxidation and photocatalysts and combination of the mentioned processes [11]. Hence, innovative, highly proficient, and low budget technologies for wastewater treatment is today's necessity [12]. Conventional methods clubbed with newer rapid technologies can enhance the reaction rate. Some rapid and significant processes such as bioremediation, adsorption, photo catalytic oxidation are being used to treat the polluted waters [13, 14].

Advanced oxidation processes (AOPs) are promising techniques used in the past decades for the complete mineralization of organic content, hazardous pollutants and non-biodegradable matter in wastewater [15]. AOPs have gained importance because of their fast reaction rates, moderately small setup, reduced toxicity and perhaps complete mineralization of pollutants, no sludge formation, and no further treatment of the product formed [16, 17]. AOP is a special class of oxidation grounded on the production of OH· radicals in the aqueous phase leading to the destruction or disintegration of their molecular structure resulting in losing their chemical identity [13]. Hence, refractory organic matter is oxidized into carbon dioxide, water and biodegradable inorganic waste thus treating the water. OH· radical is a non-selective, highly reactive, and short-lived oxidizing agent which has an oxidizing potential ranging between 2.8V(pH 0) and 1.95 V(pH 14) *vs* Saturated Calomel Electrode(SCE), a commonly used reference electrode in water treatment techniques [18, 19]. Hydroxyl methods can be introduced into the reaction mixture using different techniques such as a combination of oxidizing agents (using $O_3$ and $H_2O_2$), irradiation of UV light or ultrasound, and catalysts (such as $Fe^{2+}$) [20]. The intermediate formed during the course of the reaction depends on the substrate compound that ultimately oxidizes to $CO_2$ and $NO_3^-$. Moreover, species such as chlorinated alkenes are more susceptible to hydroxyl radical attack, as compared to saturated molecules (*i.e.* alkanes). It is true that AOP processes are highly efficient but should always be used after conventional methods have significantly degraded the biodegradable waste [21]. The application of AOP at the correct locus in a process chain can increase the efficiency of the process by oxidizing the pollutant and mineralizing it completely into non-toxic and/or inorganic byproducts [22].

Recently, field application of AOPs combined with conventional methods is much in use [23]. Selection of treatment, compatibility of all selected processes and cost effectiveness of the total setup remains a task. Certain factors that must be taken into consideration are the nature of the solution under investigation and parent contaminant to be removed; selection of primary and secondary methods [24], efficiency post AOP and cost relevance, long term sustainability and eco-friendly approach [25]. The parameters determining the efficacy of the oxidation process are pH, temperature, concentration of pollutant and hydrogen peroxide, the reaction contact time, dose of catalyst, wavelength and intensity of UV radiations [26]. From an environmental point of view, one must take cognizance that the intermediate formed during the reaction should not be more toxic than the initial pollutant under study. Thus, it is unconditionally important to know the progress of the reaction and the final product to be characterized. Although the degradation of organic pollutants by hydroxyl radical is stepwise, a plethora of intermediates is formed, with different oxidation states before the complete mineralization into carbon dioxide and water [27]. Identification and quantification of all intermediates and determination of the kinetics and mechanisms of the individual is a daunting task. Moreover, while studying actual industrial effluents, which involve complex mixtures, complexities can be many folds and some of them might be resistant to degradation and accumulate in the system [28, 29]. Hence, degradation pathways and mechanisms can be still considered in their infancy for most of the AOPs. In comparison to conventional AOPs, photochemical and photocatalytic processes have gained a huge response to the treatment of polluted waters and effluents. It is cost effective and eco-friendly option for effluent treatment in tropical and subtropical regions. Due to the wide applications of AOPs and because of their complex mechanism, the intermediate formed could be more toxic than the parent and the excess use of peroxide and precipitation of iron during the Fenton reaction and its removal needs further study [30].

Nano means 'dwarf"; Richard Feynman laid the foundation for nanotechnology. The term nano-technology was introduced by Norio Taniguchi in 1974 [31]. Nano materials (NM) are emerging fields, which are finding varied applications in today's world. Nano materials are claimed to be environment friendly, on the basis of processing efficiency, enhancing reactivity, surface area, economic benefit and low energy consumption [32]. Nano materials usually have a larger specific area and hence, have a high density at active sites, which results in improved surface reactivity. Some nanomaterials show super para magnetism, or even quantum confinement effect in a given particle size range [33]. The size dependent property of nano materials can be used for wastewater effluent treatment systems in order to improve the quality of water at discharge. Nano materials are multifaceted and hence, are effective in membrane processes [34], adsorption and separation processes [35], catalytic oxidation [36], sensing and disinfection.

Nano-particles possess favorable physical, chemical and biological characteristics due to their size range between 1nm-100nm. Several transition metals in their oxide form such as Cu, Zn [37], Fe [38, 39], Co, Sn [40], W [41], and Ti [42] have found useful applications in advanced fields of material science, chemistry and physics. These oxides have been especially useful in water treatment as they incur the low cost and provide better treatment, and are thus gaining popularity to be utilized in wastewater treatment practices. Iron oxide exists in many phases such as goethite, hematite, magnetite, akaganeite and maghemite. Amongst these hematite is the most significant and attractive, which can be synthesized by different methods like sol-gel [43], precipitation, solvothermal, electrochemical, hydrothermal [44], combustion [45], pyrolysis, micro emulsion and oxidation [46]. Presently, numerous nano-particles widely in use are divided into five classes which include:

- Nano adsorbents like metal oxides are applied usually for the removal of heavy metals.
- Nanomaterials of carbon are used for effective adsorption and conduction process.
- Graphene-based nanoparticles for environmental remediation.
- Nanotubes have been used for effective removal of pollutants by means of hybrid nano membranes, nano fibers and carbon nanotube membranes.
- Recyclable nano composites, mats, and beads in water decontamination.

Finally, some zero-valent nano sized metals show strong absorption capability and operational simplicity [47 - 49]. For the elimination of such heavy metal ions, certain novel nanometerials are introduced, such as metal-organic frameworks, g-$C_3N_4$ [50] and M Xenes [51]. Metal-organic frameworks (MOF) [52 - 54] show better efficiency over a wide pH range, they can be reused and regenerated easily. Modified composites possess increased surface area, they also have more porosity hence, the efficiency of NPs enhances.

Nanoparticles in wastewater treatment have to be used with utmost care, taking into consideration the damages it can cause to the environment [55]. Some protocols and guidelines have to be followed so that the nanoparticles achieve the purpose of water purification without harming the ecosystem [56]. With the growing awareness and understanding, from the ongoing studies about nano-technology and nano-toxicology, precaution and attention should be laid while its use and risk assessment should be done before its application [57].

Presently, benzoic acid degradation is studied using iron oxide nano particles. Benzoic acid is an aromatic compound containing carboxylic group directly attached to the benzene ring, it is white, crystalline, and weakly acidic when soluble in water. Usually, it is found in nature, in many plants, resins and animals. It is used for the production of glycol benzoates, as plasticizer in the adhesive formulation and used to manufacture alkyd resins and drilling mud additives for crude oil recovery applications. Benzoic acid is used as polymerized activator and retardant. Similarly, it is used in industries as a corrosion inhibitor, antifreeze coolant, a nucleating agent for a polyolefin, a dye intermediate, and in photographic processing as a stabilizer. Benzoic esters are used as solvents in dying carriers, disinfectant additives, penetrating agents, pesticides, and manufacturing of other compounds. The preservatives are frequently used as antimicrobial substances in varieties of foods, such as marinated fish, fruit-based fillings, jam, salad cream, soft drinks, and beer. The use of food additives has increased enormously in the last few decades [58]. As a result, it has been estimated that today about 75% of the Western diet is made up of various processed foods, each person consuming an average of 8–10 lbs of food additives per year, with some possibly eating even more. Some adverse effects after the consumption of food additives are eczema, urticaria, angioedema, exfoliative dermatitis, irritable bowel syndrome, nausea, vomiting, diarrhoea, rhinitis, bronchospasm, migraine, anaphylaxis, childhood hyperactivity [59] and other behavioural disorders [60 - 63]. Some of the usages are mentioned in the tabular form in Table **1**.

**Table 1. Common uses of benzoic acid.**

| Chemical | Usage | References |
|---|---|---|
| Benzoic acid and its salts and esters | Toothpastes(0.5%), mouthwashes and dentifrices | [64, 65] |
| Benzoic acid | Cosmetics (0.5%) (creams and lotions) | [66, 67] |
| Benzoic acid | Deodorants | [68] |
| Benzoyl peroxide | Bleach flour (0.015%-0.075%) | [66] |
| Benzoyl peroxide Benzoic acid+ salicylic acid Whitfield's ointment) | Dermatological antifungal preparations Fungicidal treatment for ringworm | [69] |

In the present study, iron oxides were used as a photocatalyst to assist the degradation of benzoic acid in the aqueous sample [70]. Oxidants such as hydrogen peroxide enhanced the oxidation process by producing hydroxyl radicals (OH⋅). Photocatalysts absorb equal or more energy from the source gap (Sunlight, UV, and Vis) to overcome the band gap and initiates the shifting of electron from the valence band to the conduction band of nano particles. Thus, an electron and hole pair ($e^-h^+$) is generated [71, 72]. Hence, for carrying out effective photocatalytic oxidation, a model photocatalyst should possess remarkable properties such as: eco-friendly, budget caring, photo-stability, and chemically and biologically inert nature [73].

## METHODS AND MATERIALS

Degradation of benzoic acid was assisted by using Fe nano particles and hydrogen peroxide as an oxidant in the presence of visible light. Optimal pH was maintained using NaOH and $H_2SO_4$. The mode of synthesis was a hydrothermal process, where iron (III) oxide was synthesized using 4.85 g nitrate nano hydrate (Fe $(NO_3)_3 \cdot 9H_2O$ and 2.73 g of KOH, which was dissolved in 10 ml respectively. With vigorous stirring, the alkali solution was added to iron nitrate salt. The fine precipitate obtained was loaded into 100 ml Teflon–lined autoclave (Fig. **1**), the solution was made up to the 75% of the total volume by adding distilled water. The autoclave was maintained at 100°C for a period of 6 hours, wrapped with aluminum foil in order to trap the heat. After 6 hours, the autoclave was allowed to cool down in ambient conditions to room temperature. A yellow colored solid was obtained which was washed several times using absolute ethanol and distilled water in order to remove excess alkali. The subsequent product was allowed to dry at 60° C for 12 hours to yield iron oxide nano particles. The product obtained was goethite (α- FeOOH). This (α- FeOOH) was further annealed at 200°C for 3 hours to yield hematite (α- $Fe_2O_3$). Annealing of goethite (α- FeOOH) to hematite (α- $Fe_2O_3$)was assisted with a color change from yellow to red which confirms the transition. The resultant nanoparticles have been characterized using XRD, SEM, and FT-IR [74].

**Fig. (1).** Autoclave used to synthesize nanopaticles by hydrothermal process.

The nano particles that were obtained from the above-mentioned process were applied to a sample for the degradation of benzoic acid in the batch reactor. The reactor comprised of a cylindrical glass vessel with a narrow opening to extract samples for studies after required time intervals during the course of the reaction. To maintain the constant temperature cold water was circulated through the jacket fitted at the wide opening. A visible light source of 8W (philips) was used to enhance the reaction rate. Variation of nanoparticle concentration in the benzoic acid sample was carried out in order to reach an optimal value. The pH was measured using digital pH meter MKVI. After every 15 minutes, the sample was drawn and centrifuged. The decrease in the absorbance of the sample was recorded using a Labmann spectrophotometer at 272 nm (Fig. **2**). The nano systems were studied for the effect of change in concentration of nanoparticles and oxidants and the effect of pH on the degradability. The concentration of nanoparticles used were 0.01, 0.03, 0.05, 0.07, 0.1 g. Application of nano particles can be applied to a wide range of pH, presently experiments were carried out at the optimized pH of 7. The diagram displays the complete process for synthesis and subsequent degradation of benzoic acid (Fig. **3**).

**Fig. (2).** Labmann Spectrophotometer.

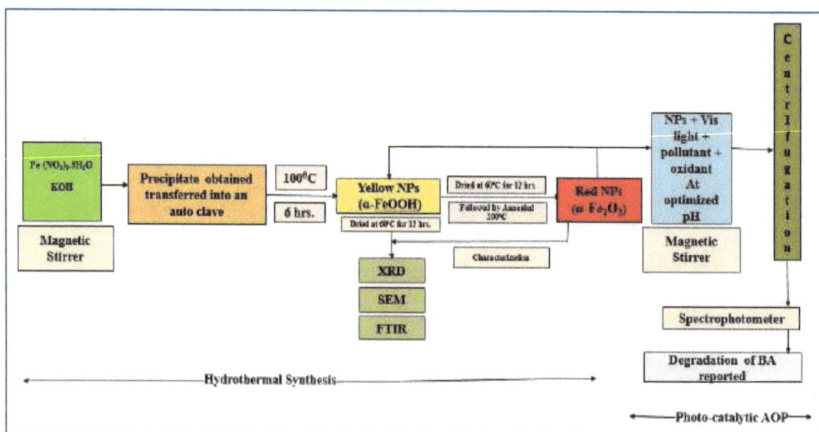

**Fig. (3).** Schematic diagram showing complete process from synthesis of Iron oxide nano particles by hydrothermal mode to photo catalytic aided AOP for the treatment of BA.

## RESULTS AND DISCUSSION

X-ray diffraction (XRD), Scanning Electron Microscope (SEM) and Energy dispersive analysis of X-rays (EDAX) were used for the characterization of the nano particles.

SEM images were used to study the morphology and size of the particles. The uniform size of iron nanoparticles were shown in Fig. (**4a and b**). α– FeOOH phases seen at ordinary temperature possess rod-like structure that gives high yield and were reproducible. However, when annealed at 200° C, α –FeOOH phase gets converted to α-$Fe_2O_3$ [38]. The colour changes from yellow to reddish brown at 200° C confirms the conversion of α –FeOOH phase to α-$Fe_2O_3$ as shown in equation (1). The removal of a water molecule from α –FeOOH, makes α-$Fe_2O_3$ nano particle more porous and slightly larger than the size. The size was found to vary from 19 nm to 25 nm [39] as calculated by the XRD data.

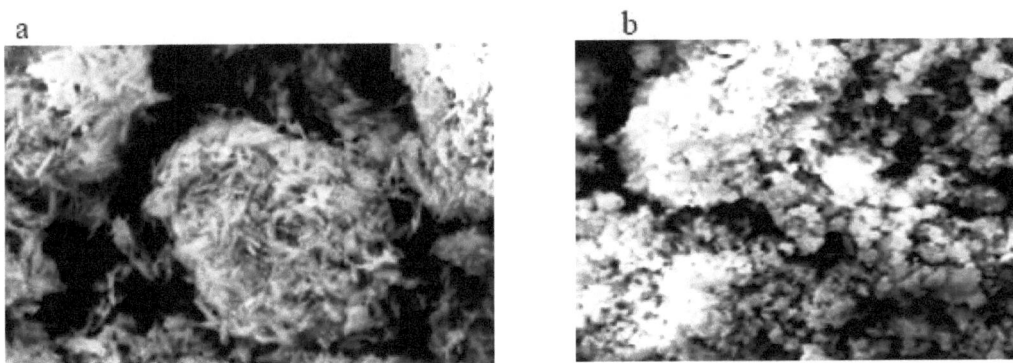

a

b

**Fig. 4. (a.)** SEM image of α –FeOOH nanoparticles obtained by the hydrolysis of $Fe(NO_3)_3.9H_2O$ at 100°C for 6 h.(b) SEM image of α -$Fe_2O_3$nanoparticles obtained after annealing at 200° C.

$$\alpha\text{–FeOOH} \xrightarrow{\text{dehydration}} \alpha\text{ -Fe}_2O_3 \tag{1}$$

For characterization of the nanoparticles formed using the hydrothermal method X-ray diffraction was observed using a Model Ragaku D 600, X-ray 40kV, 15mA, filter K-beta (x1), detector SC- 70, within a scanning range of 10.0000-90.0000 degree at room temperature. The scan speed/ duration time was 10.0000 deg/min.

The α- FeOOH nanoparticles that were synthesized using the hydrothermal method showed XRD pattern with a wide angle range of 2θ from 20° to 60° with CuK (λ=0.154 nm) radiation. 2θ range of 24.2°, 33.09°, 35.49°, 54.08°, 62.9° are in comparison to the universal JCPDS card number 87-1164. Observing the diffraction peaks of the XRD, it can be inferred that α- FeOOH exists in an orthorhombic structure with lattice constants of a=4.62 A, b=9.95 A, c=3.02 A.

The XRD pattern shows a phase transformation at higher temperatures. α- FeOOH when annealed at 200°C for a period of 3 hours shows phase transformation to α-Fe$_2$O$_3$, which exists in a rhombohedral structure. 2θ range of 24.15°,33.18°, 35.69°, 54.08°, 62.56°, with lattice parameters as a=b=5.03A, c=13.76 A were in comparison to the universal JCPDS card number 89-8104. Debye-Scherrer equation was used to find the average crystalline size of the nanoparticles (19-25 nm), (2) (Fig. **5a** and **b**).

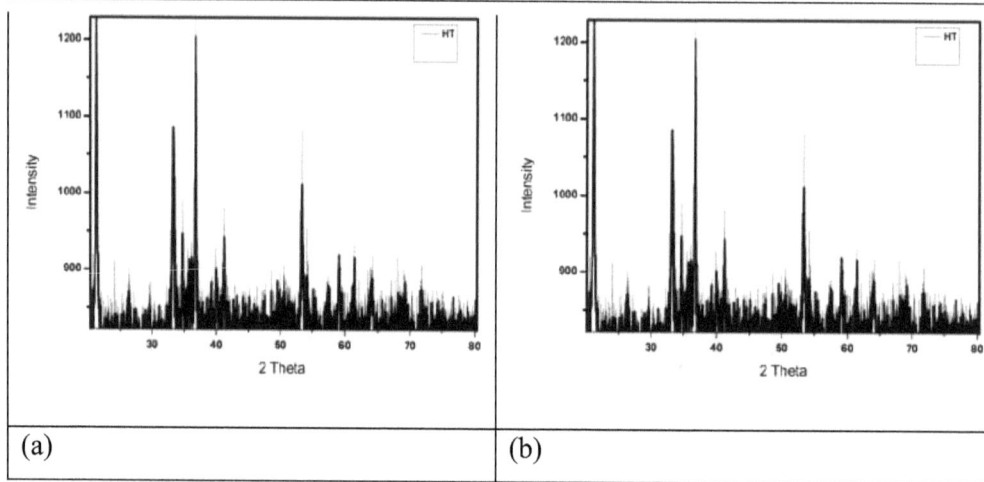

(a)    (b)

**Fig. (5).** XRD diffraction patterns of Iron nanoparticles at (**a**) at ordinary temperature and (**b**) annealed at 200° C prepared by hydrothermal process.

$$D = 0.94\ \lambda\ /\ \beta\ Cos\ \theta \tag{2}$$

Where λ is the X-ray wavelength, β is the full width of the diffraction line at the half of maximum intensity and θ the Bragg's angle [75].

ALPHA II Bruker spectrometer was used for performing FTIR. Molecular vibration developed in the infrared region of the electromagnetic spectrum works on the absorption of molecules. The absorption relates to bonds present in the molecules. As shown in Fig. (**6**), the IR radiation *versus* wavelength was taken by spectrometer measuring absorption of sample material usually from 4000-400 cm$^{-1}$. Opus-Touch software generated the report on measurement and evaluation. The IR Spectrum of α-FeOOH obtained shows the absorption at 519, 540, 564, 574, 625, 796, 896 and spectra for α- Fe$_2$O$_3$ shows absorption at 508, 515, 524, 531, 544, 554, 556, 562, 580, and 639. The high energy region should be associated with the presence of organic and water species and the low energy region corresponded to Fe-O vibration [76].

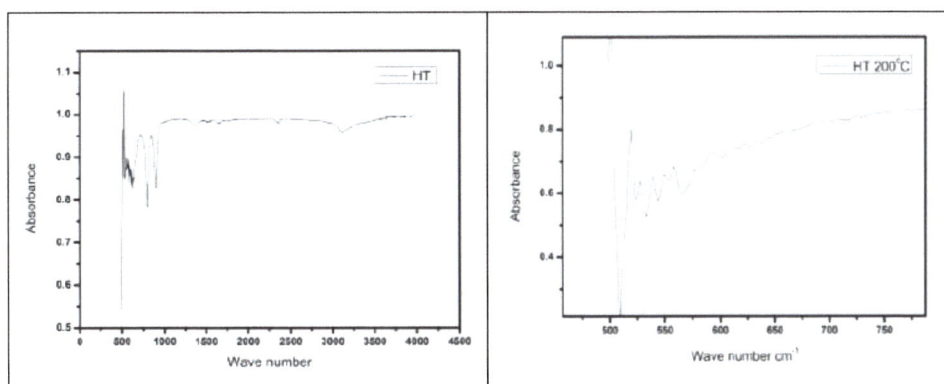

**Fig. (6).** FTIR spectra of α-FeOOH and α- $Fe_2O_3$ synthesised at ordinary temperature and 200°C respectively by hydrothermal process.

The EDAX image confirms iron oxide formation by hydrothermal process. It established the presence of Fe and O atoms in the sample as shown in Fig. (7). The image shows the higher C percentage, which may be due to organic solvent and the incomplete removal of K was found in the spectra may be due to insufficient washing of precipitate formed. The atomic percentages of Fe, O, C, and K as obtained from EDAX were 41%, 28%, 15% and 4%, respectively.

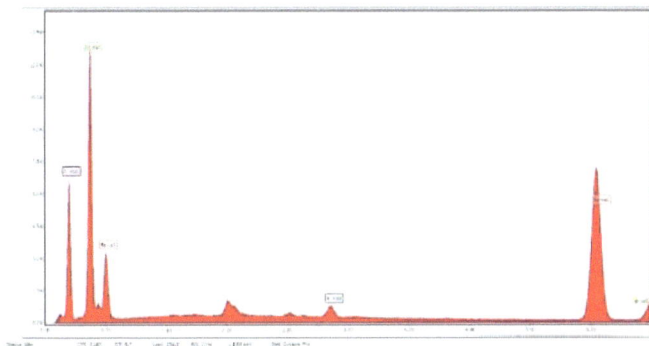

**Fig. (7).** EDAX image showing percentage of atoms present in Iron nanoparticles synthesized by hydrothermal route.

The nano particles synthesised by hydrothermal process were used as photocatalyst in AOP studies. Synthesized nanoparticles by hydrothermal process at ordinary temperature and elevated temperature were used to study the benzoic acid degradation. Goethite α-FeOOH was the phase at ordinary temperature, which transfers to α-$Fe_2O_3$ by annealing at 200°C. At optimised $H_2O_2$ and pH 7, using visible light for irradiation, the effect of variation of nano particles were studied in the batch reactor. Detail study of change in concentration with time is shown in Tables **2** and **3** and Fig. (**8** and **9**). Table **4** and Fig. (**10 a-c**) below show that percentage degradation of benzoic acid using 0.07g of α-FeOOH was 49.02% whereas the highest degradation of 90.90% was observed using 0.05g of α-Fe$_2$O$_3$.

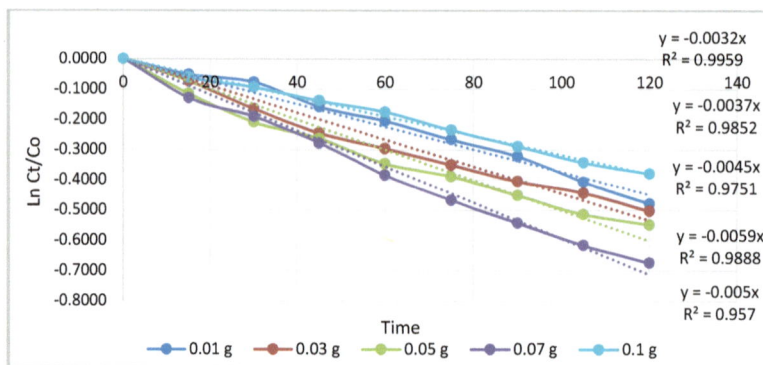

**Fig. (8).** Optimization of NP at ordinary temperature for BA at optimized [H$_2$O$_2$] = 560 ppm.

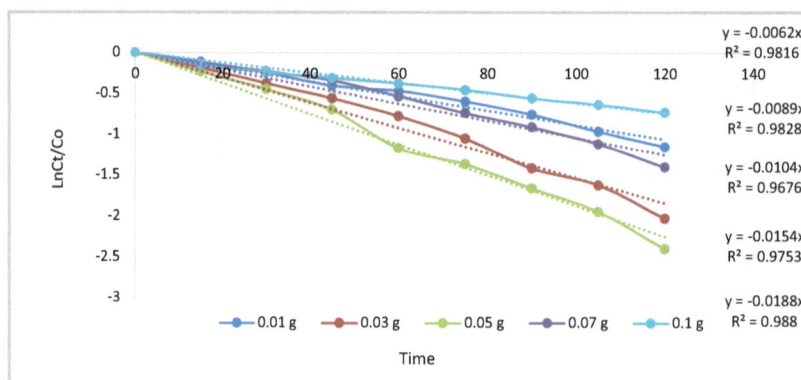

**Fig. (9).** Optimization of NP at 200° C for BA at optimized [H$_2$O$_2$] = 560 ppm

**Table 2. Optimization of NP at ordinary temperature for BA; [H$_2$O$_2$] = 560 ppm (pH = 7; [BA]$_0$ = 5.328×10$^{-4}$mg/L; λ Max= 272 nm; Temp= 25 ±2°C).**

| Time/ min | 0.01 g NP | | 0.03 g NP | | 0.05 g NP | | 0.07 g NP | | 0.1 g NP | |
|---|---|---|---|---|---|---|---|---|---|---|
| | [BA] | Ln(C$_t$ /C$_o$) | [BA] | Ln(C$_t$ /C$_o$) | [BA] | Ln(C$_t$ /C$_o$) | [BA] | Ln(C$_t$ /C$_o$) | [BA] | Ln(C$_t$ /C$_o$) |
| 0 | 0.000554 | 0.000 | 0.000524 | 0.000 | 0.000534 | 0.000 | 0.000558 | 0.000 | 0.000524 | 0.000 |
| 15 | 0.000526 | -0.0515 | 0.000486 | -0.0743 | 0.000476 | -0.1152 | 0.00049 | -0.1295 | 0.000495 | -0.0616 |
| 30 | 0.000513 | -0.0771 | 0.000444 | -0.1654 | 0.000433 | -0.2085 | 0.000461 | -0.1906 | 0.000479 | -0.0939 |
| 45 | 0.000473 | -0.2053 | 0.00041 | -0.2447 | 0.00041 | -02630 | 0.000422 | -0.2783 | 0.000459 | -0.1378 |

*(Table 2) cont.....*

| Time/ min | 0.01 g NP | | 0.03 g NP | | 0.05 g NP | | 0.07 g NP | | 0.1 g NP | |
|---|---|---|---|---|---|---|---|---|---|---|
| | [BA] | $Ln(C_t/C_o)$ | [BA] | $Ln(C_t/C_o)$ | [BA] | $Ln(C_t/C_o)$ | [BA] | $Ln(C_t/C_o)$ | [BA] | $Ln(C_t/C_o)$ |
| 60 | 0.000425 | -0.2053 | 0.000039 | -0.2962 | 0.000378 | -0.3460 | 0.00038 | -0.3840 | 0.000442 | -0.1754 |
| 75 | 0.000451 | -0.2661 | 0.000369 | -0.3504 | 0.000362 | -0.3886 | 0.00035 | -0.4670 | 0.000416 | -0.2347 |
| 90 | 0.000401 | -0.3217 | 0.00035 | -0.4043 | 0.00034 | -0.4507 | 0.000324 | -0.5424 | 0.000395 | -0.2884 |
| 105 | 0.000369 | -0.4066 | 0.000336 | -0.4431 | 0.000319 | -0.5131 | 0.000301 | -0.6159 | 0.000375 | -0.3420 |
| 120 | 0.000344 | -0.4779 | 0.000317 | -0.5024 | 0.000309 | -0.5478 | 0.000284 | -0.6738 | 0.000316 | -0.3783 |

**Table 3. Optimization of NPat 200° C for BA; $[H_2O_2]$ = 560 ppm**
**(pH = 7; $[BA]_0 = 2.642 \times 10^{-4}$ mg/L ; $\lambda$ Max= 272 nm; Temp= 25 $\pm$2°C).**

| Time/ min | 0.01 g NP | | 0.03 g NP | | 0.05 g NP | | 0.07 g NP | | 0.1 g NP | |
|---|---|---|---|---|---|---|---|---|---|---|
| | [BA] | $Ln(C_t/C_o)$ | [BA] | $Ln(C_t/C_o)$ | [BA] | $Ln(C_t/C_o)$ | [BA] | $Ln(C_t/C_o)$ | [BA] | $Ln(C_t/C_o)$ |
| 0 | 0.000237 | 0.000 | 0.000294 | 0.000 | 0.00028 | 0.000 | 0.000255 | 0.000 | 0.000255 | 0.000 |
| 15 | 0.000212 | -0.1133 | 0.000244 | -0.1848 | 0.000221 | -0.2329 | 0.000224 | -0.1315 | 0.000219 | -0.1534 |
| 30 | 0.000186 | -0.2412 | 0.000204 | -0.3751 | 0.000179 | -0.4452 | 0.000203 | -0.2279 | 0.000201 | -0.2399 |
| 45 | 0.000159 | -0.4029 | 0.00168 | -0.5586 | 0.000139 | -0.6975 | 0.000182 | -0.3412 | 0.000188 | -0.3084 |
| 60 | 0.000149 | -0.4659 | 0.000136 | -0.7746 | 0.0000871 | -1.1658 | 0.00015 | -0.5316 | 0.000175 | -0.3751 |
| 75 | 0.000131 | -0.5960 | 0.000103 | -1.0504 | 0.0000714 | -1.3649 | 0.000122 | -0.7367 | 0.000162 | -0.4540 |
| 90 | 0.000111 | -0.7563 | 0.0000714 | -1.4155 | 0.0000532 | -1.6582 | 0.000103 | -0.9092 | 0.000146 | -0.5561 |
| 105 | 0.0000908 | -0.9606 | 0.0000581 | -1.6219 | 0.0000399 | -1.9459 | 0.0000835 | -1.1178 | 0.000136 | -0.6334 |
| 120 | 0.000075 | -1.1510 | 0.0000387 | -2.0273 | 0.0000254 | -2.3979 | 0.0000629 | -1.4006 | 0.000123 | -0.7269 |

**Table 4. Comparison between percentage degradation of BA using Nanoparticles synthesized by the hydrothermal process at ordinary temperature and at elevated temperature 200°C.**

| S.No. | Nano particles | 0.01 g | 0.03 g | 0.05 g | 0.07 g | 0.1 g |
|---|---|---|---|---|---|---|
| 1. | % D HT | 37.99 | 39.49 | 42.17 | 49.02 | 31.49 |
| | k | 0.0037 | 0.0045 | 0.0059 | 0.005 | 0.0032 |
| | $R^2$ | 0.9852 | 0.9751 | 0.9888 | 0.957 | 0.9959 |
| 2. | % D HT 200 | 68.36 | 86.83 | 90.90 | 75.35 | 51.65 |
| | k | 0.0089 | 0.0154 | 0.0188 | 0.0104 | 0.0062 |
| | $R^2$ | 0.9828 | 0.9753 | 0.988 | 0.9676 | 0.9816 |

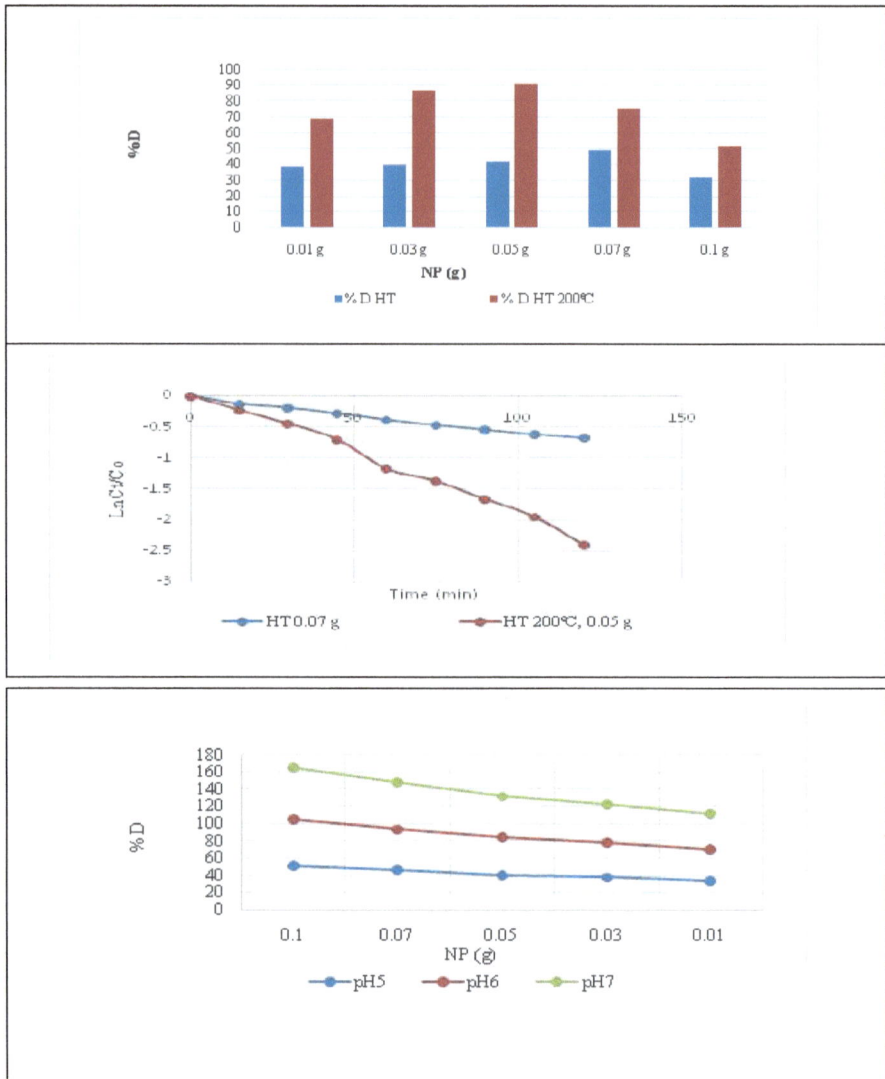

**Fig. (10).** (**a**)Graph showing % D of benzoic acid at different concentration using: (i) α-FeOOH, (ii) α-Fe$_2$O$_3$. (**b**) ln C$_t$/C$_o$ Vs t for optimized values. (**c**) Optimization of pH.

## CONCLUSION

By using a hydrothermal process, α-FeOOH and α-Fe$_2$O$_3$ nanoparticles were successfully synthesized. The slow addition of KOH in iron nitrate and the temperature of synthesis were the key factors for controlled growth and crystal structure. From the XRD data, 19-25 nm was the average crystalline size of the particles synthesized. $\alpha$-$Fe_2O_3$ was formed by the dehydration of $\alpha$-FeOOH nanowires at elevated temperatures. $\alpha$-$Fe_2O_3$ possesses a high surface area and shows better degradation of benzoic than $\alpha$-FeOOH due to mesoporous structure. The percentage degradation of benzoic acid using 0.07 g $\alpha$-FeOOH and 0.05 g $\alpha$-$Fe_2O_3$ was 49.02% and 90.90%, respectively. From the results, $\alpha$-FeOOH showed less photocatalytic property than the $\alpha$-$Fe_2O_3$ nanoparticles with the oxidant. Moreover, the recovery and reusability of the nanoparticles were subjected to their magnetic property. The degradation of benzoic acid with respect to hydrogen peroxide shows first order reaction. The highest rate constant observed was 0.0188 $min^{-1}$. Benzoic acid degradation using nano- AOP was found to be an eco-friendly, time-saving and economical method. Hence, nano aided AOP processes can be further used for the degradation of the organic compounds found in industrial wastewater effectively and efficiently.

## CONSENT FOR PUBLICATION

Not applicable.

## CONFLICT OF INTEREST

The authors declare no conflict of interest, financial or otherwise.

## ACKNOWLEDGEMENTS

The authors are grateful to the Director, Laxminarayan Institute of Technology, R.T.M. Nagpur University, Nagpur (MS) for providing the necessary facilities required during the work.

## REFERENCES

[1]     Wojnárovits L, Takács E. Radiation induced degradation of organic pollutants in waters and wastewaters. Top Curr Chem (Cham) 2016; 374(4): 50.
        [http://dx.doi.org/10.1007/s41061-016-0050-2] [PMID: 27573402]

[2]     Pendergast MM, Hoek EMV. A review of water treatment membrane nanotechnologies. Energy Environ Sci 2011; 4(6): 1946.
        [http://dx.doi.org/10.1039/c0ee00541j]

[3]     World Health Organization. Gender, climate change and health Bull World Health Organ 2013.

[4]     Beck L, Bernauer T, Beck L, Bernauer T. How will combined changes in water demand and climate affect water availability in the Zambezi river basin? Glob Environ Chang 2011.
        [http://dx.doi.org/10.1016/j.gloenvcha.2011.04.001]

[5]     Wang C, Du X, Li J, Guo X, Wang P, Zhang J. Environmental Photocatalytic Cr (VI) reduction in metal-organic frameworks: A. Appl Catal B 2016; 193: 198-216.
        [http://dx.doi.org/10.1016/j.apcatb.2016.04.030]

[6]     Zhang Y, Wu B, Xu H, *et al.* Nanomaterials-enabled water and wastewater treatment. NanoImpact 2016; 3–4: 22-39.
        [http://dx.doi.org/10.1016/j.impact.2016.09.004]

[7]     Deshpande BD, Agrawal PS, Yenkie MKN. Nanoparticles aided AOP for Degradation of p - NitroBenzoic acid. Mater Today 2020.

[8]     Salem HM FA, Eweida EA. Heavy metals in drinking water and their environmental impact on human health. ICEHM 2000; 542-56.

[9]     Pignatello JJ, Oliveros E, MacKay A. Advanced oxidation processes for organic contaminant destruction based on the fenton reaction and related chemistry. Crit Rev Environ Sci Technol 2006; 36(1): 1-84.
[http://dx.doi.org/10.1080/10643380500326564]

[10]    Deshpande B D, Agrawal P S, Yenkie M K N. Advanced oxidative degradation of benzoic acid and 4-nitro benzoic acid – A comparative study Advanced Oxidative Degradation of Benzoic Acid and 4-Nitro Benzoic Acid – A Comparative Study. AIP 2019; 210003

[11]    Deshpande BD, Agrawal PS, Yenkie MKN. AOP as a degradative tool for oxidation of 4-hydroxybenzoic acid. 2019; 020034(May): 3-9.
[http://dx.doi.org/10.1063/1.5100402]

[12]    Parmar A. Fenton process: A case study for treatment. 2014; 1(2): 23-30.

[13]    Liu P, Zhang H, Feng Y, Yang F, Zhang J. Removal of trace antibiotics from wastewater: A systematic study of nanofiltration combined with ozone-based advanced oxidation processes. Chem Eng J 2014; 240: 211-20.
[http://dx.doi.org/10.1016/j.cej.2013.11.057]

[14]    Jian M, Liu B, Zhang G, Liu R, Zhang X. Colloids and surfaces A: Physicochemical and engineering aspects adsorptive removal of arsenic from aqueous solution by zeolitic imidazolate framework-8 (ZIF-8) nanoparticles. Colloids Surf A Physicochem Eng Asp 2015; 465: 67-76.
[http://dx.doi.org/10.1016/j.colsurfa.2014.10.023]

[15]    Del Moro G, Mancini A, Mascolo G, Di Iaconi C. Comparison of UV/$H_2O_2$ based AOP as an end treatment or integrated with biological degradation for treating landfill leachates. Chem Eng J 2013; 218: 133-7.
[http://dx.doi.org/10.1016/j.cej.2012.12.086]

[16]    Sharma S, Ruparelia J, Patel M. A general review on advanced oxidation processes for waste water treatment Int Conf Curr. 8-10.

[17]    Oturan MA, Aaron JJ. Advanced oxidation processes in water/wastewater treatment: Principles and applications. A review. Crit Rev Environ Sci Technol 2014; 44(23): 2577-641.
[http://dx.doi.org/10.1080/10643389.2013.829765]

[18]    Liu K, Pérez-González A, Urtiaga a M. New concepts on UV/$H_2O_2$ oxidation. 2013; 127.(June)

[19]    Krishnan S, Rawindran H, Sinnathambi CM, Lim JW. Comparison of various advanced oxidation processes used in remediation of industrial wastewater laden with recalcitrant pollutants IOP Conf Ser Mater Sci Eng. 206
[http://dx.doi.org/10.1088/1757-899X/206/1/012089]

[20]    Hassaan MA, El Nemr A. Advanced oxidation processes for textile wastewater treatment. Int J Photochem Photobiol 2017; 2(3): 85-93.

[21]    Toor R, Mohseni M. UV-$H_2O_2$ based AOP and its integration with biological activated carbon treatment for DBP reduction in drinking water. 2007; 66: 2087-95.
[http://dx.doi.org/10.1016/j.chemosphere.2006.09.043]

[22]    An J, Zhu L, Zhang Y, Tang H. Efficient visible light photo-fenton-like degradation of organic pollutants using *in situ* surface-modified $BiFeO_3$ as a catalyst. J Environ Sci (China) 2013; 25(6): 1213-25.
[http://dx.doi.org/10.1016/S1001-0742(12)60172-7] [PMID: 24191612]

[23]    Azbar N, Yonar T, Kestioglu K. Comparison of various advanced oxidation processes and chemical

treatment methods for COD and color removal from a polyester and acetate fiber dyeing effluent. Chemosphere 2004; 55(1): 35-43.
[http://dx.doi.org/10.1016/j.chemosphere.2003.10.046] [PMID: 14720544]

[24]   Gupta A. UNIT-I Wastewater treatment , primary treatment of wastewater wastewater engineering.

[25]   Oller I, Malato S, Sánchez-Pérez JA. Combination of advanced oxidation processes and biological treatments for wastewater decontamination--a review. Sci Total Environ 2011; 409(20): 4141-66.
[http://dx.doi.org/10.1016/j.scitotenv.2010.08.061] [PMID: 20956012]

[26]   Soler J, Santos-Juanes L, Miró P, Vicente R, Arques A, Amat AM. Effect of organic species on the solar detoxification of water polluted with pesticides. J Hazard Mater 2011; 188(1-3): 181-7.
[http://dx.doi.org/10.1016/j.jhazmat.2011.01.089] [PMID: 21353387]

[27]   Hodges BC, Cates EL, Kim JH. Challenges and prospects of advanced oxidation water treatment processes using catalytic nanomaterials. Nat Nanotechnol 2018; 13(8): 642-50.
[http://dx.doi.org/10.1038/s41565-018-0216-x] [PMID: 30082806]

[28]   Beltrán FJ, Encinar JM, González JF. Industrial wastewater advanced oxidation. Part 2. Ozone combined with hydrogen peroxide or UV radiation. Water Res 1997; 31(10): 2415-28.
[http://dx.doi.org/10.1016/S0043-1354(97)00078-X]

[29]   Pandya MT. Treatment of industrial wastewater using photooxidation and bioaugmentation technology. Water Sci Technol 2007; 56(7): 117-24.
[http://dx.doi.org/10.2166/wst.2007.694] [PMID: 17951875]

[30]   Kusic H, Koprivanac N, Bozic AL. Minimization of organic pollutant content in aqueous solution by means of AOPs: UV- and ozone-based technologies. 2006; 123: 127-37.

[31]   Taniguchi N. On the basic concept of nano-technology. Proceedings of the International Conference on Production Engineering, Tokyo, Part II.

[32]   Mohd Amil Usmani MO, Khan I, Bhat AH, Pillai RS, Ahmad N, Mohamad Haafiz MK. Current trend in the application of nanoparticles for waste water treatment and purification: A review. Curr Org Synth 2017; 14(2)

[33]   Khin MM, Nair AS, Babu VJ, Murugan R, Ramakrishna S. A review on nanomaterials for environmental remediation. Energy Environ Sci 2012; 5(8): 8075.
[http://dx.doi.org/10.1039/c2ee21818f]

[34]   Das S, Sen B, Debnath N. Recent trends in nanomaterials applications in environmental monitoring and remediation. Environ Sci Pollut Res Int 2015; 22(23): 18333-44.
[http://dx.doi.org/10.1007/s11356-015-5491-6] [PMID: 26490920]

[35]   Ayati A, Ahmadpour A, Bamoharram FF, Tanhaei B, Mänttäri M, Sillanpää M. A review on catalytic applications of Au/TiO$_2$ nanoparticles in the removal of water pollutant. Chemosphere 2014; 107: 163-74.
[http://dx.doi.org/10.1016/j.chemosphere.2014.01.040] [PMID: 24560285]

[36]   Karn B, Kuiken T, Otto M. Nanotechnology and *in situ* remediation: A review of the benefits and potential risks. Environ Health Perspect 2009; 117(12): 1813-31.
[http://dx.doi.org/10.1289/ehp.0900793] [PMID: 20049198]

[37]   Kathirvelu S, Souza LD, Dhurai B. UV protection finishing of textiles using ZnO nanoparticles. 2009; 34: 267-73.

[38]   Gandha K, Mohapatra J, Hossain MK, *et al.* Mesoporous iron oxide nanowires: Synthesis, magnetic and photocatalytic properties. RSC Advances 2016; 6(93): 90537-46.
[http://dx.doi.org/10.1039/C6RA18530D]

[39]   Ou P, Xu G, Ren Z, Hou X, Han G. Hydrothermal synthesis and characterization of uniform α-FeOOH nanowires in high yield. Mater Lett 2008; 62(6–7): 914-7.
[http://dx.doi.org/10.1016/j.matlet.2007.07.010]

[40]  Wu CH, Chang CL. Decolorization of Reactive Red 2 by advanced oxidation processes: Comparative studies of homogeneous and heterogeneous systems. J Hazard Mater 2006; 128(2-3): 265-72.
[http://dx.doi.org/10.1016/j.jhazmat.2005.08.013] [PMID: 16182444]

[41]  Shahid Arshad G, Djinović P, Zavašnik J, Pintar L. Electron trapping energy states of $TiO_2$–$WO_3$ composites and their influence on photocatalytic degradation of bisphenol A. Appl Catal B Environ 2017; 209(15): 273-84.

[42]  Tomova L B D, Iliev V, Rakovsky S. Gold modified n-doped $TiO_2$ and n-doped $WO_3$/$TiO_2$ semiconductors - photocatalysts for uv-visible light destruction of 2, 4, 6-trinitrotoluene in aqueous solution. Nanosci Nanotechnology 2011; 2(12)

[43]  Hassena H. Photocatalytic degradation of methylene blue by using $Al_2O_3$/$Fe_2O_3$ Nano Composite under Visible Light. Mod Chem Appl 2016; 4(1): 3-7.

[44]  Byrappa K, Adschiri T. Hydrothermal technology for nanotechnology. 2007; 53

[45]  Aruna ST, Mukasyan AS. Combustion synthesis and nanomaterials. Curr Opin Solid State Mater Sci 2008; 12(3–4): 44-50.
[http://dx.doi.org/10.1016/j.cossms.2008.12.002]

[46]  Qu X, Brame J, Li Q, Alvarez PJ. Nanotechnology for a safe and sustainable water supply: Enabling integrated water treatment and reuse. Acc Chem Res 2013; 46(3): 834-43.
[http://dx.doi.org/10.1021/ar300029v] [PMID: 22738389]

[47]  Sheng G, Tang Y, Linghu W, *et al.* Enhanced immobilization of $ReO_4$ − by nanoscale zerovalent iron supported on layered double hydroxide *via* an advanced XAFS approach: Implications for $TcO_4$ − sequestration. Appl Catal B 2016; 192: 268-76.
[http://dx.doi.org/10.1016/j.apcatb.2016.04.001]

[48]  Jan Filip O Š, Kolařík J, Petala E, Petr M, Zbořil R. Nanoscale zerovalent iron particles for treatment of metalloids. Springer 2019.

[49]  Li X, Ai L, Jiang J. Nanoscale zerovalent iron decorated on graphene nanosheets for Cr (VI) removal from aqueous solution: Surface corrosion retard induced the enhanced performance. Chem Eng J 2016; 288: 789-97.
[http://dx.doi.org/10.1016/j.cej.2015.12.022]

[50]  Wang C, Xiao-Hong Yi PW. Powerful combination of MOFs and $C_3N_4$ for enhanced photocatalytic performance. Appl Chem Environmental 2019; 247: 24-48.

[51]  Ming V, *et al.* Correction: Recent progress in layered transition metal carbides and / or nitrides (MXenes) and their. J Mater Chem A Mater Energy Sustain 2017; 5(18): 8769.
[http://dx.doi.org/10.1039/C7TA90088K]

[52]  Kobielska PA, Howarth AJ, Farha OK, Nayak S. Metal – organic frameworks for heavy metal removal from water. Coord Chem Rev 2018; 358: 92-107.
[http://dx.doi.org/10.1016/j.ccr.2017.12.010]

[53]  Ke F, Qiu LG, Yuan YP, *et al.* Thiol-functionalization of metal-organic framework by a facile coordination-based postsynthetic strategy and enhanced removal of $Hg^{2+}$ from water. J Hazard Mater 2011; 196: 36-43.
[http://dx.doi.org/10.1016/j.jhazmat.2011.08.069] [PMID: 21924826]

[54]  Dandan L. Metal-organic frameworks for catalysis: State of the art, challenges, and opportunities. Energy chem 2019; 100005(April)

[55]  Park B, Donaldson K, Duffin R, Kelly F, Mudway I. Hazard and risk assessment of a nanoparticulate cerium oxide-based diesel fuel additive — a case study Robert Guest and Peter Jenkinson. Inhal Toxicol 2008; 20(January): 547-66.
[http://dx.doi.org/10.1080/08958370801915309] [PMID: 18444008]

[56]  Maynard AD, Warheit DB, Philbert MA. The new toxicology of sophisticated materials:

Nanotoxicology and beyond. Toxicol Sci 2011; 120(1) (Suppl. 1): S109-29.
[http://dx.doi.org/10.1093/toxsci/kfq372] [PMID: 21177774]

[57]   Cassee FR, *et al.* Exposure , health and ecological effects review of engineered nanoscale cerium and cerium oxide associated with its use as a fuel additive. Crit Rev Toxicol 2011; 41: 213-29.
[http://dx.doi.org/10.3109/10408444.2010.529105]

[58]   Ünal F, Yılmaz S, Aksoy H, Zengin N, Yüzbas D. The evaluation of the genotoxicity of two food preservatives: Sodium benzoate and potassium benzoate. 2011; 49: 763-9.

[59]   Egger HL, Angold A. Common emotional and behavioral disorders in preschool children: Presentation, nosology, and epidemiology. 2006; 4: 313-37.

[60]   Michaelsson G, Juhlin L. Urticaria induced by preservatives and dye additives in food and drugs. 1973; 525-33.

[61]   Gray J, Close K. Food intolerance and food aversion. 135-42.
[http://dx.doi.org/10.1111/j.1467-3010.1984.tb01348.x]

[62]   Tuormaa T E. The adverse effects of food additives on health: A review of the literature with special emphasis on childhood hyperactivity. 1970.

[63]   Doguc D K, Vatansev H. Int J Health Nutr 2013; (January):

[64]   Kanerva E S L. Contact allergens in toothpastes and a review of their hypersensitivity. 33(2).

[65]   Ishida H. Novel 5-aminoflavone derivatives as specific antitumor agents in breast cancer. 1996; 39(18): 3461-9.

[66]   Directory of Microbicides for the Protection of Materials: A Handbook.

[67]   Shahmohammadi M, Javadi M, Nassiri-asl M. An overview on the effects of sodium benzoate as a preservative in food products. 2016; 3(3)
[http://dx.doi.org/10.17795/bhs-35084]

[68]   Wibbertmann A, Kielhorn J, Koennecker G, Mangelsdorf I, Melber C, Melber DC. Benzoic acid and sodium benzoate. World Heal Organ Geneva 2005; 26: 1-52.

[69]   Benzoyl peroxide dermatological antifungal preparations. Willingford 2005.

[70]   Benhebal H, Chaib M, Salmon T, *et al.* Photocatalytic degradation of phenol and benzoic acid using zinc oxide powders prepared by the sol-gel process. Alex Eng J 2013; 52(3): 517-23.
[http://dx.doi.org/10.1016/j.aej.2013.04.005]

[71]   Kumar A, Pandey G. A review on the factors affecting the photocatalytic degradation of hazardous materials. 2018; (November 2017):

[72]   Horváth E, Szabó-Bárdos H. Photocatalytic oxidation of oxalic acid enhanced by silver deposition on a $TiO_2$ surface. J Photochem Photobiol Chem 2003; 154(2–3): 195-201.

[73]   Bora LV, Mewada RK. Visible/solar light active photocatalysts for organic effluent treatment: Fundamentals, mechanisms and parametric review. Renew Sustain Energy Rev 2017; 76: 1393-421.

[74]   Pattanayak B C. Synthesis and characterization of alumina / iron oxide mixed nanocomposite. 2010.

[75]   Tharani K, Nehru LC. Synthesis and characterization of iron oxide nanoparticle by precipitation method. Int J Adv Res Phys Sci 2015; 2(8): 47-50.

[76]   Aliahmad M, Nasiri Moghaddam N. Synthesis of maghemite ($\gamma$-$Fe_2O_3$) nanoparticles by thermal-decomposition of magnetite ($Fe_3O_4$) nanoparticles. Mater Sci Pol 2013; 31(2): 264-8.
[http://dx.doi.org/10.2478/s13536-012-0100-6]

CHAPTER 6

# Wastewater Purification Using Nano-Scale Techniques

**Bhavna D. Deshpande[1], Pratibha S. Agrawal[1], M.K.N. Yenkie[1] and S.J. Dhoble[2]**

[1] *Department of Applied Chemistry, Laxminarayan Institute of Technology, R.T.M. Nagpur University, Nagpur, India–440010*

[2] *Department of Physics, R.T.M.Nagpur University, Nagpur, India-440033*

**Abstract:** This paper presents an exhaustive study of modern methods used to purify water with the support of nanomaterials. For deriving maximum benefits from nanotechnology, the environmental sustainability of the nano-particles must be assessed. Nanoparticles possess useful characteristics contributing to water treatment and the removal of numerous pollutants. Materials such as zeolites, chitosan, MWCNT, nano-composites ($Fe_3O_4/TiO_2$, $GO/FeO \cdot Fe_2O_3$, *etc.*), nano-oxides (ZnO, $TiO_2$, $Al_2O_3$, $Fe_2O_3$, $Fe_3O_4$, *etc.*) and MOF (MOF-808, Cu-terephthalate, $CoFe_2O_4$ /MIL-100(Fe), UiO-66-NHC(S) NHMe, *etc.*) have been included in the study including their apparent functionality in treating contaminated water streams. Additionally, known methods to synthesize these nano particles from diverse sources have been studied. The review highlights the removal of pollutants (non-biodegradable, heavy metals, inorganics, and organics) by adsorption using photo nano adsorbents. Devoid of any recognized standards, the performance of the nanomaterials in wastewater treatment needs further research. With the further advancement of nano technology, ideological guidelines along with general cons and future challenges affecting humans and the ecosystem have been reported to provide further scope for research in this domain.

**Keywords:** Advance oxidation process, Adsorbent, Nano-materials, Non-biodegradable, Oxidation, Photo-catalyst, Water treatment.

## INTRODUCTION

The man's hunt for information, knowledge, and facts has led him to imagine and evolve new components. With the new inventions, the dimensions have been reduced, and the efficiency has increased several folds.

* **Corresponding author Bhavna D. Deshpande:** Department of Applied Chemistry, Laxminarayan Institute of Technology, R.T.M. Nagpur University, Nagpur, India–440010; Tel: +91-9881370448; E-mail: dnd.bhavna@gmail.com

R. M. Belekar, Renu Nayar, Pratibha Agrawal and S. J. Dhoble (Eds)
All rights reserved-© 2022 Bentham Science Publishers

A few decades ago, the technology was micro and macro-based, where micro-energy and microparticles were used by engineers in microchips, micrometers, microcells, and microprocessors. It is evident that miniaturization of devices from the micron to nanoscale improves experimental efficacy, although the proper demarcation between the two remains unclear [1]. Microparticles are microscopic in size, varying between 0.1 and 100 μm. Microparticles such as pollen, powdered sugar, sand, dust, and flour, which we come across daily [2]. Printer heads, sensors, and integrated circuits are examples of micro-scale products. Microparticles available commercially include ceramics, glass, polymers, and metals. They have been employed in wide applications such as pharmaceuticals, cosmetics, imaging, coating, electronics, and printing media, and have shown wide application in wastewater treatment. Micro-particles may have more than two components to exhibit different properties [3]. Due to their high porosity, non-toxicity, and high surface area to volume ratio, calcium carbonate micro-particles are used in various industrial applications, such as material filling, biomedical, the food industry, and environmental studies [4]. Cross-linked poly microparticles were prepared using an emulsifier-free and a single–step swelling polymerization process. Various organic pollutants are readily adsorbed by polystyrene microparticles. The application of a particle depends upon its size and shape [5]. Polymeric microparticles infused into ceramic, alumina, silica carbide, and titanium oxide were used for the decontamination of potable water [6]. Palladium microparticle exhibits catalytic activities, and it reduces dissolved oxygen in water and nitrobenzene to aniline [7]. Silver-microparticles complexed with chitosan were prepared using crosslinking agents to probe the behavior of pesticides such as methyl parathion (MP) [8]. Hence, magnetic alginate microparticles were used for the purification of α-amylases [9], providing a means for soil and water pollution remediation [10].

Today, the scale has been further pulled down to nano, increasing the strength, chemical reactivity, and surface area and at the same time reducing weight. Micro and nanoscale phenomena are widely used to overcome traditional limits on materials, systems, and technologies. The size and characteristics of particles determined by their growth mechanism define their applications. The nano era has reached every nook and corner of the world. Technologies are blending with nanotechnology, creating a change in fascinating ways. Nanoparticles have shown wider applications in various fields as compared to microparticles, which have limited applications. Micro- and nano-particulates have been used in the manufacturing sector, the electronic world, such as LED bulbs, tubes, TV, radios, and detectors on a pre-clinical basis as new drug-delivery devices, and in the water treatment process. Hybrid microparticles increase the adsorption capacity and hence, are more effective. The synthesis of micro and nanoparticles requires a stable chemical environment. The particle size of both micro and nano depends upon pressure, temperature, and concentration. Researchers have made an evolution in bringing out various applications of nanoparticles. The thought of being able to live in a world free from environmental issues, diseases, species extinction, starvation, and poverty is everyone's dream; nanotechnology has helped to achieve that goal in various fields. This review highlights the applications of nanotechnology in the purification of water, their synthesis, and their toxicity to the environment and humans. The review suggests that there exists enough room for further work in terms of the impact and risk of nanoparticles on the ecosystem.

## GENERAL APPLICATION OF MICRO AND NANOPARTICLES

Nanomaterials have shown immense potential in the fields of science and technology. They have shown immense potential in capturing solar energy like solar cells, medicines, and weather monitoring. It has paved the way for a variety of businesses, companies, commercial and industrial products, aerospace, nuclear, biomedical, electronics, energy, and metallurgical engineering. Calcium nano/microparticles have various applications in different fields such as paper, plastics, paints and coatings, medicines, the environment, catalysts, chemicals, and food industries, and wastewater treatment. The properties like high surface area, porosity, biocompatibility, and ability to alternate from micro to nano make it a versatile compound to be applied in various fields [4]. Fig. (**1**) depicts the wide applications of nanoparticles.

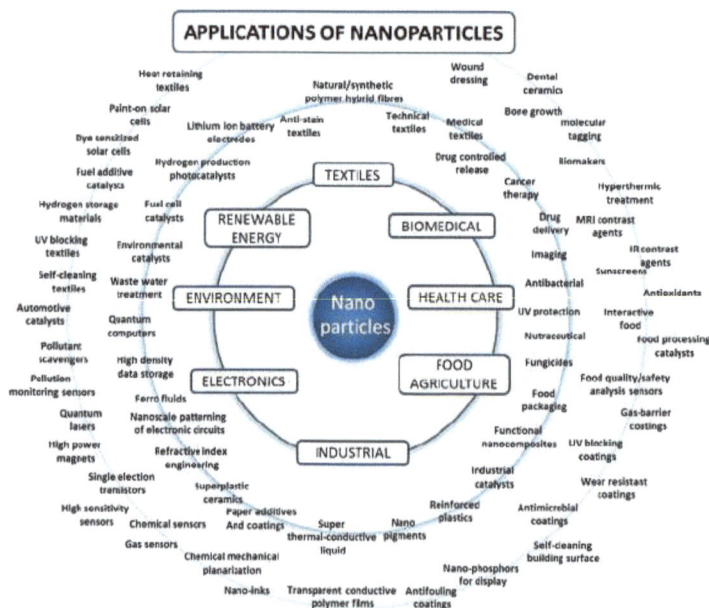

**Fig. (1).**  Nano-particles in various fields (on line source chandanashaw.svbtle.com).

General applications of micro and nanoparticles are under following headings.

*Technology*-CDs and DVDs have storage in micrometers, providing good data storage, but the storage space increases many folds when the scale is reduced to nanometres. A nanoscale film works as an antifog, antimicrobial, resistant to UV or IR, anti-reflective, self-cleaning, water-resistant, and repellent on computers, camera displays, automobile windows, and eyeglasses. Ultra-high-definition displays on television and computer screens use quantum dots to produce more vibrant colors and are energy efficient. Nanotechnology has revolutionized abundant electronic merchandise, processes, and their uses.

*Environment*-Nanoparticles are used to provide clean energy sources such as solar cells. Plastic solar cells are a cheap and alternative source of energy. These cells, used on buses and building roofs, can reduce electricity consumption. Nano-paints can reduce pollution, keep the house cool in the summer, and these paints are much more durable (chip resistant). They also help in decomposing air pollutants into safer compounds, thus keeping the air around us clean.

*Materials*-Nanoparticles have surprised us with the production of special stain-resistant clothes. Nano-fibres create a cushion of air around the fabric, hence used for preparing astronaut suits. Nanoscale additions in fabrics can make them lightweight, resistant to wrinkling, bacterial growth, and staining. These properties benefit the manufacturing of body armors for defense personnel, police force, anti-bomb squad, and space suits. Smack-free ketchup bottles are specially designed to provide a non-stick coating to food packaging [11].

*Health Care*- Nanoparticles assist in locating cancer cells, tumors, brain research, neuro-electronic interphase, and nerve cells in living beings and therefore are widely used as drugs and delivery devices. They are widely applied in biosensors, chemical, and biological sensors. Further areas benefiting from nanotechnology are: metal oxide nanoparticles like $ZnO$ and $TiO_2$ are widely used in cosmetics and skincare products for their sunscreen active properties. These nanoparticles are non-irritant towards the skin, hence photo-stable in nature [12, 13]. Photo-protective properties of $ZnO$ and $TiO_2$ can be enhanced by coating aluminum oxide, silicon dioxide, or silicon on them. Cotton fabrics are made UV protected by coating them with nano $ZnO$ and $TiO_2$ [14 - 16]. Noble metals like silver, gold and inorganic nanoparticles improve antimicrobial activity when added to $ZnO$ and $TiO_2$ [17, 19]. Hair growth is enhanced by using lotion-containing nanoparticles called ethosomes. The proteins derived from stem cells are encapsulated in liposome nanoparticles, which help prevent the aging of the skin, by merging with the membranes of skin cells to deliver the proteins. $Al_2O_3$ nanoparticles are widely used in plastic, rubber, and ceramic for reinforcing and toughening the products [20]. Silver nanoparticles are widely used in antibacterial mechanisms, along with clinical application [21, 22]. $CeO_2$ nanoparticles have proven to be an eco-friendly fuel additive, as they reduce fuel consumption, carbon dioxide emissions, and particulate emissions too [23, 24]. Lightweight automotive bodies can enhance fuel efficiency considerably. Nano-scale addition to producing metals or alloys has improved their properties, making them stiff, durable, resilient, and noncorrosive. Eco-friendly ways to produce ethanol for fuel from wood chops, perennial grasses, sugarcane, and corn stalks by nano-bioengineering of enzymes can save the environment largely. Nano-technology has also entered the series of high-grade household equipment, such as stain removers, degreases, air purifiers, filters, specialized self-cleaning paints, and sealing products. In case of oil spills, scientists have developed 'paper towels', which could absorb oil twenty times its weight, which were made out of potassium manganese oxide tiny wires for water clean-up purposes [25]. In aircraft, air filters contain nanoscale adsorbents that allow "mechanical filtration" and the charcoal layer acts as a deodorant. In recent years, steel industries are using electroless plating made of nano copper-tungsten-silicon electrodes successfully. Iron oxide (IO) nanoparticles are widely used in magnetic data storage [26, 27], bio-sensing [28], and drug-delivery. SPIONs are ideal for biomedical applications due to their manageable sizes, relatively long half-life, and low agglomeration [29, 30]. Nanotechnology has also brought immense progress in medicine, therapies, and medical equipment, and has offered solutions from prevention to treatment. Researchers are working to create needleless vaccines and anti-flu vaccines. Numerous applications of nanoparticles have made them inevitable in today's world. With its growing usage, its presence may affect both environment and ecosystem directly or indirectly [31, 32]. Hence, nanoparticles with unique properties and characteristics have shown their potential and applications worldwide for environmental remediation [33].

## NANOTECHNOLOGY FOR WATER PURIFICATION

Escalating demands for clean water throughout the world are due to the overuse of freshwater resources. Most aqueous bodies undergo a natural purification process by using sunlight to break down the organic molecules into simpler molecules. Natural sensitizers and semiconductor colloids can accelerate this process using solar energy. However, this natural process gets disturbed due to human interference. Due to industrialization and a population explosion, the manufacturing and production sectors have gone up in the past few years, leading to the scarcity of natural resources. Various unconventional water sources, such as stormwater, brackish water, rainwater, and contaminated water, can be used to meet the requirements of freshwater supply. Several physical, biological, chemical, physico-chemical, commercial, and non-commercial techniques have been developed to treat water in different ways. The freshwater resources are largely consumed in the pharmaceutical and textile industries, food and beverages, cosmetic and consumer products. Numerous non-biodegradable, aromatic compounds enter the ecosystem through the wastewater, affecting the living organisms. These pollutants have toxicological effects in high as well as low concentrations. Hence, before the onset of alarming conditions, urgent steps need to be taken for the development of new technologies and clean-up processes. Many processes are studied for the removal of organic hazardous wastes, such as adsorption of toxic matter on adsorbents that transfer the pollutant from one form to another. However, the photocatalytic process has a major advantage over such processes, which does not require further treatment; hence, they are easily disposable. Photocatalysts are reusable, and can be recycled as they are self-regenerating [34].

Advanced oxidation processes (AOPs) have widely been used in recent years to mineralize organic pollutants, into non-toxic, biodegradable, and simpler products. Photo-catalysis using heterogeneous nanoparticles is practiced for water treatment, and has gained encouraging results in the destruction of hazardous compounds. Dyes are the main pollutants emerging from textile industries, paints, glue, ink, and coloring industries, AOP has provided immense potential in treating such pollutants. AOPs are based on the production of highly reactive species generated *in situ* by different means. The various means to generate such highly reactive species include ozonation, Fenton oxidation, $UV/H_2O_2$, electrochemical oxidation, photocatalysis, and their combinations. The organic compounds (benzoic acids, their derivatives, phenols, phthalic acid and its derivatives, halogenated compounds, and their derivatives) have been destroyed in batch scales and field studies using homo and heterolytic photocatalysts [35 - 37]. Amongst this photocatalyst, mild conditions with primary reagents such as oxygen are used. The source of light used may be UV/Vis/solar. Recently, visible and solar sources have been used as renewable and cost-effective means for photo-degradation. These nano-sized photocatalysts have emerged as a boon for water treatment. Some water contaminants treated by photocatalysts are shown in Table **1**.

CECs are the newly observed substances in an environment whose adverse effects are yet to be understood completely. A low concentration of these compounds may lead to serious problems in living beings. EDCs are compounds that affect the endocrine system of living being adversely. It may lead to dreadful diseases like cancer, gene mutation, and genetic deformation in due course of time. Photocatalysis also possesses a bactericidal and germicidal potential that ultimately destroys internal components.

The nanoparticles used to improve water quality work on three key parameters [42]: catalytic or photocatalytic degradation of refractory organic compounds, removal of pollutants by adsorption, and disinfection of microbial organisms in contaminants. It is evident that nanotechnology has significant effects in exploring photocatalysis. The properties of nanoparticles can be further enhanced using doping, coupling, capping, and sensitizing techniques. Semiconductor composites with organic and inorganic compounds can also enhance photocatalytic or optical activity. Photocatalysts serve as a bridge between mechanisms and the significance of photoinitiated processes in semiconductors [34].

**Table 1. Main Water Pollutants Treated *via* Photocatalysis[38-41].**

| Type of Pollutant | Examples |
|---|---|
| Contaminants of Emerging Concern (CECs) | Additives (polybrominated diphenyl ethers) Antibiotics (amoxicillin, ampicillin, metronidazole) Disinfectants (haloacetic acids, trihalomethanes) Dyes (methylene blue, methyl orange, rhodamine B) Pharmaceuticals (carbamazepine, diclofenac, ibuprofen) Preservatives (dimethylphenols, parabens, salicylic acid). |
| Endocrine Disrupting Compounds (EDCs) | Compounds (EDCs) Alkylphenols (phenol, 4-methylphenol, 4-n-heptylphenol) Bisphenol A Heavy metals ($Cr^{6+}$, $As^{5+}$, $Hg^{2+}$, $Cu^{2+}$, $Pb^{2+)}$ Organotins (monobutyltin, dibutylin, tributyltin) Pesticides (atrazine, chlorpyrifos, diazinon) Polycyclic aromatic hydrocarbon (phenanthrene, fluoranthene) Phthalates (dimethyl phthalate, di(2-ethylhexyl)phthalate) Steroid hormones (estrone, 17α-ethinylestradiol, 17β-estradiol), Aromatic compounds (alcohols, aldehydes), hydrocarbons (cyclohexane) |
| Pathogenic germs (Disinfection) | Bacillus subtilis, Escherichia coli, Micrococcus lylae, Salmonella typhi, Staphylococcus aureus |
| Cyanotoxins | Microcystins, Cylindrospermopsin, Nodularins, and Anatoxin-a, as main cyanobacteria families studied |

A semiconductor upon irradiation undergoes a photo-induced process which results in the promotion of an electron from the VB (creating a hole and generating an $e^-/h^+$ pair) to the CB. The irradiation used may be UV/Vis/ solar, which matches the wavelength within the energy range of the semiconductor bandgap and hence interacts with the solid. Charges so produced migrate to the surface of the material from the bulk. Different reactions are studied during the photoirradiation of semiconductors, ultimately leading to the mineralization of pollutants. In an aqueous medium, dissolved oxygen adsorbed on photocatalyst surface is reduced by electrons giving superoxide radical anions, $O_2^-$, the holes oxidizes water and hydroxyl anions producing hydroxyl radicals $HO^-$. Moreover, protonation of $O_2^-$ produces hydroperoxyl radicals $HOO^-$.

The recombination of radicals/charges leads to the scavenging effect that reduces the efficiency of photocatalysts.

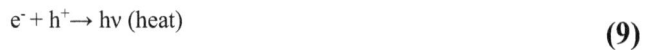

$$(O_2)_{ads} + e^- \rightarrow O_2^{\cdot -} \tag{1}$$

$$H_2O + h^+ \rightarrow HO\cdot + H^+ \tag{2}$$

$$HO^- + h^+ \rightarrow HO\cdot \tag{3}$$

$$O_2^{\cdot -} + h^+ \rightarrow HOO\cdot \tag{4}$$

$$HO\cdot / HOO\cdot / O_2^{\cdot -} + P \rightarrow \text{Intermediates} \rightarrow CO_2 + H_2O \tag{5}$$

$$HO\cdot + H^+ + e^- \rightarrow H_2O \tag{6}$$

$$HOO\cdot + HOO\cdot \rightarrow H_2O_2 + O_2 \tag{7}$$

$$HOO\cdot + H^+ + e^- \rightarrow H_2O_2 \tag{8}$$

$$e^- + h^+ \rightarrow h\nu \text{ (heat)} \tag{9}$$

Where P-pollutant.

Thermodynamically, oxidant radicals are produced only when the reaction potential falls within the band gap between VB and CB of the photocatalyst [43, 44]. A photocatalyst's efficiency can be influenced by wide band gaps that depend on its crystalline phase (as in the case of $TiO_2$, a commonly used photocatalyst), adsorption capacity, size effect, and lattice structure defects [45].

UV/Vis/solar radiations have been effectively used to activate semiconductors to generate the charged particles which act as carriers, that helps to induce electronic transitions in oxides and sulfides such as $TiO_2$, ZnO, $SnO_2$, $SrTiO_3$, $WO_3$, $Fe_2O_3$,

CdS, ZnS and zero-valent iron [46 - 49]. $TiO_2$ [50] possesses antimicrobial properties. However, silver [51], titanium [52], and zinc [49] nanoparticles also possess disinfecting properties. These materials are used as photocatalysts with suitable band gap energies to absorb UV/Vis/solar radiation. Silver, titanium, and zinc are stable towards photo-corrosion, are economical, non-hazardous, and chemically photostable. Studies show that $TiO_2$, ZnO, $SnO_2$, $Fe_2O_3$, are found to be active catalysts for degrading a wide variety of organic compounds under UV light. Moreover, a study conducted on naphthalene showed that both the anatase and rutile structure of $TiO_2$ have high photocatalytic reactivity. 4-chlorophenol showed complete removal (99.20%) using Fe/N/S-doped $TiO_2$, which was eight-fold more as compared to $TiO_2$ Degussa P-25, under photocatalyst dose at neutral pH. The reaction showed a first-order rate constant. Fe/N/S-doped $TiO_2$ showed stability until four cycles of reuse [53]. In another study, UV irradiated nanocomposite $Fe_3O_4$–$TiO_2$–Ag was used to degrade 4-chlorophenol in an aqueous medium by 97% in 165 min. The efficiency of the photocatalyst after five cycles was 94%. The kinetic rate followed zero-order kinetics [54]. Extensive work on $TiO_2$, ZnO, and $SnO_2$ shows that its effectiveness remains at its maximum using UV radiation for dyes, phenols, and pesticides and for toxic organic pollutants. A number of halogenated hydrocarbons such as polychlorinated dibenzo-p-dioxins, chlorinated alkanes, organic dyes such as crocein orange G, methylene blue, methyl red, Congo red, toxic metals were effectively mineralized using $TiO_2$ [38, 55]. The effect of pH and electron acceptors on the degradation of dyes was studied using P25 (Degussa) [56]. In another study, o-methyl benzoic acid *via* UV radiation at pH 3 was completely decomposed in two hours using $TiO_2$ aqueous suspension [57]. In another study, GO/$TiO_2$ showed 57% degradation of salicylic acid under sunlight [58]. The studies were also conducted on the removal of $Pt^{2+}$, $Au^{3+}$, $Rh^{3+}$, $Cr^{6+}$ and $Ag^+$ ions as contaminants using $TiO_2$ aqueous suspension under UV light [59, 60]. The electronic structure of $TiO_2$ determines its photocatalytic properties, when a quantum of light is absorbed by a semiconductor, a free electron and an electron-hole are formed which migrate or recombine in a semiconductor.

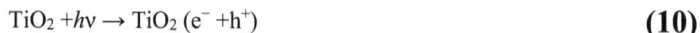

$$TiO_2 + h\nu \rightarrow TiO_2 \ (e^- + h^+) \tag{10}$$

These photocatalysts oxidize non-biodegradable organic pollutants into non-toxic organic or inorganic products. (Fig. **2**) shows the general mechanism of degradation of such toxic organic compounds through nano photocatalysts. However, removal of residual material, reaction time, compounds used for precipitation, low yield, and workforce oppose their wide applications. Therefore, before applying any water treatment process, the following conditions must be considered: (1) flexibility of the system and ultimate efficacy, (2) recycling and reuse of nanomaterials used; and (3) eco-friendly and cost-effectiveness [52, 61]. To overcome the above drawbacks, and to increase the efficiency of $TiO_2$, modified forms are preferred. These forms reduce the bandgap furthermore and show unique features for photocatalytic mineralization. Modification can be done either by doping, to nullify the probability of recombination of charges generated during photocatalysis, or by the addition of noble metals such as gold, platinum, silver, or palladium that trap the electron effectively [62, 63]. In another approach, a combination of semiconductor oxide and $TiO_2$ improves the photocatalytic property by increasing the time period of electrons on its surface. $WO_3$, $SnO_2$, $ZrO_2$, $ZnO_2$, $CeO_2$, the coupling has been studied by researchers. In a study, a nano-composite of $WO_3$-ZnO in the presence of ultrasonic irradiation were used to degrade brilliant blue dye up to 90% in 40 min [64]. Recently, ternary nano-composites $CuO/TiO_2/ZnO$ (with molar ratio of 0.5:1:1) have been reported for improved degradation of methylene blue. Under optimum conditions, 100% and 98% degradation of methylene blue was observed in 2hrs using UV and Vis light with the rate constants of 0.045 and 0.025 min-1 respectively [65]. Another ternary nano-composites $TiO_2/SnO_2/WO_3$ (with molar ratio 80:10:10) synthesized by sol-gel and hydrothermal process showed high photocatalytic activity for degrading 1,2- dichlorobenzene. The results were appreciable in the case of the hydrothermal mode using vis light as compared to the sol-gel process [66]. N-doped $TiO_2/WO_3$ and $TiO_2/SnO_2$ have been used to increase the photocatalytic activity of 2,4,6 trinitrotoluene (TNT) and 4-chlorophenol respectively. In another study, a ternary nanocomposite of Fe-doped $TiO_2/rGO$ was used to degrade rhodamine B, with its initial concentration of 20 mg/L at 6 pH, after the solar irradiation of 120 min 91% removal was achieved. Moreover, complete mineralization was observed after 300 min. Further addition of oxidants such as $H_2O_2$ enhanced the efficiency of the reaction leading to the complete mineralization [67].

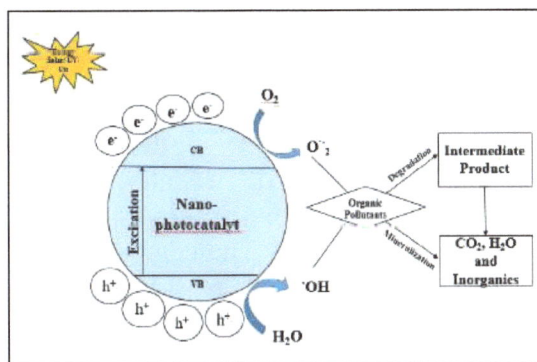

**Fig. (2).** General mechanism of degradation of toxic organic compound through nano-photocatalysts.

In another study, the rate constants of TNT under visible light photo-oxidation catalysis follow the order: Au/N-doped $WO_3/TiO_2$> Au/$WO_3/TiO_2$> Au/N-doped $TiO_2$> N-doped $WO_3$/ $TiO_2$>$WO_3$/ $TiO_2$> N-doped $TiO_2$>> Au/$TiO_2$>$TiO_2$. UV irradiation showed better results than visible light. In comparison to all the possible oxidations, the rate constant of TNT catalyzed by Au/N-$WO_3$ / $TiO_2$ under UV light was 4 times higher than that carried out with $TiO_2$ alone. This increase in performance with noble metal may be attributed to the formation of the Schottky barrier at the metal-semiconductor interface, hence promoting the trapping of electrons effectively [67, 68].

In a study, MgO nanoparticles were tested for the removal of azo [Reactive blue (RB 19) taken as a model] and anthraquinone [Reactive red (RR 198) taken as a model] dyes. Optimized dosage, contact time, and pH obtained were 0.2g, 5 mins, and pH8. The adsorption capacity of RB 19 and RR 198 onto MgO was 166.7 and 125 mg g-1 respectively. MgO is a non-toxic, simple-to-synthesize, and cost-effective material. These unique and novel properties make it a promising and feasible alternative for dye removal [69].

$Fe_3O_4$ hollow nanospheres offer an effective sorbent for red dye (with an adsorption capacity of 90 mg g−1) [70]. The synthesis of iron oxide nanoparticles by thermal evaporation and co-precipitation process showed the photodegradation of Congo red dye. The maximum removal efficiency was 96% at a size of 100 nm [71, 72]. Magnetic $Fe_3O_4$ is used for water purification for the removal of arsenate, cadmium, nickel, and arsenite. It is also used to remove hardness, alkalinity, decolorization, and desalination of industrial effluents like pulp mills, dyes, and textiles [73].

Nano hematite has shown effective removal of heavy metals such as Pb, Cd, Cu, and Zn species from solutions. Adsorption increased unexpectedly with the increase in nanoparticle concentration. It was observed that with 0.5 g/L nanohematite, 100% of Pb and Zn species were adsorbed, whereas, 94% and 89% of Cd and Cu species were adsorbed efficiently [74]. Heavy metals, Pb(II), and Cu(II) removal were studied at the optimized pH of 5 using maghemite at varying temperatures. The maghemite ($\gamma$- $Fe_2O_3$) was synthesized using spray pyrolysis, the maximum Langmuir adsorption capacity reported was 68.9 mg/g for $Pb^{2+}$ at 45 °C and 34.0 mg/g for $Cu^{2+}$ at 25°C [75]. Recently, a study was carried out to degrade rhodamine B (a dye used in the printing and textile industries) using $Fe_2O_3$ doped with $In_2O_3$ in the $H_2O_2$ /UV. The degradation efficiency was 94% at pH 4 [76]. In a recent approach, $\alpha$- $Fe_2O_3$ (coral-like) synthesized by co-precipitation methods, effective degraded methylene blue (95%), methyl orange (94%), Methyl red (76%), and Bromo green (94%) at pH 7 under UV/ $H_2O_2$ system. The system was reusable with a 10% loss after five cycles [77].

Photocatalyst degradation of rhodamine B dye was further studied using titanium ($TiNPs-Fe_2O_3$) and silver ($AgNPs-Fe_2O_3$) nanoparticles in a slurry reactor with UV and vis irradiation. The heterogeneous photocatalyst $Fe_2O_3$ reduces the bandgap from 2.2 eV to 2.0 eV for $TiNPs-Fe_2O_3$ and 1.8 eV for $AgNPs-Fe_2O_3$ respectively. The decrease in the bandgap and increase in the surface area shows remarkable photocatalytic action of $Fe_2O_3$ under UV and Vis light. The percentage degradation of silver-iron composite showed best results under UV light (94.1%, k=0.0222 min-1) than under visible light (58.36%, k=0.007 min-1) however, under same conditions $TiO_2$- P25 catalyst showed (61.5%, k=0.0078 min-1) and (44.5%, k=0.0044 min-1) respectively [78].

While comparing various adsorbents such as $TiO_2$, $Al_2O_3$, and MgO, it was reported to have maximum sorption of heavy metals in the order $Cu^{2+}$ (149.1) >$Ni^{2+}$ (149.9) > $Pb^{2+}$ (148.6) > $Cd^{2+}$ (135 $mgg^{-1}$), followed by $Al_2O_3$ nanoparticles as: $Cd^{2+}$ (118.9) > $Cu^{2+}$ (47.9) > $Pb^{2+}$ (41.2) > $Ni^{2+}$ (35.9 mg $g^{-1}$). For example, $TiO_2$ was found to have minimum adsorption, 120.1, 50.2, 39.3, and 21.7 mg g-1 for $Cd^{2+}$, $Cu^{2+}$, $Ni^{2+}$, and $Pb^{2+}$, respectively [79]. The order of adsorption of metals depends upon its electronegativity but the difference in experimental techniques, and conditions, source of adsorbents, type of metals, and their charge to radius ratio can vary the adsorption criteria. MgO showed the greatest metal adsorption capacity compared to $Al_2O_3$ and $TiO_2$. The cations sorbed from a single-component solution show better results as compared to the multiple-component solutions [79]. Researchers have observed that with the increase in the concentration of $TiO_2$, available surface positions increase, leading to better removal of pollute. Cd adsorption increases as the concentration of nanoparticles increases from 0.01g/L to 0.1 g/L from 84.3% to 99.8% [80].

Removal of organic waste from water could be achieved using transition metal oxides by adsorption and catalytic combustion at low temperatures. One such azo-dye commonly used in textile industries was 90% removed by adsorbing it on $MnO_2$ at pH 7.6 [81]. In another study, organic dyes such as crystal violet, janus green, methylene blue, thionine, neutral red, congo red, and reactive blue were removed from raw water by using magnetite nanoparticles loaded with tea waste (MNLTW) as adsorbent. It was observed that the adsorption capacity of MNLTW was better than other adsorbents for neutral red, crystal violet, congo red, and methylene blue at different pHs [82]. MNLTW shows high adsorption to cationic dyes as shown in Table **2**.

In another study, waste from iron ores was used as starting material for forming magnetite NP (MNPs) which were further used for treating contaminated water. The adsorption of dyes was maximum at pH <7 for anionic dye and at pH> 7 for cationic dyes.

The adsorption capacities of MNPs for methylene blue were reported at 70.4 mg $g^{-1}$ while for Congo red was reported 172.4 mg $g^{-1}$ [83].

Carbon nanotubes (CNTs) and composite membranes are effectively used in wastewater treatment, reuse of brackish water, and desalination of seawater [84, 85]. CNTs have proved to be a sustainable technology because they offers easy processing, low operation cost, capital, low energy consumption, relatively low environmental pollution, high stability, and efficiency. Membrane technology, both polymeric and inorganic, has shown effective applications in the lab and industrial scales. However, these membranes show selectivity, permeability, and fouling properties that limit their applications. Hence, to increase the overall usability and enhance their performance, some modifications are made, which increase the mechanical strength, chemical inertness, and antifouling property [86]. CNTs possess two allotropes, SWCNTs and MWCNTs that are widely used and reported to adsorb a large variety of organic, inorganic, and heavy metal pollutants from wastewater. The adsorption mechanism of CNTs is complex and the adsorption capacity of adsorbents varies depending upon their physical and chemical nature. Open CNTs offers larger active sites than capped CNTs. The adsorption sites available in CNTs include internal sites, interstitial channels, outer surfaces, and grooves. The adsorption reaches equilibrium much faster inside the tubes and interstitial channels as they are exposed directly to the adsorbing material. The activity of the adsorption site reduces with the presence of impurities such as soot, carbon-based contaminants, and catalytic properties [87]. Its potential can be enhanced by modifying the parent adsorbent by combining it with more suitable polymer or chelating agents or functional groups like -OH, -CO, and -COOH that increase its surface area [88]. While studying such adsorbents it was reported that sorption capacities of the MWCNTs (multi-walled carbon nanotubes) were larger than powder activated carbon and granular activated carbon by 3 to 4 times for the adsorption of heavy metals such as Pb(II), Cu(II), and Cd(II), as shown in Table **1**. Volatile organic compounds were better adsorbed on MWCNTs than carbon black in aqueous solutions, which was reported by Li *et al*. Four water-soluble dyes, acridine orange, ethidium bromide, eosin bluish, and orange G showed high sorption capacity and selectivity with caged MWCNTs [89]. While studying the effectiveness of SWCNT (single-walled carbon nanotubes) and activated carbon in the removal of 17R-Ethinyl estradiol (EE2) and bisphenol A (BPA) from wastewater, studies showed relatively higher adsorption of both BPA and EE2 on SWCNT than on activated carbon. Carbon nanotubes showed higher adsorption of trihalomethanes than activated carbon, on the other hand, SWCNT lasts longer than activated carbon due to its rigid structure and resistance to strong oxidants [90]. Research shows the adsorption affinity of naphthalene, phenanthrene, and pyrene on fullerenes, SWCNT, and multiwalled carbon nanotubes MWCNTs, increases in order naphthalene< phenanthrene < pyrene. Order of phenanthrene adsorbed on various nanotubes was in order: SWCNT> MWCNTs > fullerene. High PAHs (polycyclic aromatic hydrocarbons) adsorption capacity indicates environmental risks and their effects when released into the environment. SWCNT had the highest Kf value for phenanthrene $(Kf)=10^{2.52}=331(mg\ g^{-1})/(mg/L)1/n$, close to activated carbon, 273 $(mgg^{-1})/(mg/L)1/n$ [91] which suggests successful removal of PAHs from wastewater and drinking water [92]. CNTs can also be modified using metal oxides iron oxide [93], manganese dioxide [94] and aluminum oxide [95]. Nanoparticles their modified versions, modifications with chelating agents, carbon composites, nanofibres, and nanorods and wires [96], zeolites [97], zero-valent iron (ZVI) [98], nanoflowers [99], and silver nanoparticles [11, 51, 100], *etc.* are used to treat water successfully.

Noble metals (silver and gold) are generally non-toxic to humans. These inorganic nanoparticles possess novel, and enhanced physical, chemical, and biological properties, which enable them as transporters for drug and gene delivery, and in anti-fouling coatings. Silver has been used for ages to build resistance and delay spoilage. In ancient days, medicines, liquids, and water used to be stored in Ag-coated bottles to protect the content from microbial action [101]. Also, because of its antibacterial nature, it is used in dentistry (dental resins) [100]. $Fe_3O_4$ attached silver nanoparticles have been used for water treatment [102], which can be easily removed from the system using a magnetic field to avoid further pollution in the ecosystem [103]. A number of investigators have noticed the application of silver for water filtration. The performance of silver nanoparticles coated with polyurethane was checked with the water having a bacterial load of 106 CFU/ml and no bacterial growth was observed, hence this technique could be used for domestic purposes. Synthesis of these nanoparticles is simple, cost-effective, and an eco-friendly approach [104]. Commercially (Aqua pure) silver nanoparticles surface coated membranes are widely used, which removes 99.9% of pathogen and halogenated pollutants present in water [105]. Pesticides present in water can also be effectively removed using silver nanoparticles [106]. Silver nanoparticles are applicable to purify water polluted with heavy metals. Its synthesis can be carried out by chemical, physical, or biological methods. Chemical reductions are generally carried out using sodium borohydride, sodium citrate, sodium ascorbate, and elemental hydrogen for the synthesis [42, 49, 107 - 109]. In the course of silver nanoparticle synthesis, silver ions get reduced to silver atom (Ag0), atoms agglomerates to form oligomeric clusters which finally combine to form nanoparticles [62, 110]. The strength of the reducing agent decides the size of nanoparticles, mono-dispersed sodium borohydride leads to smaller particle size, whereas weaker reducing agents lead to the formation of larger particle size, due to the slow rate of reduction [111]. Heavy metals find their way through silver nanoparticles efficiently by chemisorption of metal cations. Metals such as mercury and cadmium are effectively removed with silver nanoparticles. Silver nanocomposites with chitosan and alginate have been used to disinfect water and to remove pesticides effectively [112 - 114].

Polycyclic aromatic compounds (PAHs) such as anthracene and naphthalene get adsorbed on fullerenes effectively [115]. Another fullerene, which has been synthesized and characterized as amphiphilic polyurethane nanoparticles, in aqueous solutions has sorbed PAHs and increased their bioavailability [116, 117]. Chlorinated alkenes [tetrachloroethylene (PCE) and trichloroethylene (TCE)] in aqueous solutions showed high sorption capacity on Sodium dodecyl sulfate (SDS) incorporated into aluminum layered sulfate (SDS) and double hydroxides (LDHs) than organoclay [118]. The sorption capacity of some heavy metals ions and organic compounds are given in Tables **2** and **3** respectively.

Table 2. Sorption capacity of heavy metal ions on various adsorbents.

| Researcher | Sorption Metal | Nano Particle / Size | Capacity of Sorption |
|---|---|---|---|
| Nataå lia Moreno *et al.*, 2001 [119] | (pH 6.8) Zn(II) Cu(II) Cd(II) Fe(II) Mn(II) Pb(II) | NaP1 zeolite | Concentration decrease 174 to 0.2 mg Zn/L 10g/L to <0.1 mg Cu/L 400 to <0.1 µg Cd/L. 10g/L<0.1 mg Fe/L 74 to 6 mg Mn/L, 10 g/L<0.1 µg Pb/L |
| Deliyanni *et al.*, 2003 [120] | As(V) pH 12 | Akaganeite [b-FeO(OH)] Nano crystals | 75% As(V) |
| Alvarez *et al.*, 2003 [121] | Cr Ni Zn Cu Cd | synthetic zeolites | 43.6 mg/g 20.1mg/g 32.6mg/g 50.5mg/g, 50.8 mg/g |
| Li *et al.*, 2003 [122] | Pb(II) (10 mg/g) | MWCNTs | 97.08 mg/g |
| Qi & Su 2004 [123] | Pb(II) | Chitosan (40–100 nm) | 398 mg/g |
| Peng *et al.*, 2005 [124] | Pb(II) pH 5 | CNTs-iron oxide composite | 0.51 mmol g$^{-1}$ |
|  | Cu(II) | CNTs-iron oxide composite | 0.71 mmol g$^{-1}$ |
| Crini 2005 [125] | Pb(II) Cu(II) Cd(II) Cu(II) As(II) Cu(II) | Cross linked starch gel Alumina/ chitosan composite | 433 mg/g 135mg/g 150mg/g 164mg/g 230mg/g 200mg/g |
| Peng *et al.*, 2005 [126] | As(V) | Cerium oxide high surface area (189 m$^2$/g) carbon nanotubes (CeO$_2$-CNT) | 10 mg/g |

*(Table 2) cont.....*

| Researcher | Sorption Metal | Nano Particle / Size | Capacity of Sorption |
|---|---|---|---|
| Peng *et al.*, 2005 [126] | As(V) + divalent cations Ca and Mg | (CeO$_2$-CNT) | 10 to 82 mg/g |
| Lazaridis *et al.*, 2005 [127] | Cr(VI) pH 5.5 | Nanocrystalline Akaganeite | 80 mg Cr g$^{-1}$ |
| Bhakat *et al.*, 2006 [128] | As(V) | Modified calcined bauxite | 99% |
| Lu *et al.*, 2006 [75] | As(III) pH 7 | α-Fe$_2$O$_3$ nanoparticles | 95mg/g |
| Lu *et al.*,2006 [129] | As(V) pH 7 | α-Fe$_2$O$_3$ nanoparticles | 47mg/g |
| Wang *et al.*, 2007 [130] | Pb(II), pH 5 | Acidified MWCNT | 85 mg/g |
| Kabbashi *et al.*, 2009 [69] | Pb(II), pH 5 | CNTs | 102.04 mg/g |
| Li *et al.*, 2010 [132] | Cu(II), pH 5 | CNTs- immobilized by calcium alginate | 67.9 mg/g |
| Engates *et al.*, 2011 [80] | Cd(II) | TiO$_2$ 0.01g/L to 0.1 g/L | 84.3% to 99.8%. |
| Shipley 2013 [74] | Pb(II) pH 8 Cd(II) Cu(II) Zn(II) | Hematite | 100% 94% 89% 100% |
| Dave *et al.*, 2014 [133] | As(III) | γ-Fe$_2$O$_3$ nanoparticle | 67.02mg/g |
| Nicomel *et al.*, 2015 [134] | As(III) pH 4 | PEG-MWCNTs | 83mg/g |
| Lee *et al.*, 2015 [135] | As(V) pH 7 | Ti-loaded basic yttrium carbonate (Ti-BYC) | 348.5mg/g |
| Li *et al.*, 2017 [136] | Cr(VI) pH 7 Pb(II) Cu(II) Co(II) | Porous Fe$_2$O$_3$ | 175.5 mg/g 97.8 mg/g 66.2mg/g 60.4mg/g |
| S Rajput *et al.*, 2017[75] | Pb(II) | γ- Fe$_2$O$_3$ | 69 mg/g (45°C) |
| | Cu(II) | γ-Fe$_2$O$_3$ | 34 mg/g (25°C) |

*MWCNT multiwalled carbon nanotubes

**Table 3. Sorption capacity of organic compounds on various adsorbents.**

| Researcher | Sorption Metal | Nano Particle / Size | Capacity of Sorption |
|---|---|---|---|
| Mangun *et al.*, 2001 [137] | benzene, toluene, pxylene and ethylbenzene | nanoporous activated carbon fibers (ACFs) 1.16 nm | 171 to 483 mg/g. |
| Peng *et al.*, 2003 [138] | 1,2-dichlorobenzene | CNTs | 30.8 mg/g. |
| B. Fugetsu *et al.*, 2004 [89] | Acridine orange Ethidium bromide Eosin bluish, Orange G | 1.0 mg of the caged MWCNTs | 0.44 µmol 0.43 µmol 0.33 µmol 0.31 µmol |

(Table 3) cont.....

| Researcher | Sorption Metal | Nano Particle / Size | Capacity of Sorption |
|---|---|---|---|
| B. Fugetsu *et al.*, 2004 [89] | Ethidium bromide | MWCNTs>caged CNFs >caged ACTC control vesicles. | |
| | Eosin bluish | MWCNTs>caged CNFs >caged ACTC> control vesicles | |
| | Aridine orange (10.0 µM with the absorbance of 0.38) | MWCNTs CNFs Activated carbon MWCNTs> caged ACTC> caged CNFs | 0.037 (absorbance decreases) 0.042 0.071 |
| J. Li, Jinbo Fei, *et al.*, 2008 [81] | Congo red (100 mg/L$^{-1}$) pH 7.6 | $MnO_2$ (0.03g) | 90% |
| Zhang *et al.*, 2008 [139] | Phenol | $Fe_3O_4$ | 85% |
| Zhang *et al.*, 2009 [140] | Phenol aniline | $Fe_3O_4$ | 42.79% phenol, 40.38% aniline. |
| G. Moussavi *et al.*, 2009 [69] | RB 19 pH 8 | MgO (0.2g) | 166.7 mg/g |
| A.Afkhami *et al.*, 2010 [141] | Brilliant cresyl blue Thionine Janus Green B | $Y-Fe_2O_3$ modified by sodium dodecyl sulphate | 93.3% 98.4% 94.5% |
| H.Zhu *et al.*, 2010 [142] | Methyl Orange | Chitosan/kaolin/ $Y-Fe_2O_3$composite | 71% |
| H.Y. Zhu *et al.*, 2010 [142] | Methyl Orange | $Y-Fe_2O_3$/chitosan nanocomposite | 98.25% |
| Giri. S.K *et al.*, 2011 [83] | Methylene blue pH <9.2 | Magnetite nanoparticles (MNPs) | 70.4 mg/g |
| | Congo red pH >6.2 | Magnetite nanoparticles (MNPs) | 172.4 mg/g |
| Z. Zhang *et al.*, 2011 [143] | Methylene blue Cresol red | $Fe_3O_4$@C | 90% 20% |
| I.Stephen *et al.*,2011 [144] | Methylene Blue Rhodamine B | $Fe_2O_3$ (coated with c-glutamic acid) $Fe_2O_3$ (modified with humic acid) | 100% 98.5% |
| G.R Chaudhary *et al.*, 2012 [145] | Coomassie brilliant blue R-250 | $Fe_2O_3$ nanoparticles | 98% |
| P. Panneerselvam *et al.*, 2012 [146] | Rhodamine B | $Fe_3O_4$ | 92.2% |
| Yang *et al.*, 2012 [147] | 1-naphthylamine, 1-naphthol and naphthalene | GO/FeO·$Fe_2O_3$ composites | Adsorption naphthalene < 1-naphthol < 1-naphthylamine |

*(Table 3) cont.....*

| Researcher | Sorption Metal | Nano Particle / Size | Capacity of Sorption |
|---|---|---|---|
| Le *et al.*, 2012 [148] | Diazinon | $Fe_3O_4$/hydroxyapatite nanocomposite | 75% (photodegradation) |
| T. Madrakian, *et al.*, 2012 [82] | Crystal violet pH10￼Janus green￼Methylene blue￼Thionine￼Neutral red pH6￼Congo red￼Reactive blue 19 pH3 | Magnetite nanoparticles loaded tea waste (MNLTW)(0.01g-0.05g) | 113.64 mg/g￼129.87 mg/g￼119.05 mg/g￼128.21 mg/g￼126.58 mg/g￼82.64 mg/g￼87.72 mg/g |
| Kakavandi *et al.*, 2013 [149] | Aniline | $AC\text{-}Fe_3O_4$ | 90.91mg/g |
| G.R. Chaudhary *et al.*, 2013 [150] | Acridine orange￼Direct Red | $\gamma\text{-}Fe_2O_3$ nanospindles | 99.7%￼100% |
| Jing *et al*, 2013 [151] | Quinoline | $Fe_3O_4/TiO_2$ composites | 90% |
| G.R Chaudhary *et al.*, 2013 [152] | Congo red | $Fe_2O_3$ | 98% |
| Zerjav *et al.*,2017 [153] | Bisphenol A | $TiO_2\text{-}WO_3$/UV | 87% |
| Gao *et al.*, 2017 [154] | Rhodamine B￼Methylene Blue￼Methyl Orange | $TiO_2$ w f* /UV￼$TiO_2\text{-}WO_3$ w f */UV￼$TiO_2$ w f* /UV￼$TiO_2\text{-}WO_3$ w f* /UV￼$TiO_2$ w f */UV￼$TiO_2\text{-}WO_3$ w f* /UV | 97.2%￼99.8%￼92.4%￼96.6%￼89.6%￼96.9% |
| Li *et al.*, 2017 [136] | Humic acid￼Methyl Blue | Porous $Fe_2O_3$ | 159.4 mg/g￼425.9 mg/g |
| A. F J. Villaluz *et al.*, 2019 [53] | 4- chlorophenol | Fe/N/S-doped $TiO_2$ | 99.20% |
| B. kakavandi *et al.*, 2019 [44] | tetracycline(TC) | $TiO_2$/(MAC@T)/UV/US | 93% |

wood fiber- wf

Heavy metals enter the ecosystem from various anthropogenic origins and reach the soil, fresh, and groundwater systems as shown in Fig. (**5**). Industrial effluents, which, are poured untreated into the water bodies are the main cause of heavy metal pollution. Heavy metals have a density of more than 5g per cubic centimeter. Hence, nanoparticles are involved in the removal of such heavy metals from the environment. Studies have shown metal sorption on synthetic zeolite is ten times greater than natural zeolite because of the higher H+ exchange capacity. Sorption of metals on synthetic zeolites is pH dependent, sorption increases slightly at higher pH. High charges on cations and low hydration energies decrease the sorption capacity of Ni, Zn, and Cd. The maximum adsorption shown by synthetic zeolites for Cr, Ni, Zn, Cu, and Cd is 43.6 mg g-1, 20.1mg g-1, 32.6mg g-1, 50.5mg g-1, and 50.8 mg g-1 respectively. The results show that synthetic zeolite could be used to purify wastewaters produced by Ni, Cr, and acid Zn during the electroplating process, and with the zeolite dosage of 2.5g/l, 66%, 86%, and 91% of metal zinc, nickel, and chromium were removed, respectively [121], shown in Table **2**.

NaP1 zeolites are used to remove heavy metals from acid mine wastewaters [119]and metal electroplating wastewaters [121]. The metals such as Cr(III)(II), Zn(II), Cu(II), and Cd(II) were successfully removed using NaP1 zeolite. It was studied that using a high dosage of zeolites at pH 6.8, the concentration of metals such as Zn, Cu, and Cd was reduced substantially, from 174 to 0.2 mg Zn/L, 74 to 6 mg Mn/L, and from 400 to 0.1 mg Cd/L. Metals such as Co, Ni, and Sr showed a considerable reduction with the same concentrations [119]. Toxic metal ions [155], anions [156], and radionuclides [157] are effectively removed by self-assembled monolayers on mesoporous supports (SAMMS). For organic solutes in aqueous solutions, carbonaceous nanomaterial serves as a high capacity and selective sorbent.

Nano-alumina modified with 2, 4-dinitrophenylhydrazine showed adsorption capacity for some heavy metals such as Pb(II), Cr(III), and Cd(II) ions [158]. Efforts were made by many to develop eco-friendly, effective, and efficient methods to reduce these pollutants from water [159, 160]. Removal of these heavy metals was explored by a number of metal oxides, and salts including, ferric hydroxides [161], $TiO_2$ [162], akaganeite [127], and $Na_2SO_3/FeSO_4$ [78].

Selenium, commonly results from mining activities, hence the drainage water from the mines contains a high selenium concentration. Low pH helps in selenium adsorption and high temperature enhance its removal. Many types of research have reported that nano-magnetite has shown effective, spontaneous, endothermic, removal of $SeO_3^{2-}$. While comparing the adsorption performances of nano magnetite and nano iron-on selenium, $SeO_3^{2-}$ was more adsorbed on magnetite, while $SeO_4^{2-}$ showed better results with nano iron. With an initial concentration of 100 µg-Se/L, after adsorption, the final effluent concentration was found to be below 5 µg-Se/L. Hence, selenium can be effectively removed from the aqueous phase using nano-magnetite [27].

Graphene Oxide modified with EDTA increases the adsorption capacity of GO for the removal of toxic heavy metals such as Pb(II), Cu(II), Ni(II), and Cd(II)from waste water. The optimized adsorption was observed at 6.8 pH.

Studies show in Table **4** that the highest adsorption capacity for Pb(II) increases in the order: EDTA-GO> GO > carbon nanotubes. The highest value of Pb(II) was 479 ± 46 mg/g on EDTA-GO [164].

**Table 4. The table showing capacity of Pb(II) adsorption on different adsorbents [165-171].**

| Adsorbents | BET Surface($m^2$/g) | Capacity for Pb(II)(mg/g) |
|---|---|---|
| GO | 430 | 328 ± 39 |
| EDTA-GO | 623 | 479 ± 46 |
| EDTA-RGO | 730 | 204 ± 26 |
| AC | 1070 | 80−120 |
| SWCNT | 145−1200 | 30−80 |
| Acidified SWCNT | 77−237 | 90 |
| Activated Carbon cloth. | 1689 | 210 |
| Activated Carbon Fibres | 1375 | 40−360 |
| Zirconium phosphate | - | 398 |
| Calcined phosphate | - | 155 |
| Modified zeolite | - | 123 |

Zero-valent metals, semiconductor materials, and metallic nanoparticles have been used as nanocatalysts for treating water contaminants such as chlorinated organic compounds, PBCs, dyes, and pesticides, inorganic ions including nitrates and perchlorates, and refractory organic compounds. The properties of nanoparticles are accelerated orfacilitated by the mode of synthesis and the polymers selected for synthesis [172]. Owing to flexibility and multiple actionabilities, nanoscale zerovalent iron (nZVI) particles represent promising solutions for water purification. Due to its particle size, it has been considered an outstanding electron donor in aquatic environments. nZVI reacts with hydrogen ions, dissolved oxygen, nitrate, and sulphate to form redox couples which are readily scattered in an environment as electron acceptors [173].

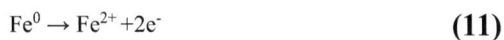

$$Fe^0 \rightarrow Fe^{2+} + 2e^- \tag{11}$$

nZVI nanoparticles can be synthesized by reducing ferric chloride with sodium borohydride. However, ferric chloride possesses highly hygroscopic and acidic properties; it increases chloride levels in the final products. To overcome this drawback, iron sulphate replaces ferric chloride. Besides the above process, nZVI can also be synthesized by reducing goethite with hydrogen gas and heat, followed by iron pentacarbonyl decomposition under an inert atmosphere using an organic solvent. nZVI can also be synthesized by chemical vapuor deposition technique or vacuum sputtering if required in large quantities. For low cost, a simple and quick production electrolysis method can be used [174, 175]. nZVI nanoparticles have shown success in the removal of chlorohydrocarbons from contaminated water and support the conversion of carcinogenic and hazardous hexavalent chromium to non-hazardous trivalent chromium [176, 177].

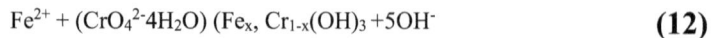

$$Fe^{2+} + (CrO_4^{2-}4H_2O) \; (Fe_x, Cr_{1-x}(OH)_3 + 5OH^-) \tag{12}$$

Moreover, arsenic (III, IV) is effectively removed from groundwater. Journals report that arsenic accumulates in the lungs, skin tissues, liver, and kidney causing a number of diseases [178]. nZVI nanoparticles with activated carbon have been used effectively for the elimination of arsenic and heavy metals such as $Co^{2+}$, $Pb^{2+}$, $Ni^{2+}$, and dyes for purification of water [179 - 181].

nZVI reacts very fast with metals (and less with organic pollutants). The nZVI particles help to remove a broad range of heavy metals and metalloids. The reaction between NZVI and water evokes an alkaline reaction that includes two mechanisms: (i) direct reduction of metal ions by NZVI and (ii) primary adsorption of metal ions onto the NZVI surface followed by their subsequent reduction [172].

The reaction mechanism involved between nZVI, and metals can be a basic reaction of iron and water [182, 183].

$$Fe(0)_{(s)} + 2H_2O_{(aq)} \rightarrow Fe\,(II) + H_2 + 2OH^- \tag{13}$$

$$Fe\,(II) + 2H_2O \rightarrow Fe\,(OH)_2 + 2H^+ \tag{14}$$

The widely known interactions between NZVI and metals are reduction, adsorption oxidation/re-oxidation, co-precipitation, precipitation [184].

Superparamagnetic iron oxide nanoparticles ($Fe_3O_4$) are used to treat water; they show effective elimination of organic pollutants and heavy metal ions by chemical or physical adsorption. These are generally preferred because of their non-toxic nature and chemical inertness. The mode of synthesis of $Fe_3O_4$ nanoparticles plays an important role in its effective application. The morphology, size, magnetic properties, and surface to mass ratio depends upon the type of synthesis. Coprecipitation, solvothermal, microemulsion, hydrothermal, sonochemical are some methods by which $Fe_3O_4$ nanoparticles can be synthesized. Amongst all these processes, coprecipitation is the simplest and highly efficient process. Under alkaline conditions, a stoichiometric mixture of ferric and ferrous precursors is turned into superparamagnetic nanoparticles [185]. For high crystallinity, large size, and controlled morphology hydrothermal synthesis is used [186].

In addition, the sonochemical method of magnetic nanoparticle synthesis with an average size of 10 nm involves irradiation of iron (II) acetate in water, with a high-intensity ultrasonic probe under 1.5 atm at 25°C [187]. Magnetic nanoparticles can be synthesized by the oil-in-water emulsion method, in which $FeSO_4/Fe(NO_3)_3$ mixture are iron sources, polyether phosphates are used as surfactants, and cyclohexane as an oil phase. The nanoparticles formed can possess different shapes, such as rods, needles, and hollow spheres. This method can control the size of nanoparticles by controlling the aqueous phase [188]. $Fe_3O_4$ nanoparticles are widely used because of their low cost, greater stability, strong adsorption ability, and easy separation for water purification. An increase in industrialization has led to increased heavy metal pollution in water. These adversely affect humans, flora, and fauna, and hence, their removal is mandatory. The efficiency of magnetic $Fe_3O_4$ nanoparticles can be increased by treating them with chelating agents. The heavy metals trapped can be successfully removed by an external magnetic field. $Cu^{2+}$ ions were removed from iron oxide modified with a chelating agent such as 1,6-hexadiammine, adsorption capacity was found to be 25 mg g$^{-1}$ [189].

In another study, 95% of $Cu_2^+$ and $Cd_2^+$ were removed by encapsulating magnetic nanoparticles with carbon. In furtherance, modification of magnetic nanoparticles with gum Arabic increases the adsorption capacity from 17.6mg g$^{-1}$ (unmodified) to 38.5 mg g$^{-1}$ [190]. When modified by Chitosan, the adsorption capacity of $Pb^{2+}$, $Cu^{2+}$, and $Cd^{2+}$ increases appreciably [191, 192]. Magnetic nanoparticles modified with acetylcholinesterase and octadecyl trichlorosilane have effectively removed chlorothalonil and phosphorous pesticides [193].

Textile industrial effluent contains red dye, which can be removed using a hollow nanosphere of $Fe_3O_4$ with an adsorption capacity of 90 mg g$^{-1}$ [70]. The hollow nanosphere of $Fe_3O_4$ successfully removes organic pollutants, water from oil refineries, and heavy metals [194, 195]. Nanocomposites of $Fe_3O_4$ with $MnO_2$ and $TiO_2$ have been used to remove As(V) [196] and photo catalytically degrade methyl orange dye respectively [49]. $TiO_2$ nanoparticles are used for large-scale water purification because of their excellent chemical steadiness, high reflective index, and photo-catalytic ability. These particles can remove heavy metals, methyl orange dye photolytically, and enhance anti-bacterial properties, helpful in the removal of bacteria from contaminated water. A number of processes are known for the synthesis of $TiO_2$, such as chemical and physical vapour deposition, the sol-gel method, micelle, and inverse micelle method, solvothermal, microwave process, and the sonochemical method [197]. In the sol-gel process [198], $TiO_2$ nanoparticles are synthesized using titanium (IV) alkoxide as a precursor, subjected to acid-catalyzed hydrolysis followed by condensation. $TiO_2$ can be synthesized using $TiCl_4$ as the source, ammonia as precipitating agent, cyclohexane, poly(oxyethylene)5 nonylephenol, and poly(oxyethylene)9 nonylephenol ether as emulsifiers in a reverse microemulsion system. The amorphous particles formed were further converted into anatase by heating from 200°C to 750°C and the rutile phase by heating above 750°C [199]. $TiO_2$ nanoparticles are synthesized using the hydrothermal method with the particle size ranging from 7 nm to 25 nm [200]. In the hydrothermal method, titanium alkoxide in an acid ethanol-water solution is used. For the better-controlled size and morphology of nanoparticles, the solvothermal method is preferred over the hydrothermal method. Amongst the processes described in the literature for the synthesizes of $TiO_2$ by solvothermal, one is titanium(IV) isopropoxide used as the source had been mixed with toluene in the weight ratio of 1-3:10, heated at 250° C for 3 h [201], and another is the hydrolysis of $Ti(OC_4H_9)_4$ with linoleic acid [202]. The size of nanoparticles can be further reduced below 10 nm by the chemical vapuor deposition technique. In this technique, titanium precursors under an inert atmosphere (helium or oxygen) undergo pyrolysis [203] to attain the desired size.

Metal-organic frameworks (MOFs), recently explored crystalline porous materials that are hybrids of inorganic-organic materials, have been used in bio-applications and water purification. Due to their outstanding properties such as high crystallinity, porosity, and surface area, modifiable pore size, variable and huge tailorable design structures, MOFs have led to varied applications related to energy and the environmental. Potential applications of MOF nanofibers can be observed in water treatment, air pollutants, heterogeneous catalysis, and gas storage and separation [204]. Hence, MOFs as heterogeneous catalysts possess specific advantages over other porous materials such as activated carbon, clays, polymers, silica, and zeolite [205]. These are synthesized under mild conditions from metals (single, chain, or cluster) and organic linkers. About 20,000 metal-organic frameworks (MOFs) are known so far, which show potential applications in water purification for heavy metal removal (Table **5**). In a study, Pb (II) and Hg (II) were effectively removed from aqueous solution using (MIL100-Fe and MOF 808) MOF–polymer [206]. The adsorption capacity of arsenic (V) on Fe-BTC a MOF was higher than 96% at 4 pH. This adsorption capacity was found seven times more than $Fe_2O_3$ NPs (50 nm size) [207]. However, As(V) and As(III) showed moderate adsorption on ZIF-8 (zeolitic imidazolate frameworks) up to 49 mg/g and 60 mg/g respectively at neutral pH [208]. The hybrid nanocomposite of copper terephthalate MOF–Graphene oxide [Cu(tpa).GO] was used to remove heavy metals from Sungun mine drainage. Sorption data for $Cu_2^+$, $Zn_2^+$ and $Pb_2^+$ with Cu(tpa).GO was better than graphene oxide (GO) and Cu(tpa) adsorbent alone [209]. The kinetics of metal ions adsorption showed a pseudo-second-order rate. In another study, As(V) exhibited an adsorption capacity of 21 mg/g and 25 mg/g on MIL-53(Fe) [210] and on MOF-808 [211] respectively, and 106 mg/g on MOF MIL-53(Al) at pH 8 [212].Adsorption of As(V) on UiO-66 showed unexpected results of 303mg/g at pH 2 [213]. List of heavy metals adsorbed on numerous MOFs is discussed below in Table **4**. Multi-walled carbon nanotube modified MIL-53 (Fe) was effectively used to remove tetracycline antibiotics, such as tetracycline hydrochloride (TCN), oxytetracycline hydrochloride (OTC), and chlortetracycline hydrochloride (CTC) which fits pseudo-second-order through Langmuir equation of adsorption isotherm. At pH 7, the adsorption capacity of TCN (364.37 mg g-1), OTC(325.59 mg g-1), and CTC(180.68 mg g-1) on MWCNT/MIL-53 (Fe) was better than 3.34 times than that of single MWCNT [214]. Recently, ZIF-67/PAN filter has shown high efficiency for removing organic pollutants from the wastewater by activation of peroxymono sulfate (PMS), which generates $SO_4^-$. Acid yellow 17 (AY), methylene blue (MB), tetracycline (TC), and bisphenol A (BPA) were effectively removed during the ZIF-67/ PAN filter [215]. In a study, MIL-96, synthesized using varying solvents was used as adsorbent and catalyst. MIL-96 can be reused for the catalytic degradation of water pollutants such methyl orange after the $CO_2$ adsorption by gas phase was completely done. Maximum adsorption on MIL-96-Et for p-HBA was 521 mg/g [216]. In similar studies, MOF-11-Co/ PMS (peroxymon sulfate) was used to degrade pharmaceutical pollutants and personal care products. sulfachloropyradazine (SCP) and para-hydroxybenzoic acid (p-HBA) removal were rapid in less than 30 mins at 25°C and in 5 min at 45°C respectively. Lewis bases in the structure of bio- MOF-11 helps to promote electron transfer that help in the activation of PMS, promoting the degradation of pollutants [217]. Binary MOFs, such as HKUST-1 and UiO-66 was used for the removal of methylene blue under a wide range of pH. The adsorptive capacity of single MOFs (454 and 107 mg/g) and activated carbon (303 mg/g) was less as compared to the binary MOF (526 mg/g). The binary MOFs with reduced surface area show better adsorption than pristine MOFs, fast kinetic was observed and fits pseudo 1st and pseudo 2nd order models [218].

**Table 5. Sorption of heavy metals on metal organic framework (MOF).**

| Researcher | Sorption Metal/pH | Nano Particle | Capacity of Sorption / % |
|---|---|---|---|
| Fang *et al.*, 2010 [219] | Cd (II)<br>Hg(II) | PCN-100 | 1.6 Cd(II) per formula<br>1.4 Hg(II) per formula |
| Ke *et al.*, 2011 [220] | Hg(II) | Thiol-HKUST-1 | 714 mg/g |
| Zhu 2012[207] | As (V) pH 4 | Fe-BTC | 96% |
| Zou *et al.*, 2013 [221] | Cd(II) pH 7<br>Pb(II) | HKUST-1MW@$H_3PW_{12}O_{40}$ | 32 mg/g<br>98 mg/g |
| M R Sohrabi *et al.*, 2013 [222] | Hg(II) pH 5.5 | SH@$SiO_2$/Cu(BTC)$_2$ | 210 mg/g |
| Li *et al.*, 2014 [212] | As (V) pH 8 | MOF MIL-53(Al) | 106 mg/g |
| Jian *et al.*, 2014 [208] | As (III) pH neutral<br>As (V) pHneutral<br>As (III) | ZIF-8<br>ZIF-8<br>$CoFe_2O_4$/MIL-100(Fe) | 60 mg/g<br>40 mg/g<br>143.6 mg/g |
| Tahmasebi *et al.*, 2014 [223] | Cd(II) pH 10<br>Cr(III)<br>Pb(II) | TMU-5<br>TMU-5<br>TMU-5 | 43 mg/g<br>123 mg/g<br>251 mg/g |
| Vu *et al.*, 2015 [210] | As (V) pH 5 | MIL-53(Fe) | 21 mg/g |
| Li *et al.*, 2015 [211] | As (V) pH 4 | MOF-808 | 25 mg/g |
| Wang *et al.*, 2015 [224] | Cd (II) pH 6 | $Cu_3(BTC)_2$-$SO_3H$ | 89 mg/g |
| Saleem *et al.*,2015 [225] | Cd(II)<br>Cr(III)<br>Hg(II) | UiO-66-NHC(S)NHMe | 49 mg/g<br>117 mg/g<br>99% |
| Rahimi and Mohaghegh 2015 [209] | Cd(II) pH 7<br>Pb(II) | Cu-terephthalate | 100mg/g<br>80 mg/g |
| Maleki *et al.*, 2015 [226] | Cr(VI) pH 7 | Cu-BTC | 48 mg/g |
| Li *et al.*, 2015 [211] | Cr(VI) | Silver triazolato MOF-1-$NO_3$ | 93% |
| Zhang *et al.*, 2015 [227] | Cr(VI) | ZJU-101 | 245 mg/g |
| Li *et al.*, 2015 [211] | Cr(VI) pH 5 | ZIF-67 | 18 mg/g |
| Abbasi *et al.*, 2015 [228] | Hg(II) pH 6 | Co(II)MOF | 70% |
| F. Luo *et al.*, 2015 [229] | Hg(II) pH 5 | Zn(hip)(L)(DMF)(H2O) | 333 mg/g |
| L.J. Huang *et al*, 2015 [230] | Hg(II) pH 3 | $Fe_3O_4$@$SiO_2$@HKUST-1 | 264 mg/g |
| K.Folen *et al.*, 2016 [231] | As (III) pH 7<br>As (V) pH 7 | $Fe_3O_4$@MIL-101(Cr) | 121.5mg/g<br>80 mg/g |
| Wang *et al.*, 2016 [213] | As (V) pH 2 | UiO-66 | 303 mg/g |

(Table 5) cont.....

| Researcher | Sorption Metal/pH | Nano Particle | Capacity of Sorption / % |
|---|---|---|---|
| A. Chakraborty *et al.*,2016 [232] | Cd (II) pH 6 Pb(II) Hg(II) | AMOF-1 | 41 mg/g 71 mg/g 78 mg/g |
| Zhang *et al.*, 2016 [227] | Cd (II) pH 6 Pd(II) | HS-mSi@MOF-5 | 98 mg/g 312g/g |
| Yang *et al.*, 2016 [233] | Cr(IV) pH 2 | $Fe_2O_4$@MIL-100Fe | 18 mg/g |
| Wang *et al.*, 2016 [213] | Cr(IV) acidic pH | A chitosan MOF(UiO-66) composite | 94 mg/g |
| Rivera *et al.*, 2016 [234] | Pb(II) | MOF-5 | 659 mg/g |
| M. Mon *et al.*, 2016 [235] | Hg(II) pH 7 | Bio MOF | 900 mg/g $HgCl_2$, 166 mg/g $CH_3HgCl$ |
| E. Rahimi *et al.*, 2017[209] | Cu(II) Zn(II) Pb(II) | *Cu(tpa).GO | 243 mg/g 131 mg/g 88 mg/g |
| M. Mon *et al.*, 2017 [236] | Hg(II) | {CuII 4[(S,S)-methox]$_2$} . $5H_2O$ | 99.5% |
| T.T. Zheng *et al.*, 2017 [237] | Cu(II) pH 6.7 | Cd-MOF-74 | 189.5 mg/g |
| Huang *et al.*, 2018 [238] | Pb(II) pH 5.2 | ZIF-67 | 1348.42 mg/g |
| Alqadami *et al.*, 2018 [239] | Pb(II) pH 5.8 Pb(II) pH 5.1 | *AMCA-MIL-53(Al) ZIF-8 | 390 mg/g 1119.80 mg/g |
| J. Wen *et al.*,2018 [240] | Pb(II) Hg(II) Cd(II), 14 Cr(III) Cr(VI) As(III) As(V) | MOF | 8.40-313mg/g 0.65-2173 mg/g 3.63–145 mg/ g 14.0-127 mg/ g 15.4-145 mg/ g 49.5-123 mg/ g 12.3-303 mg/ g |

*Cu(tpa).GO copper terephthalate MOF-graphene oxide; {CuII 4[(S,S)-methox]$_2$}. $5H_2O$ methox is bis[(S)-methionine]oxalyl diamide; *AMCA-MIL-53(Al) -amide citric anhydride

Cadmium shows effective adsorption on different frameworks. $Cu_3(BTC)2$-$SO_3H$ MOF at pH 6 displays an adsorption capacity of 89 mg/g [213]. The maximum uptake of Cd(II) on AMOF-1 and UiO-66-NHC(S)NHMe was found to be 41 mg/g and 49 mg/g. The mode of synthesis varies depending upon the particle size required for the adsorption process. HKUST-1-MW@H3PW12O40 was an extra-stabile MOF with a capacity of 32 mg/g. Higher removal capacity of Cd(II) was noticed with increasing metal ion concentration and an increase in temperature between 273-323K [221].

Investigation shows chromium (III), (IV), and (V) can be effectively adsorbed on various MOF synthesized through different modes. TMU-30 an isonicotinic N-oxide-based MOF was highly effective for removing Cr(VI) from water, the adsorption capacity was 145mg/g achieved in a short time over a wide range of pH(2-9) [241]. In a study, green, safe, and fast MOR-1-HA (UiO-66 based, amino-functionalized MOF with alginic coating) was synthesized which displayed a maximum adsorption capacity of 280 mg/g. The MOF not coated with alginic acid had a lesser adsorption capacity. The pH variation between 2 to 8 did not alter adsorption capacity, but at pH 1, an evident drop was observed. Photocatalytic reduction and separation using MOF have been progressed along with the conventional means [242, 243].

The maximum adsorption capacity of Pb(II) on zirconium-based, UiO-6--NHC(S)NHMe, and a silica-coated thiolated MOF-5 HS-mSi@MOF-5 was reported 232 mg g$^{-1}$ [225]and 312 mgg$^{-1}$ at optimum pH 6 respectively [227]. In another study, the magnetic composite MIL-53 (Al@100aBDC) with 50% or more $-NH_2$ moieties reported higher uptake, about 492 mg/g of Pb(II) [244]. Removal of Pb(II) was studied using lanthanide based MOF Dy(BTC)(H$_2$O)(DMF)1.1, which achieved equilibrium in 10 min. Its uptake was moderate at 5 mgg$^{-1}$ at pH 7. However, MnO$_2$-MOF recorded an unexpected 917 mgg$^{-1}$ adsorption capacity of Pd(II) [245]. An interesting fact was reported while studying the adsorption of Pb(II) on MOF-5, in which maximum adsorption capacity reached 659 mgg$^{-1}$ at 45°C. This value kept on changing with the change in temperature, hence no trend was observed. This is attributed to the structure of MOF, which possesses both acidic and basic sites. The adsorption value (reported at pH 5, 450 mgg$^{-1}$) at pH 4, and pH 6, was 750 mgg$^{-1}$ and 660 mgg$^{-1}$ respectively, showing an increase with both high and low pH [246].

Mercury has been highly adsorbed on a number of MOFs, hence a means of better environmental remediation. Sulfur modified MOF-FJI-H12 (created from Co(II) metal ions and 2,4,6-tri(1-imidazolyl)-135-triazine linkers) reported maximum uptake of 440 mgg$^{-1}$ at 7 pH. No increase in adsorption was reported with the increase of pH from 3 to 6 [247]. Another novel post synthetically modified MOF that shows effective Hg(II) adsorption has been Cr-MIL-101-AS. This includes compactly packed thiol groups along with alkenyl function. 99% of Hg got adsorbed with the initial concentration of 10 ppm [248].

The composites of g-C$_3$N$_4$/MOF proved to be effective remediation for removing heavy metals from the wastewater. Cr(VI) and organic pollutants (personal care products and pharmaceutical) were removed under sunlight. g-C$_3$N$_4$/MOF are binary photocatalysts which has wide applications in degrading dyes such as Rhodamine B (ZIF-8/g-C$_3$N$_4$, g-C$_3$N$_4$/UiO-66 [249], g-C$_3$N$_4$/MIL-125(Ti), g-C$_3$N$_4$/MIL-100(Fe) [250], Methylene blue (g-C$_3$N$_4$/ UiO-66 [251], g-C$_3$N$_4$/NH$_2$-MIL-88B(Fe) [252], g-C$_3$N$_4$/MIL-88B(Fe) [253], Bisphenol A (BPA) g-C$_3$N$_4$/MIL-101(Fe) [254], Dimethyl chlorophosphate (DMCP) g-C$_3$N$_4$/CuBTC [255 - 258]. Some heavy metals are highly toxic to the entire ecosystem whereas some have less toxicity but do compete for the pollution. A list of some metals is given in Table **6**.

Table 6. List of heavy metals with less toxic nature.

| Researcher | Sorption Metal/pH | Nano Particle | Capacity of Sorption / % |
|---|---|---|---|
| F. Zou *et al.*, 2012 [221] | Cu(II) pH 7 | Cu-terephthalate MOF | 225 mg/g |
| Fang Zou *et al.*, 2013 [221] | Fe(III) pH 7<br>Mn(II) pH 7<br>Zn(II) pH7 | Cu-terephthalate MOF | 115 mg/g<br>175 mg/g<br>150 mg/g |
| Zhang *et al.*, 2015 [227] | Al(III) pH 6<br>Fe(III) pH 6 | 3D Co(II) MOF | 90%<br>100% |
| Cheng *et al.*, 2015[259] | Ag(I) | MIL-53(Al) | 183 mg/g |
| N. Bakhtiari *et al.*, 2015 [260] | Cu(II) pH 5.2 | MOF-5 | 290 mg/g |
| A. Maleki *et al.*, 2015 [226] | Ni(II) pH 5 | Chitosan-MOF | 60 mg/g |
| E. Rahimi *et al.*, 2015 [209] | Co(II) pH 10 | TMU-5 | 63 mg/g |
| Conde Gonzalez *et al.*, 2016 [261] | Ag(I) pH 6-7 | HKUST-1 | 100% |
| Zhang *et al.*, 2016 [227] | Cu(II) pH 4 | ZIF-8 | 800 mg/g |
| T. Zheng *et al.*, 2017 [237] | Cu(II) pH 6.7 | Cd-MOF-74 | 190 mg/g |

Recently, chiral catalysis over MOFs has shown wide applications in pharmaceutical and related products. MOFs can be modified using functional groups to improve selectivity (such as chemoselectivity, wettability selectivity, chiral selectivity) and enhance activity. MOFs are multifunctional catalysts with numerous active sites that enable different substrates and multi-component coupling reactions. Photo-assisted MOFs accelerate the reaction by producing the heat around the active site as compared to the traditional heating methods, leading to an increase in process efficiency [262]. With tremendous advancement and applications, MOF catalysts face great challenges. For tailoring targeted MOFs, its intrinsic mechanism plays an important role that remains unclear. Some MOF is synthesized using high-priced metal precursor, organic and inorganic ligands that involve several steps and high-energy consumption. Hence, affordable, large-scale production with a high yield remains a challenge. The application of MOFs in practical needs stability and the ability to retain the activity during the complete course of the reaction. The stability of MOFs thermally, chemically, and mechanically still remains questionable. Moreover, with low thermal stability than catalysts such as zeolites, their regeneration remains unclear.

As MOFs are mostly micro-porous, mass transfer remains a problem for large products. MOFs show poor conductivity, limiting their applications.

## MATERIAL USED FOR NANO PARTICLES SYNTHESIS

Nanoparticles can be prepared using different raw materials. Some natural raw materials, such as chitosan, alginate, and gelatine, show biocompatibility, biodegradability, and antibacterial activities and are non-toxic, hence they are commonly used for nano synthesis. Noble metals, magnetic materials, and quantum dots also possess special characteristics, that enable the synthesis of nanomaterials. Synthetic polymers used for synthesis increase the efficiency and affectivity of nanomaterials and their applications. Some natural and synthetic materials are mentioned below.

### Natural Materials

Some natural raw materials used for the synthesis of nanoparticles are as follows:

*Chitosan*- Chitosan is a polysaccharide composed of poly-2-acetamido-2-d-oxy-D-glucose [263], which possesses the highest chelating ability than natural polymers, exhibiting β (1→4) linkage. Chitin forms the shells of all crustaceans [264]. Chitosan, obtained by the deacetylation of chitin. The presence of amino polysaccharides enables chitosan to undergo several chemical modifications [265]. Chitosan exhibits varied applications owing to its biocompatibility, non-toxic, cationic nature, biodegradability, and antibacterial activity. Further, chitosan microparticles exhibit enhanced properties that enable them for drug delivery applications [264].

*Alginate*- Brown seaweeds are the source of alginate, an unbranched copolymer of D-mannuronic and L-guluronic acid linked by β (1→4) linkage. It shows remarkable applications in the pharmaceutical world due to its unique properties of low immunogenicity, high biocompatibility, and sol-gel transition initiated in the presence of multivalent cations. The synthesis of alginate microparticles requires mild reaction conditions. It remains an important precursor for proteins, drugs, and cell microencapsulation synthesis.

*Gelatine*- Gelatine is the denatured form of animal-derived collagen. Gelatine micro particles are commercially used for diagnostic detection of HIV-1, syphilis, and measles and have been studied for sustained release application [266].

*Noble metals*-Noble metals (silver and gold) are generally regarded as harmless to humans, hence they are used for drug and gene delivery, and in anti-fouling coatings. Due to surface Plasmon resonance (SPR) property noble metals are used for imaging [267] and photoacoustic imaging [268] as well. In photo-acoustic imaging, signals are detected *via* ultrasound. Free electrons of metals form a cloud in the conduction band, which surrounds the surface of the material, producing a reflective surface.

As the electrons exhibit wave nature, the cloud vibrates, absorbing the same frequency of light, and producing an absorbance signal. The thickness of the material may alter the signals. Signals for nanomaterials may be altered due to the aggregation and adsorption of molecules on the material surface. Noble metal biosensors have been developed using this unique property [269].

*Magnetic Materials*- Superparamagnetic iron oxide nanoparticles (SPIONs), usually <15 nm in diameter, are the most commonly employed magnetic materials for various applications. SPION was initially synthesized by precipitation and microemulsion methods [270], but a high-temperature precursor decomposition method [271] has been developed that yields small, uniform nanoparticles. Large nano and microsized particles, show ferromagnetic properties and produce permanent magnetization aligned with the external field; SPIONs with small sizes display superparamagnetic behaviour and have a single magnetic domain. These magnetic properties find their application for cell separation, drug delivery, biomolecule manipulation [272], and magnetic resonance imaging (MRI) [273]. It is also used for therapeutic purposes, such as curing tumours and improving the efficiency of chemotherapy [272].

*Quantum Dots*- These are small, <10 nm in diameter, spherical shape, fluorescent nanocrystals made from semiconductors. Commonly used commercially available quantum dots are synthesized through high-temperature decomposition [274]; or by precipitation route [275]. Cadmium selenide has been used as a core, surrounded by a zinc sulphide shell, for the synthesis of commercial quantum dots. Quantum dots have shown a large number and a broad range of biological applications, including single-molecule biophysics [276], optical barcoding [277], molecular [276], cellular [278], and *in vivo* imaging [279].

Quantum dots display a unique property of large absorption cross-sections, which makes them brighter than organic dyes and fluorescent proteins. This property enables them to have more stability against photo bleaching, and show narrower and more symmetric emission spectra. The major limitation of quantum dots has been their larger size than organic dyes, causing steric hindrance [280].

## Synthetic Polymers

Poly(lactic acid) (PLA), Poly(glycolic acid) (PGA), and Copolymers (PLGA) Poly (ε-caprolactone) (PCL), Poly(ortho ester) (POE), polyacrylamide are synthetic polymers used for controlled drug delivery because of their high permeability, lack of toxicity, biodegradability, and biocompatibility [281].

## SOME PROCESS OF SYNTHESIS OF NANOPARTICLES FOR WATER PURIFICATION

There are various processes of nanoparticles synthesis. The processes are widely divided into physical, chemical, and biological methods. The process of synthesis, the physical conditions (temperature, stirring, milling, grinding, annealing), and the precursor used, decide the properties of the nanoparticle formed, which in turn decides its further application.

### Physical Methods

Physical methods include mechanical methods such as high energy ball milling and melt mixing, and vapour methods include physical vapour deposition, laser ablation, sputter deposition, and ion implantation. Grinding micro or macro scale particles using a ball mill, a planetary ball mill (360 rpm), a vibratory mill (1200 rpm), and a tumbler mill produces nanoparticles in a simple way. Simple ball mills turn Co, Cr,W, Al, Fe, Ag materials into nanoparticles. A molten stream of Cu-B and Ti form $TiB_2$ nanoparticles by melt mixing method. single-walled carbon nanotubes are synthesized by laser beam technique. The laser pyrolysis method can be used for $Al_2O_3$, WC, $Si_3N$ nanosynthesis. Physical synthesis approaches through phase transformation, begins with a grouping of two stages: nucleation and growth. Phase transformation begins with the aggregation of numerous small particles and is continuous until the transformation is completed. Nucleation starts with nuclei that act as templates for crystal growth. It determines the time for the formation of a new phase. Nucleation shows sensitiveness towards impurities in the system. Nucleation can be homogenous or heterogeneous. Some physical processes used for the synthesis of nanoparticles, which are effectively used in water treatment techniques, are discussed below.

***Physical vapour Deposition (PVD)*** This process involves the generation and condensation of vapor phase species *via* thermal evaporation, sputtering, or laser ablation.

***Inert gas condenser:*** To synthesis ultra-fine metal nano-particles, metallic sources are evaporated in inert gas; the process is carried out inside an ultra-vacuum chamber filled with helium. The morphology, size, and yield of nano-particles should be controlled.

***Evaporation technique:*** ET involves thermal and electron-beam evaporation methods.

***Thermal evaporation:*** The target material is heated to its melting point and the substrate is placed just opposite to the source. A high current evaporates the target material. The crystal monitor placed close to the substrate provides an approximate measure of how fast and how much the material gets deposited on the substrate [282].

***Electron beam evaporation:*** In this technique, a high DC voltage is applied to a tungsten filament that causes the emission of electrons. These emitted electrons are accelerated and excite the target material to produce vapours. These travel to the substrate. On reaching the surface, they condense to form a thin coating [283].

***Sputtering:*** In this technique, source materials at the cathode are sputtered by high-energy ions of inert gas atoms, most commonly argon, produced by a high voltage DC glow discharge. These sputtered atoms ejected by high voltage form a thin film coating on the substrate kept opposite to the target at the anode plate. This process is nonthermal and works at a low vacuum ($10^{-3}$) torr. In spite of this advantage, the technique is highly costly since a wide amount of energy is released during the emission. Controlling parameters is a difficult task [284].

***Spray pyrolysis:*** In this process, liquid or gas is forced through the opening at high pressure and burnt. The solid produced is air classified to recover oxide particles from by-product gases. Pyrolysis results in aggregation and agglomeration, but is simple, cost-effective, and produces a high yield. The spray pyrolysis was used to prepare $TiO_2$ [285] and copper nanoparticles [286].

***Plasma processing methods:*** This process involves a chemical and physical reaction between solid surfaces and particles in contact with the plasma. It has been used in plasma etching, thin film deposition, protective coating, surface hardening, and ion implantation. The arc discharges technique works on the plasma technique.

***Arc discharges technique:*** This process involves vaporizing materials by an arc produced by applying a very high voltage (50-100V) across the electrode. At 6000°C the noble gas ionizes producing a plasma that evaporates metal atoms and condenses them on the water-cooled substrate. The impurities are later removed by heating the substrate. Dichalcogenides nanotubes are synthesized during this method.

***Laser ablation:*** In this process, a high-power laser (Nd: YAG) is used to ablate a solid metal rod in an inert chamber. Laser ablation results in plasma that evaporates and condenses the metal atoms on the substrate, cooled by water. The impurities are removed by heating the substrate. This method assures a high deposition rate of 2-3 g/min, but the quality of material deposited is inferior and involves high temperatures to carry out deposition [287]. Silver nanoparticles were synthesized by laser ablation [288]. The size and morphology of nanoparticles depend upon the wavelength of the laser used, duration of laser pulses, and ablation, and the liquid medium was taken for synthesis [289, 290]. Physical methods dominate over chemical methods of synthesis as they are free from chemical contamination and the size of nanoparticles can be better controlled.

**Chemical Synthesis Technique:**

There are a number of chemical processes for nanoparticle synthesis. For the large-scale synthesis of inorganic nanoparticles, a wet chemical approach is used. Several methods of wet chemical synthesis are colloidal synthesis: Micro-emulsion; Sol-gel; spray pyrolysis methods. This technique enables the production of very small and controlled particles size, but a large amount of liquid is needed for the method, and the ultimate yield is lower.

*Chemical Vapour Deposition (CVD):*This process has been used in the semiconductor industry to produce thin films. The desired deposits are decomposed on the substrate surface by exposing them to the volatile precursors. The process can be studied in two ways: RF-Plasma Enhanced CVD and Microwave Plasma Enhanced CVD. High purity substrate, large-scale deposition, high productivity at low and controlled temperature makes the process economically viable.

*Precipitation method:* The process involves the mixing of anionic and cationic solutions, with constant shaking, which leads to nucleation, growth of particles, and the solid formed is precipitate. This settles at the bottom without sufficient force of gravity. The precipitate needs rigorous washing with solvent followed by calcination at the given temperature [291].

*Sol-gel Synthesis:* Sol-gel isa wet chemical process that has been widely used in material science and ceramic engineering and was synthesized by M. Ebelman in 1845. In this process, precursors used undergo hydrolysis and poly-condensation to form a colloid that forms a stable sol. Gelation proceeds *via* poly-condensation or polyesterification reaction [292]. Later gel ages to form a solid mass, which causes phase transformation, and Ostwald ripening, which also causes contraction of gel. Drying leads to a change in the gel structure. Dehydration carried at a high temperature protects it from rehydration. Hydrolysis, condensation, growth of particles, and agglomeration are the processes that completely explain the processes involved in the sol-gel. Sol-gel has shown its wide application in drug delivery systems, in preparing superior strength fibers, films, and scratch and abrasion-resistant materials. It has been used to manufacture UV protection gel, fire retardant gel, and anti-corrosion coatings. The process is easy, economical, procures are low cost and easily available, less energy consuming, and less polluting the environment [292].

***Combustion Method:*** Combustion is known for its self-propagation high-temperature synthesis (SHS). This process finds its way into various industries because of its slow cost, and effective processing for producing nanoparticles. The initial reaction helps to classify the combustion process as solution-combustion or gas-phase solution combustion. In the former process, synthesis occurs in an aqueous medium, whereas in the latter process, nanoparticle synthesis occurs in the presence of flame. Studies show that the materials produced by the latter combustion process are not economical [294].

***Conventional SHS:*** Condensed phase combustion- Synthesis of nanoparticles by this process has been a challenging task because the particle size of initial solid reactants is in the order of 10-100 μm and the process proceeds at a very high temperature. Nevertheless, several methods follow this approach for synthesis: (i) SHS, followed by intensive milling; (ii) SHS + mechanical activation (MA); (iii) SHS, followed by chemical treatment; boron, aluminum, silicon, nitrides nanoparticles were prepared by this processes. (iv) SHS with additives; alkali metals when added to transition metal oxide reduce to metal particles (v) carbon combustion synthesis (CCS) has been shown to produce noble techniques for micro and nanoparticle synthesis of higher oxides. The material synthesized by this method comprises battery electrode material, diesel emission removal catalysts, multiferroics, ferroelectrics, fuel cell components, and hard/soft ferrites [294]. These products are more economically viable and better than those prepared by conventional methods. The conventional approach failed to produce oxides synthesized from this process.

***Solution combustion synthesis-*** The solution combustion process is widely used for water and air purification due to its simplicity, rapidness, and versatility. The process involves the reaction between oxidizers and fuel. The mode of propagation depends upon the precursor used and the conditions used for the process. This process yields nano oxides and allows even and constant doping of rare-earth impurities in one step. Nano-sized ZnO/carbon composites showed relatively higher specific capacitance than micro-sized ZnO powder [295]. $WO_3$ synthesized from this process was used to remove 90% of methylene blue from the aqueous solution [296]. Fuel plays an important role in this synthesis; generally, urea and glycine are used for the uniform production of oxides. Organic compounds used as fuels are ammonium salts of acetate and tartrate, serine, asparagine, methylcellulose, and alanine [297].

***Colloidal Synthesis:*** A solution-phase or wet process synthesis route, where different ions form insoluble precipitates when mixed under controlled temperature and pressure. This method is economical, flexible, and scientifically simple to implement. This method has been used to produce nanoparticles and nano-composites of a varied range of materials.

***Green Synthesis of nano particles:*** Green nanoparticles are prepared by simple processes. Commonly, the plant products are boiled at a given pH. A known amount of precursors are added to produce the desired nanomaterial. Noble metal nanoparticles such as gold and silver are synthesized using this mode are economical, biodegradable, and eco-friendly. Studies have shown that nanoparticles obtained from plants and fungi possess different shapes and sizes. It was observed that gold nanoparticles obtained from the leaves of geranium plants are rod and disk-shaped whereas, nanoparticles prepared from Colletotrichum sp. fungi, are spherical. Hence, the source of nanoparticles decides their physical appearance. The study shows that nanoparticles synthesis can be from the bacterial medium as well as fungal extracts [298 - 301]. Green synthesis has proved to be a step to purifying polluted water without using harmful, toxic, and expensive substances to prepare nanoparticles.

***Solvothermal process:*** This process involves reactions with water and solvents other than water, such as organic solvents (n-butyl alcohol). The hydrothermal method refers to a process with water as a solvent where the crystals grow at a high temperature and pressure in an auto clave (a thick-walled steel vessel). Nano-size titanium dioxide, graphene, and carbon nanotubes are effectively synthesized by the solvothermal process.

Chemical processes have numerous advantages over other processes as they are simple, work on simple, inexpensive instruments, require low temperature, a large amount of material can be obtained, size and shape of particles can be controlled. The process needs simple assembly and facilitates doping effectively.

## CHARACTERIZATION OFNANO PARTICLES

Characterization of microparticles is simple, inexpensive, and less time-consuming. Coated particles are simple to read without microscopic aid, a variety of substances can be studied, it is a highly sensitive method, and less time is required to obtain results. Factors such as surface charge and morphology, size, shape, chemical composition, and aggregation tendency play a vital role in deciding the nature of particles. Slight modifications to these properties result in substantial fluctuations. The methods for characterization of micro/nanoparticles are electron microscopy (*i.e.*, SEM, TEM), x-ray diffraction (XRD), determines the crystal shape, unit cell dimensions, measures of sample purity, and identifies of an unknown mineral through dynamic light scattering (DLS), zeta-potential measurement, X-ray photoelectron spectroscopy (XPS) [302] analysis, atomic force microscopy (AFM) [303, 304]and optical light or fluorescence microscopy. Moreover, differential scanning calorimetry (DSC) has been used to assess changes in polymer thermal properties, infrared spectra (IR or FTIR) has been used to analyze interactions between the functional groups of multiple materials, and surface tension among them has been used to evaluate particle aggregation. Infrared spectroscopy is a valuable technique for environmentalists for monitoring air quality, water quality, and soil analysis.

## GENERAL CONS OF NANOTECHNOLOGY [305]

Nanotechnology has entered diverse fields. Enormous studies reveal their importance, usage, and applications that have brought a revolution in the fields of science and technology. Nevertheless, along with the advantages, there are certain disadvantages, and the lesser evil precedes the greater one.

- *Job insecurity-* Developments in science have led to job insecurities in conventional farming as in well as the manufacturing industry. With nano-robots that work effectively, efficiently, and continuously as compared to humans, employment remains a challenge. In developed countries, these nanorobots are abundantly used, saving time and money.
- *Traceability-* The nano-robots are puppets; they are handled by others. Robot technology can be a boon if used for the wellbeing of humans or curse causing immense destruction. However, the traceability of robots might cause immense problems if lost.
- *Precision-* Nanotechnology enhances the degree of precision. Atomic weapons with laser technology are more precise, accessible, powerful, and destructive, never the less particle dispersed can cause havoc due to its nano size.
- *Non collectability-* The nano-scale particles are difficult to recollect after their use. Either its fate ends up in water or in soil pollution. Problems arise by inhaling these particles, which often may lead to respiratory troubles, exposure to the body may affect the skin, its penetration in the skin may affect the lined cell and tissues. Workers at paint factories might suffer from lung diseases due to the inhaled nanoparticles.
- *Specialized and expensive-*Nanoparticle synthesis requires specialized laboratory and analytical grade chemicals for effective results. Different syntheses require different specifications, for characterization makes it a costly affair. Its production at a large scale might reduce its cost, but infrastructure cost rises enormously, which probably makes nano products a costly affair.

## CHALLENGES: EFFECT OF NANO-MATERIALS ON HUMAN BEINGS AND ECOSYSTEM USED FOR WATER PURIFICATION

The adverse effect of chemicals on the world environment is toxicology. Toxicology depends upon the concentration, dosage, and time of administration. Despite the success of nanoparticles and microparticles as biomaterials, the greatest challenge remains regarding their fate in the body [306]. The size, surface charge, and degree of agglomeration of nanoparticles in an aqueous medium and the factors affecting these are vital to studying the toxicology of nanoparticles. The study of some factors such as pH, surface chemistry, and ionic strength is imperative to examine the dispersion of nanoparticles [307]. Though the minimum amount of these particles is administered during therapy or drug administration, the surface-to-volume ratio remains significant, which means it can alter the chemical interface within the body. Sometimes, due to their small size, they get absorbed by cells and organs and are often too large to be cleared by the kidneys. Hence, knowing the biological behavior of micro/nano-size particles remains vital to their further applications. Further, the mobility of nanoparticles in water bodies may vary depending upon their properties, such as their chemical and physical behavior, solubility in water, and binding capabilities with other components as shown in Fig. (3). Surface charges of nanoparticles along with these factors may cause immobilization.

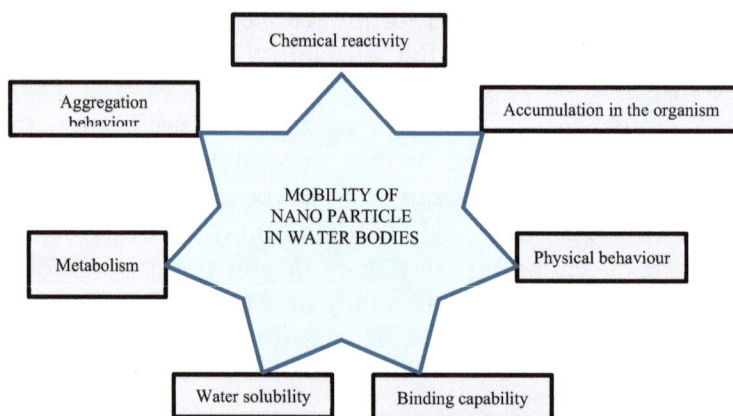

**Fig. (3).** Properties of nanoparticles in water bodies.

Nanoparticles are the tiniest particles produced naturally in a number of ways. They are derived from industrial activities, solid waste incineration, fuel combustion, traffic emissions, power plants, and military explosions and shooting ranges, hence their presence in the atmosphere remains permanent and inevitable. Both fabricated and naturally occurring nanoparticles affect adversely human health. Nanoparticles' presence everywhere allows them to enter the human system through ingestion, inhalation, and permeation, causing serious damage to cells, tissues, and organs. This causes mutations and unrepairable damage in human beings.

Although some nanomaterials have established biocompatibility, the toxicity of carbon nanostructures and quantum dots needs forethought. Carbon nanostructures are synthesized using metallic catalysts, which can incite cellular responses [308]. Unmodified CNs can penetrate the cell membrane [309] and form granulomas in tissue [310]. This adverse effect can be moderated to yield hydrophilic surfaces by acidic oxidation. CNs have shown direct toxicity, through the generation of reactive oxygen species (ROS), which can be controlled using surface coating [311].

Nano traces are evident in water bodies and soil, due to the disposal of treated water into the environment. Entry into the human system through the food chain by aquatic organisms remains unavoidable. The effect of photocatalytic nano $TiO_2$ on the environment and different organisms including fish, plants, invertebrates, algae, and bacteria was studied in past. $TiO_2$ nanoparticles are widely used in consumer products such as paints, pharmaceutical preparations, food additives, and cosmetics for their ability to confer opaqueness and whiteness [312, 313]. The cytotoxicity of $TiO_2$ nanoparticles is still in controversy. Some data shows the potential ability of nanoparticles to induce tumors. Carbon nanotubes and $TiO_2$ at high concentrations cause respiratory tract cancer in animal models [314]. In some literature studies, $TiO_2$ nanoparticles are reported non-toxic [315], wherein some have reported them toxic, which may induce oxidative stress, genotoxicity, and immunotoxicity [316 - 318]. In another study, $TiO_2$ does not release toxic ions as released by ZnO, and quantum dots. $TiO_2$ generally used in sunscreen creams has been reported into liver, spleen, and lymph with the size varying between 150 nm to 500 nm [319]. General effects on aquatic organisms reported were respiratory distress, behavioral changes, and pathological changes intestine and gills. Time exposure also plays an important role to evaluate toxicity. More the exposure time, the higher the toxicity. Emission of nanoparticles may come from point sources, like wastewater plants, landfills, or factories, or from nonpoint sources, as material containing nanoparticles, its washing. Nanoparticles show different behavior in water bodies, their movement in water depends upon their water solubility; their agglomerates, which undergo redox reaction, physically bind them to the components in water. A study carried on rats showed $MnO_2$ nanoparticle's presence in lungs, olfactory organs, and brain, after inhalation for seven days causing oxidative stress and neurological diseases. Labors in the mining, welding, steel, and glass industry get exposed to fumes and dust of nanoparticles causes progressive damage to the central nervous system [320]. It was reported that copper nanoparticles orally induced can adversely affect and cause injury in kidney, liver, and spleen of mice under experimentation than micro copper particles [321]. Studies have shown that nanoparticles possess different chemical compositions for people suffering from Crohn's disease, and colon cancer or ulcerative colitis. The NPs of carbon, zirconium, silver, silicon, stainless steel have been detected on colon mucosa [322].

Various processes are involved in wastewater treatment. The removal of nanoparticles takes place through interaction with microorganisms that degrade them. Due to their large specific surface area, nanoparticles form large clusters. In the case of non-aggregation, naturally, surfactants are added to remove nanoparticles by sedimentation, filtration, or flotation [323].

It is very difficult to judge the fate, toxicity, and transport of nanoparticles used for water purification. Recent studies have shown the toxic effect of nanoparticles on the environment as well as on humans. Because of their large surface area to mass ratio, they are more biologically and chemically active than normal particles. Nanotoxicology shows that particles less than 300 nm can easily get through the cell membrane, though their effects are unclear. Silver nanoparticles suppress cellular growth and multiply rapidly, causing cell death. An experiment shows that titanium dioxide nanoparticles cause DNA and chromosomal damage and inflammation too. ZnO NPs are abundantly used in industries; hence, consumers and workers are in constant exposure to these particles. These particles have shown wide applications in personal care products, dyes, cosmetics, and electronics [320]. Generally, the size of ZnO NPs used for personal care products is less than 100 nm, which means they are not able to stratum cornea layer, unlike the sprays that might find a way to vital organs. Several studies reveal the oxidative damage, genotoxicity, and cytotoxicity caused by ZnO. It has been reported that human epidermal cells exposed to 0.8 µg/ml of 30 nm-sized ZnO NPs had DNA damages [324]. Zinc oxide and titanium dioxide are toxic to the human brain and lung cells. Aluminum, Barium, and Strontium nanoparticles enter the environment through high altitude spraying technique, hence relentless contact remains unavoidable. Their presence in water bodies and in the soil gradually enters food chains, leading to their accumulation in the body.

MOFs are extensively used in biomedical, pharmaceutical, and drug delivery applications; hence, the raw materials used for their synthesis are required to be non-toxic. The most appropriate metals used as metal nodes are Ca, Cu, Mn, Mg, Zn, Fe, Ti, and Zr, but the data on their biological fate is not enough. Hence, the organic linkers used, reduce the risk of accumulation in the body thereby reducing the risk of adverse influence on humans and the environment. Such MOFs are synthesized, which have large spherical voids and high BET surface area. For medical applications, MOFs synthesis, recommends instability of the chemical used to avoid their accumulation followed by self-degradation in the body.

It is a fact that NP usage in varied sectors has increased unconditionally over the years. With so many practical applications, whether intentionally or unintentionally, human exposure to these nano materials is indispensable.

## IDEOLOGIES FOR ECO-FRIENDLY USE OF NANOTECHNOLOGY

Twelve designing ideologies of green engineering were suggested by Anastas, P.T., and Zimmerman, J.B., which were later redrafted after the Sandestin conference into 9 points [325 - 327]. The objective of these principles was to optimize the use of nanomaterials with maximum benefits including the use of eco-friendly, non-toxic raw materials, reusable, recyclable, material with less landfill, cost-effective, and commercially viable. Some of its important principles were conservation of nature, minimum depletion of resources, protecting human health, and applying such solutions, which are cognizant of local geography and heritage and inherent rather than circumstantial (Use of non-toxic, non-hazardous raw material after and before).

1. Prevention instead of treatment (Lesser amount of material should be used rather than treating it later).

2. Design for separation (Surface properties and magnetic properties may be used for the separation and purification of nanomaterials to minimize energy and material consumption).

3. Maximum efficiency (System should be designed simple, sustainable, and have maximum efficiency in areas like mass, energy, space, and time).

4. Output-pulled *versus* input pushes (Minimising the input to check output for commercial viability).

5. Conserve Complexity (Search for recycling, reuse, and reproducible opportunity, embedded entropy, and complexities should be considered).

6. Durability rather than immortality (Design should be sturdy and long-lasting allowing changes when required).

7. Meet need, minimize excess (design as per requirement and need, optimize the use of nanomaterial, avoid multiple usages of the same system).

8. Minimise material diversity (reduce variability and raw materials (sources) of a given nanomaterial).

9. Integrated material and energy flow (Systems must be integrated and interconnected with material and energy flow).

10. Design for the commercial afterlife (The system should be of practical usage and have commercially viable).

11. Renewable rather than depleting (Look for renewable, easily disposable nanomaterial, with minimum landfill, after its use).

Besides the above environmental principles, the cost comparison of various nanomaterials must be considered before their usage. The manufacturing cost of the nanomaterial may come down in the near future, but the present high cost may be the main obstacle to its application in water treatment. Cost, in turn, depends upon the volume of water to be treated and upon the lifetime capacity, reproducibility, and reusability of nanomaterials, rarely tested to exhaustion. Moreover, the prices, if calculated for the small-scale bases and research purposes, will be higher than those calculated for commercial and practical purposes in bulk. Despite the high cost, nanomaterials are gaining interest in the water treatment sector for a number of reasons, such as the relatively low quantity required, unique properties of nanomaterials leading to new applications, excess usability has shown decreasing trends in the cost of nano-related products, and cut down on capital investment for localized or centralized set-up.

The large-scale treatment plant can provide low-cost treatment, however, the initial capital cost might be high for developing countries. Small point-of-use type of system has low infrastructure costs but require high operational cost which can be comparable with highly advanced point-of-use treatment systems. As the technology grows and applications escalate, the prices of raw materials and nanoparticles fall. With the recycling and reuse ability of nanomaterials, their applicability to purify water increased [325].

In order to attain a balance between the consumption of freshwater in domestic usage, industrial setups, and commercial use, and its reuse to ensure its availability to generations to come, novel and advanced water treatment technologies are required. To ensure the high quality of water, nano-engineering has offered innovative methods of water treatment that can be adopted depending upon their specific applications. For cost-effectiveness, many nanomaterials can be integrated with conventional methods and existing treatments. Due to the unique properties of nanomaterials such as high surface to mass ratio, high reaction rate, reusability, along with the removal of contaminants, it has been preferred above all other treatment processes.

However, with numerous advantages, the deposition and emission of nanoparticles in the environment and their retrieval have been a challenge to all. Its traces in water and soil are a cause of concern. Another important aspect of the nanomaterial to be studied is, whether it can be adapted on a large treatment processes. Their compatibility with conventional processes at large-scale needs to be studied further.

In order to extract maximum benefits from nanotechnology increased responsiveness and thought have to be developed from the ongoing studies on nanotechnology. Utmost care should be taken while deciding the process, the raw material for synthesis, as well as the accidental release of NPs. The complexities are due to nano size, which triggers their interactions with biological materials, causing risk. Interdisciplinary sciences should come together to reflect on its challenges, reduce the ill effect of nanomaterials, and suggest innovative ways for its application with fewer hazards.

## CONCLUSION

The extraordinary and exclusive properties of nano-materials have laid the way for water and wastewater treatment. Nano-materials are categorized as nano-adsorbents, nanotech membrane, and nano photocatalysts. Although these processes have not been used in large-scale wastewater treatment plants as per the literature study, it is an effectively competent, eco-friendly, and efficient approach. This technique produces less waste, is easy to handle, and effective method. Amongst all nanoparticles studied Ag, $Fe_3O_4$, $TiO_2$, Zero-valent nanoparticles are largely used to remove pathogens, toxic organic compounds, pesticides, and heavy metals from the wastewater, moreover, the catalytic properties of nanoparticles inactivate the bacteria present in the water. Carbon nanotubes are toxic and water-insoluble until functional groups are introduced which makes them biocompatible and water-soluble. Modified nanoparticles have more acceptability, due to their less toxicity to humans and low environmental hazards. In the near future nanoparticles would be an important component to treat industrial, commercial and public wastewater. Efforts are being made to synthesize cost-effective and environmentally friendly nanoparticles.

The sustainability of nanoparticles depends on the extent to which nanomaterials are safe. Nanotechnology should stick to the rules laid out by the pollution control board and show ecological responsibility for a safe and healthy atmosphere.

Natural nanoparticles synthesized by natural nano-materials such as chitosan, gelatine, alginate, and peptides showing antimicrobial properties should be used instead of those containing heavy metals and their oxides. Raw materials used should be cost-effective and easily locally available. Green nano-synthesis is a step to meet all the drawbacks set by various nano-materials. Green nanoparticles are prepared by using plants and their parts. It involves three ideas: primarily the use of less energy, recycle of nanoparticles, and use of eco-friendly and environment caring nanomaterials, which are effective, and efficient pollutant removers. Their mode of synthesis is simple and cost-effective. The objective of green nanotechnology is to reduce health hazards, provide an economic way to treat polluted water, and replace the present metal nanoparticles with environmentally friendly nanoparticles synthesized by plants.

Hence, nanoparticles can be considered as a means to treat wastewater and help to resolve water-related problems for a growing population in the future, provided some safety measures are followed and the toxic effect of nanoparticles is taken into consideration before usage.

## CONSENT FOR PUBLICATION

Not applicable.

## CONFLICT OF INTEREST

The author declares no conflict of interest, financial or otherwise.

## ACKNOWLEDGEMENTS

The authors are grateful to the Director, Laxminarayan Institute of Technology, R.T.M. Nagpur University, Nagpur(MS) for providing the necessary facilities required during the work.

## REFERENCES

[1]     Ferrari M. Cancer nanotechnology: Opportunities and challenges. Nat Rev Cancer 2005; 5(3): 161-71.
        [http://dx.doi.org/10.1038/nrc1566] [PMID: 15738981]

[2]     Leppard GG, Mavrocordatos D, Perret D. Electron-optical characterization of nano- and micro-particles in raw and treated waters : An overview 2000; 6: 1-8.

[3]     Luo G, Du L, Wang Y, Wang K. Recent developments in microfluidic device-based preparation, functionalization, and manipulation of nano- and micro-materials. Particuology 2019; 45: 1-19.
        [http://dx.doi.org/10.1016/j.partic.2018.10.001]

[4]     Boyjoo Y, Pareek VK, Liu J. Carbonate particles and their applications. J Mater Chem A Mater Energy Sustain 2014; 2(35): 14270-88.
        [http://dx.doi.org/10.1039/C4TA02070G]

[5]     Suri S, Ruan G, Winter J, Schmidt CE. Materials Used for Microparticle Synthesis. Third Edit.. Elsevier 2007; 2000.

[6]     Ribeiro AR, Nunes OC, Pereira MFR, Silva AMT. An overview on the advanced oxidation processes applied for the treatment of water pollutants defined in the recently launched Directive 2013/39/EU. Environ Int 2015; 75: 33-51.
[http://dx.doi.org/10.1016/j.envint.2014.10.027] [PMID: 25461413]

[7]     Huang DL, Zeng GM, Feng CL, *et al.* Mycelial growth and solid-state fermentation of lignocellulosic waste by white-rot fungus Phanerochaete chrysosporium under lead stress. Chemosphere 2010; 81(9): 1091-7.
[http://dx.doi.org/10.1016/j.chemosphere.2010.09.029] [PMID: 20951406]

[8]     Hashimoto K, Irie H, Fujishima A. TiO$_2$ photocatalysis : A historical overview and future prospects 2006; 44(12): 8269-85.

[9]     Safaríková M, Roy I, Gupta MN, Safarík I. Magnetic alginate microparticles for purification of α-amylases. J Biotechnol 2003; 105(3): 255-60.
[http://dx.doi.org/10.1016/j.jbiotec.2003.07.002] [PMID: 14580797]

[10]    Yoshizuka K, Lou Z, Inoue K. Silver-complexed chitosan microparticles for pesticide removal. React Funct Polym 2000; 44(1): 47-54.
[http://dx.doi.org/10.1016/S1381-5148(99)00076-0]

[11]    Bacon R. Growth, structure, and properties of graphite whiskers. J Appl Phys 1960; 31(2): 283-90.
[http://dx.doi.org/10.1063/1.1735559]

[12]    Mitchnick MA, Fairhurst D, Pinnell SR. Microfine zinc oxide (Z-cote) as a photostable UVA/UVB sunblock agent. J Am Acad Dermatol 1999; 40(1): 85-90.
[http://dx.doi.org/10.1016/S0190-9622(99)70532-3] [PMID: 9922017]

[13]    Nohynek GJ, Dufour EK, Roberts MS. Nanotechnology, cosmetics and the skin: Is there a health risk? Skin Pharmacol Physiol 2008; 21(3): 136-49.
[http://dx.doi.org/10.1159/000131078] [PMID: 18523411]

[14]    El-Hady MM, Farouk A, Sharaf S. Flame retardancy and UV protection of cotton based fabrics using nano ZnO and polycarboxylic acids. Carbohydr Polym 2013; 92(1): 400-6.
[http://dx.doi.org/10.1016/j.carbpol.2012.08.085] [PMID: 23218312]

[15]    Farouk A, Sharaf S, Abd El-Hady MM. Preparation of multifunctional cationized cotton fabric based on TiO$_2$ nanomaterials. Int J Biol Macromol 2013; 61: 230-7.
[http://dx.doi.org/10.1016/j.ijbiomac.2013.06.022] [PMID: 23811163]

[16]    Kathirvelu S, Souza LD, Dhurai B. UV protection finishing of textiles using ZnO nanoparticles 2009; 34: 267-73.

[17]    Selvam S, Rajiv Gandhi R, Suresh J, Gowri S, Ravikumar S, Sundrarajan M. Antibacterial effect of novel synthesized sulfated β-cyclodextrin crosslinked cotton fabric and its improved antibacterial activities with ZnO, TiO$_2$ and Ag nanoparticles coating. Int J Pharm 2012; 434(1-2): 366-74.
[http://dx.doi.org/10.1016/j.ijpharm.2012.04.069] [PMID: 22627018]

[18]    Hebeish AA, Abdelhady MM, Youssef AM. TiO$_2$ nanowire and TiO$_2$ nanowire doped Ag-PVP nanocomposite for antimicrobial and self-cleaning cotton textile. Carbohydr Polym 2013; 91(2): 549-59.
[http://dx.doi.org/10.1016/j.carbpol.2012.08.068] [PMID: 23121944]

[19]    Dastjerdi R, Montazer M. A review on the application of inorganic nano-structured materials in the modification of textiles: Focus on anti-microbial properties Colloids Surf B Biointerfaces 2010; 79(1). 5-18.
[http://dx.doi.org/10.1016/j.colsurfb.2010.03.029] [PMID: 20417070]

[20]    Fröhlich E. The role of surface charge in cellular uptake and cytotoxicity of medical nanoparticles. Int J Nanomedicine 2012; 7: 5577-91.
[http://dx.doi.org/10.2147/IJN.S36111] [PMID: 23144561]

[21]   You C, Han C, Wang X, *et al.* The progress of silver nanoparticles in the antibacterial mechanism, clinical application and cytotoxicity. Mol Biol Rep 2012; 39(9): 9193-201.
[http://dx.doi.org/10.1007/s11033-012-1792-8] [PMID: 22722996]

[22]   Doll TA, Raman S, Dey R, Burkhard P. Nanoscale assemblies and their biomedical applications. J R Soc Interface 2013; 10(80): 20120740.
[http://dx.doi.org/10.1098/rsif.2012.0740] [PMID: 23303217]

[23]   Park B, Donaldson K, Duffin R, Kelly F, Mudway I. Hazard and Risk Assessment of a Nanoparticulate Cerium Oxide-Based Diesel Fuel Additive — A Case Study Robert Guest and Peter Jenkinson. Inhal Toxicol 2008; 20(January): 547-66.
[http://dx.doi.org/10.1080/08958370801915309] [PMID: 18444008]

[24]   Cassee FR, *et al.* Exposure, health and ecological effects review of engineered nanoscale cerium and cerium oxide associated with its use as a fuel additive Crit Rev Toxicol 2011; 41: 213-29.
[http://dx.doi.org/10.3109/10408444.2010.529105]

[25]   Riley L, Thomson E. investigators IMF chief economist MIT community 2008; 52(28)

[26]   Lee J, Kwak SY. Mn-Doped Maghemite ($\gamma$-$Fe_2O_3$) from Metal-Organic Framework Accompanying Redox Reaction in a Bimetallic System: The Structural Phase Transitions and Catalytic Activity toward NOx Removal. ACS Omega 2018; 3(3): 2634-40.
[http://dx.doi.org/10.1021/acsomega.7b01865] [PMID: 31458548]

[27]   Wei X, Bhojappa S, Lin L, Viadero RC Jr. Performance of Nano-Magnetite for Removal of Selenium from Aqueous Solutions. Environ Eng Sci 2012; 29(6): 526-32.
[http://dx.doi.org/10.1089/ees.2011.0383]

[28]   Shrivastava S, Jadon N, Jain R. Next generation polymer nanocomposites based electrochemical sensors and biosensors: A review. Trends Analyt Chem 2016; 82: 55-67.
[http://dx.doi.org/10.1016/j.trac.2016.04.005]

[29]   Wang X, Yang L, Chen ZG, Shin DM. Application of nanotechnology in cancer therapy and imaging. CA Cancer J Clin 2008; 58(2): 97-110.
[http://dx.doi.org/10.3322/CA.2007.0003] [PMID: 18227410]

[30]   Cordova G, Attwood S, Gaikwad R, Gu F, Leonenko Z. Nano biomed eng magnetic force microscopy characterization of superparamagnetic iron oxide nanoparticles. Nano Biomed Eng 2014; 6(1): 31-9.
[http://dx.doi.org/10.5101/nbe.v6i1.p31-39]

[31]   Chae K, Wang S, Hendren ZD, Wiesner MR, Watanabe Y, Gunsch CK. Effects of fullerene nanoparticles on EscherichiacoliK12 respiratoryactivity in aqueous suspension and potential use for membrane biofouling control. J Membr Sci 2009; 329(1–2): 68-74.
[http://dx.doi.org/10.1016/j.memsci.2008.12.023]

[32]   Kim MJ, Davies SHR, Baumann MJ, Tarabara VV, Masten SJ. Effect of ozone dosage and hydrodynamic conditions on the permeate flux in a hybrid ozonation-ceramic ultrafiltration system treating natural waters. J Membr Sci 2008; 311(1–2): 165-72.
[http://dx.doi.org/10.1016/j.memsci.2007.12.010]

[33]   Amin MT, Alazba A, Manzoor U. A review of removal of pollutants from water / wastewater using different types of nanomaterials. Adv Mater Sci Eng 2014; 2014

[34]   Beydoun D, Amal R, Low G, Mcevoy S. Role of nanoparticles in photocatalysis. J Nanopart Res 2000; 1(4): 439-58.
[http://dx.doi.org/10.1023/A:1010044830871]

[35]   Hunge YM, Yadav AA, Liu S, Mathe VL. Sonochemical synthesis of CZTS photocatalyst for photocatalytic degradation of phthalic acid. Ultrason Sonochem 2019; 56(March): 284-9.
[http://dx.doi.org/10.1016/j.ultsonch.2019.04.003] [PMID: 31101264]

[36]   Deshpande BD, Agrawal PS, Yenkie MKN. Nanoparticles aided AOP for Degradation of p -

NitroBenzoic acid. Mater Today 2020.

[37] Deshpande B D, Agrawal P S, Yenkie M K N. Advanced oxidative degradation of benzoic acid and 4-nitro benzoic acid – A comparative study Advanced Oxidative Degradation of Benzoic Acid and 4-Nitro Benzoic Acid – A Comparative Study. AIP 2019; 210003(August)

[38] Belver C, Bedia J, Go A, Pen M. Semiconductor photocatalysis for water purification. 2019.
[http://dx.doi.org/10.1016/B978-0-12-813926-4.00028-8]

[39] Fagan R, McCormack ED, Suresh C Pillai. A review of solar and visible light active TiO$_2$ photocatalysis for treating bacteria, cyanotoxins and contaminants of emerging concern. Mater Sci Semicond Process 2015.

[40] Likodimos Vlassis, Dionysiou Dionysios D, Falaras P. Clean water: Water detoxification using innovative photocatalysts. Rev Env Sci Biotechnol 2010; 9: 87-94.

[41] Lang Chen L-NS, Tang J, Cheng P. Heterogeneous photocatalysis for selective oxidation of alcohols and hydrocarbons. Appl Catal 2018.

[42] Pankaj Attri G. Nanoparticles for water purification,' particles for water purification. 2014; January

[43] Kisch H. Characterization of solid materials and heterogeneous photocatalysis and water photoinitiators for polymer applied homogeneous. 2015.

[44] Kakavandi B, Bahari N, Rezaei Kalantary R, Dehghani Fard E. Enhanced sono-photocatalysis of tetracycline antibiotic using TiO$_2$ decorated on magnetic activated carbon (MAC@T) coupled with US and UV: A new hybrid system. Ultrason Sonochem 2019; 55(March): 75-85.
[http://dx.doi.org/10.1016/j.ultsonch.2019.02.026] [PMID: 31084793]

[45] Hiroaki Tada MN, Jin Q, Iwaszuk Anna. Molecular-scale transition metal oxide nanocluster surface-modified titanium dioxide as solar-activated environmental catalysts. Am Chem Soc, 2014; 118(23): 12077-86.

[46] Chong MN, Jin B, Chow CWK, Saint C. Recent developments in photocatalytic water treatment technology: A review. Water Res 2010; 44(10): 2997-3027.
[http://dx.doi.org/10.1016/j.watres.2010.02.039] [PMID: 20378145]

[47] Mohamed E. Removal of organic compounds from water by adsorption and photocatalytic oxidation. 2011; 31-44.

[48] Benhebal H, Chaib M, Salmon T, *et al.* Photocatalytic degradation of phenol and benzoic acid using zinc oxide powders prepared by the sol-gel process. Alex Eng J 2013; 52(3): 517-23.
[http://dx.doi.org/10.1016/j.aej.2013.04.005]

[49] Mondal K, Sharma A. Photocatalytic oxidation of pollutant dyes in wastewater by TiO$_2$ and ZnO nano-materials – A Mini-review Indian Inst Tecnol 2016; 36-72.

[50] Ireland JC, Klostermann P, Rice EW, Clark RM. Inactivation of Escherichia coli by titanium dioxide photocatalytic oxidation. Appl Environ Microbiol 1993; 59(5): 1668-70.
[http://dx.doi.org/10.1128/aem.59.5.1668-1670.1993] [PMID: 8390819]

[51] Sushma D, Richa S. Use of Nanoparticles in Water Treatment : A review. Int Res J Environ Sci 2015; 4(10): 103-6.

[52] Zeng YY, Xue Y, Liang S, *et al.* Removal of fluoride from aqueous solution by TiO$_2$ and TiO$_2$ – SiO$_2$ nanocomposite. Chem Spec Bioavail 2017; 29(1): 25-32.
[http://dx.doi.org/10.1080/09542299.2016.1269617]

[53] Finella A, Villaluz J A, De Luna M D G, Colades J I, Garcia-segura S, Lu M. Removal of 4-chlorophenol by visible-light photocatalysis using ammonium iron(II) sulfate-doped nano-titania. Process Saf Environ Prot 2019; Ii

[54] Shojaie A, Fattahi M, Jorfi S, Ghasemi B. Synthesis and evaluations of Fe$_3$O$_4$ – TiO$_2$ – Ag nanocomposites for photocatalytic degradation of 4 - chlorophenol (4 - CP): Effect of Ag and Fe

compositions. Int J Ind Chem 2018; 9(2): 141-51.
[http://dx.doi.org/10.1007/s40090-018-0145-4]

[55]  Ollis DF. Comparative aspects of advanced oxidation processes. Emerg Technol Hazard Waste Manag 1993; III(1): 18-34.

[56]  Muruganandham M, Swaminathan M. Decolourisation of Reactive Orange 4 by Fenton and photo-Fenton oxidation technology. Dyes Pigments 2004; 63(3): 315-21.
[http://dx.doi.org/10.1016/j.dyepig.2004.03.004]

[57]  Wang KH, Hsieh YH, Wu CH, Chang CY. The pH and anion effects on the heterogeneous photocatalytic degradation of o-methylbenzoic acid in $TiO_2$ aqueous suspension. Chemosphere 2000; 40(4): 389-94.
[http://dx.doi.org/10.1016/S0045-6535(99)00252-0] [PMID: 10665404]

[58]  Hunge YM, *et al.* Enhanced photocatalytic performance of ultrasound treated GO / $TiO_2$ composite for photocatalytic degradation of salicylic acid under sunlight a. Photocatalysis International Research Centre, Tokyo University of Science. Ultrason Sonochem 2019; 104849.

[59]  Yonezawa Y, Kometani N, Sakaue T, Yano A. Photoreduction of silver ions in a colloidal titanium dioxide suspension. J Photochem Photobiol Chem 2005; 171(1): 1-8.
[http://dx.doi.org/10.1016/j.jphotochem.2004.08.020]

[60]  Sawhney KCL. Treatment of hazardous organic and inorganic compounds through aqueous-phase photocatalysis: A review. Ind Eng Chem Res 2004; 43(24): 7683-96.
[http://dx.doi.org/10.1021/ie0498551]

[61]  Oller I, Malato S, Sánchez-Pérez JA. Combination of advanced oxidation processes and biological treatments for wastewater decontamination--a review. Sci Total Environ 2011; 409(20): 4141-66.
[http://dx.doi.org/10.1016/j.scitotenv.2010.08.061] [PMID: 20956012]

[62]  Horváth E, Szabó-Bárdos H Czili, Attila . Photocatalytic oxidation of oxalic acid enhanced by silver deposition on a $TiO_2$surface. J Photochem Photobiol Chem 2003; 154(2–3): 195-201.

[63]  Ayati A, Ahmadpour A, Bamoharram FF, Tanhaei B, Mänttäri M, Sillanpää M. A review on catalytic applications of Au/$TiO_2$ nanoparticles in the removal of water pollutant. Chemosphere 2014; 107: 163-74.
[http://dx.doi.org/10.1016/j.chemosphere.2014.01.040] [PMID: 24560285]

[64]  Hunge YM, Yadav AA, Mathe VL. Ultrasound assisted synthesis of $WO_3$-ZnO nanocomposites for brilliant blue dye degradation. Ultrason Sonochem 2018; 45: 116-22.
[http://dx.doi.org/10.1016/j.ultsonch.2018.02.052] [PMID: 29705304]

[65]  Taufik R S A, Albert A. Sol-gel synthesis of ternary CuO/$TiO_2$/ZnO nanocomposites of for enhanced photocatalytic performance under UV and visible irradiation J Photochem chem 2017; 344: 149-62.

[66]  Nadarajan R, Wan Abu Bakar WA, Ali R, Ismail R. Effect of structural defects towards the performance of $TiO_2$/$SnO_2$/$WO_3$ photocatalyst in the degradation of 1,2-dichlorobenzene. J Taiwan Inst Chem Eng 2016; 64: 106-15.
[http://dx.doi.org/10.1016/j.jtice.2016.03.044]

[67]  Tomova L B D, Iliev V, Rakovsky S. Gold modified n-doped $TiO_2$ and n-doped $WO_3$/$TiO_2$ semiconductors - photocatalysts for UV-visible light destruction of 2, 4, 6-trinitrotoluene in aqueous solution Nanosci Nanotechnology 2011; November 2(12)

[68]  HanCao Y, Huang Shaolong, Yub Yanlong, Yan Yabin, Lv Yuekai. Synthesis of $TiO_2$-N/$SnO_2$ heterostructure photocatalyst and its photocatalytic mechanism J Colloid Interface Sci 2017; 486: 176-83.

[69]  Moussavi G, Mahmoudi M. Removal of azo and anthraquinone reactive dyes from industrial wastewaters using MgO nanoparticles. 2009; 168: 806-12.

[70]  Iram M, Guo C, Guan Y, Ishfaq A, Liu H. Adsorption and magnetic removal of neutral red dye from

aqueous solution using $Fe_3O_4$ hollow nanospheres. J Hazard Mater 2010; 181(1-3): 1039-50.
[http://dx.doi.org/10.1016/j.jhazmat.2010.05.119] [PMID: 20566240]

[71]     Farghali AA, Bahgat M, El Rouby W, Khedr MH. Decoration of MWCNTs with $CoFe_2O_4$ Nanoparticles for Methylene Blue Dye Adsorption 2012; 41: 2209-25.
[http://dx.doi.org/10.1007/s10953-012-9934-0]

[72]     Saharan P, Chaudhary GR, Mehta SK, Umar A. Removal of Water Contaminants by Iron Oxide Nanomaterials 2014; 14(1): 627-43.

[73]     Hariani PL, Faizal M, Setiabudidaya D. Synthesis and properties of $Fe_3O_4$ nanoparticles by co-precipitation method to removal procion dye. 2013; 4(3)
[http://dx.doi.org/10.7763/IJESD.2013.V4.366]

[74]     Shipley H J, Engates K E, Grover V A. Removal of Pb ( II ), Cd ( II ), Cu ( II ), and Zn ( II ) by hematite nanoparticles : Effect of sorbent concentration , pH , temperature , and exhaustion. 2013; 1727-36.
[http://dx.doi.org/10.1007/s11356-012-0984-z]

[75]     Salini Rajput D, Lok PSingh, Charles U Pittman Jr. Lead ($Pb^{2+}$) and copper ($Cu^{2+}$) remediation from water using superparamagnetic maghemite ($\gamma$-$Fe_2O_3$) nanoparticles synthesized by Flame Spray Pyrolysis (FSP). J Colloid Interface Sci 2017; 492(15): 176-90.
[http://dx.doi.org/10.1016/j.jcis.2016.11.095]

[76]     Nan Guo JH, Liu H, Fu Y. Preparation of $Fe_2O_3$ nanoparticles doped with $In_2O_3$ and photocatalytic degradation property for rhodamine B. Optik (Stuttg) 2020; 201: 163537.
[http://dx.doi.org/10.1016/j.ijleo.2019.163537]

[77]     Nishesh K S, Gupta Kumar, Ghaffari Yasaman, Bae Jiyeol. Synthesis of coral-like $\alpha$-$Fe_2O_3$ nanoparticles for dye degradation at neutral pH. J Mol Liq 2020; 301(1): 112473.

[78]     Muraro PCL, Mortari SR, Vizzotto BS, *et al.* Iron oxide nanocatalyst with titanium and silver nanoparticles: Synthesis, characterization and photocatalytic activity on the degradation of Rhodamine B dye. Sci Rep 2020; 10(1): 3055.
[http://dx.doi.org/10.1038/s41598-020-59987-0] [PMID: 32080290]

[79]     Mahdavi S, Jalali M, Afkhami A. Heavy metals removal from aqueous solutions using $TiO_2$, MgO, and $Al_2O_3$ nanoparticles. Chem Eng Commun 2013; 6445(December): 448-70.
[http://dx.doi.org/10.1080/00986445.2012.686939]

[80]     Engates KE, Shipley HJ. Adsorption of Pb, Cd, Cu, Zn, and Ni to titanium dioxide nanoparticles : Effect of particle size, solid concentration, and exhaustion 2011; 386-95.

[81]     Fei BJ, Cui Y, Yan XH, *et al.* Controlled Preparation of $MnO_2$ Hierarchical Hollow Nanostructures and Their Application in Water Treatment. Adv Mater 2008; 20(1): 452-6.
[http://dx.doi.org/10.1002/adma.200701231]

[82]     Madrakian T, Afkhami A, Ahmadi M. Adsorption and kinetic studies of seven different organic dyes onto magnetite nanoparticles loaded tea waste and removal of them from wastewater samples. Spectrochim Acta A Mol Biomol Spectrosc 2012; 99: 102-9.
[http://dx.doi.org/10.1016/j.saa.2012.09.025] [PMID: 23058993]

[83]     Giri SK, Das NN, Pradhan GC. Colloids and Surfaces A : Physicochemical and Engineering Aspects Synthesis and characterization of magnetite nanoparticles using waste iron ore tailings for adsorptive removal of dyes from aqueous solution. Colloids Surf A Physicochem Eng Asp 2011; 389(1–3): 43-9.
[http://dx.doi.org/10.1016/j.colsurfa.2011.08.052]

[84]     Zhang Y, Wu B, Xu H, *et al.* Nanomaterials-enabled water and wastewater treatment. NanoImpact 2016; 3–4: 22-39.
[http://dx.doi.org/10.1016/j.impact.2016.09.004]

[85]     Hirlekar R, Yamagar M, Garse H, Vij M. Carbon Nanotubes and its application : A Review. Asian J Pharm Clin Res 2009; 2(4): 17-27.

[86]    Ali S, Rehman SAU, Luan HY, Farid MU, Huang H, Huang H. Challenges and opportunities in functional carbon nanotubes for membrane-based water treatment and desalination. Sci Total Environ 2019; 646(November): 1126-39.
[http://dx.doi.org/10.1016/j.scitotenv.2018.07.348] [PMID: 30235599]

[87]    Ihsanullah , Abbas A, Al-Amer AM, *et al.* Heavy metal removal from aqueous solution by advanced carbon nanotubes: Critical review of adsorption applications. Separ Purif Tech 2016; 157: 141-61.
[http://dx.doi.org/10.1016/j.seppur.2015.11.039]

[88]    Bolisetty S, Peydayesh M, Mezzenga R. Sustainable technologies for water purification from heavy metals: Review and analysis. Chem Soc Rev 2019; 48(2): 463-87.
[http://dx.doi.org/10.1039/C8CS00493E] [PMID: 30603760]

[89]    Bunshi Fugetsu A. Caged multiwalled carbon nanotubes as the adsorbents for affinity-based elimination of ionic. 2004; 38(24): 6890-6.

[90]    Pan BO, Lin D, Mashayekhi H, Xing B. Adsorption and hysteresis of bisphenol A and 17 r -ethinyl estradiol on carbon nanomaterials. 2008; 42(15): 5480-5.

[91]    Richard RG, Waiters W. Equilibrium adsorption of polycyclic aromatic hydrocarbons from water onto activated carbon 1984; 186: 395-403.

[92]    Yang K, Zhu L, Xing B. Adsorption of polycyclic aromatic hydrocarbons by carbon nanomaterials. Environ Sci Technol 2006; 40(6): 1855-61.
[http://dx.doi.org/10.1021/es052208w] [PMID: 16570608]

[93]    Chen C, Hu J, Shao D, Li J, Wang X. Adsorption behavior of multiwall carbon nanotube/iron oxide magnetic composites for Ni(II) and Sr(II). J Hazard Mater 2009; 164(2-3): 923-8.
[http://dx.doi.org/10.1016/j.jhazmat.2008.08.089] [PMID: 18842337]

[94]    Moghaddam H Khademzadeh, Pakizeh Majid. Experimental study on mercury ions removal from aqueous solution by MnO$_2$/CNTs nanocomposite adsorbent. J Ind Eng Chem 2015; 21(25): 221-9.

[95]    Liang J, Liu J, Yuan X, *et al.* Facile synthesis of alumina-decorated multi-walled carbon nanotubes for simultaneous adsorption of cadmium ion and trichloroethylene. Chem Eng J 2015; 273(1): 101-10.
[http://dx.doi.org/10.1016/j.cej.2015.03.069]

[96]    Xu Y, Wen W, Wu JM. Titania nanowires functionalized polyester fabrics with enhanced photocatalytic and antibacterial performances. J Hazard Mater 2018; 343(5): 285-97.
[http://dx.doi.org/10.1016/j.jhazmat.2017.09.044] [PMID: 28988054]

[97]    Ikhlaq A, Brown DR, Kasprzyk-Hordern B. Mechanisms of catalytic ozonation: An investigation into superoxide ion radical and hydrogen peroxide formation during catalytic ozonation on alumina and zeolites in water. Appl Catal B 2013; 129: 437-49.
[http://dx.doi.org/10.1016/j.apcatb.2012.09.038]

[98]    Tosco T, Petrangeli M, Cruz C, Sethi R. Nanoscale zerovalent iron particles for groundwater remediation : A review. J Clean Prod 2014; 77: 1-12.
[http://dx.doi.org/10.1016/j.jclepro.2013.12.026]

[99]    Li M, Jiang Y, Ding R, Song D, Yu H, Chen Z. Hydrothermal synthesis of anatase TiO$_2$ nanoflowers on a nanobelt framework for photocatalytic applications. J Electron Mater 2013; 42(25): 1290-6.
[http://dx.doi.org/10.1007/s11664-013-2593-0]

[100]   Pal S, Tak YK, Song JM, Pal S, Tak YK, Song JM. Does the antibacterial activity of silver nanoparticles depend on the shape of the nanoparticle ? A study of the gram-negative bacterium escherichia coli. 2007.
[http://dx.doi.org/10.1128/AEM.02218-06]

[101]   Castellano JJ, Shafii SM, Ko F, *et al.* Comparative evaluation of silver-containing antimicrobial dressings and drugs. Int Wound J 2007; 4(2): 114-22.
[http://dx.doi.org/10.1111/j.1742-481X.2007.00316.x] [PMID: 17651227]

[102]   Gong P, Li H, He X, *et al.* Preparation and antibacterial activity of Fe$_3$O$_4$ @ Ag nanoparticles. Nanotechnology 2007; 18(28): 285604.
[http://dx.doi.org/10.1088/0957-4484/18/28/285604]

[103]   Rai M, Yadav A, Gade A. Silver nanoparticles as a new generation of antimicrobials. Biotechnol Adv 2009; 27(1): 76-83.
[http://dx.doi.org/10.1016/j.biotechadv.2008.09.002] [PMID: 18854209]

[104]   Jain P, Pradeep T. Potential of silver nanoparticle-coated polyurethane foam as an antibacterial water filter. 2005; 3-7.

[105]   Nair AS, Pradeep T. Extraction of Chlorpyrifos and Malathion from Water by Metal Nanoparticles 2007; 7(7): 1-7.

[106]   Bootharaju MS, Pradeep T. Understanding the degradation pathway of the pesticide, chlorpyrifos by noble metal nanoparticles. pubsacsorg/Langmuir Underst 2012; 28: 2671-9. Available from: pubs.acs.org/Langmuir

[107]   Andrea Tao P, Sinsermsuksakul Prasert. Polyhedral silver nanocrystals with distinct scattering signatures. 2006; 4597-601.

[108]   Pillai ZS, Kamat PV. What factors control the size and shape of silver nanoparticles in the citrate ion reduction method? J Phys Chem B 2004; 108(3): 945-51.
[http://dx.doi.org/10.1021/jp037018r]

[109]   Hong H, Gong M, Park C. A facile preparation of silver nanocolloids by hydrogen reduction of a silver alkylcarbamate complex. Bull Korean Chem Soc 2009; 30(11): 2669-74.
[http://dx.doi.org/10.5012/bkcs.2009.30.11.2669]

[110]   Kapoor NS, Lawless D, Kennepohl P, Meisel D, Serpone NJ. Reduction and aggregation of silver ions in aqueous gelatin solutions. Langmuir 1994; 10(12): 3018-22.

[111]   Brus L. Noble metal nanocrystals: Plasmon electron transfer photochemistry and single-molecule Raman spectroscopy. Acc Chem Res 2008; 41(12): 1742-9.
[http://dx.doi.org/10.1021/ar800121r] [PMID: 18783255]

[112]   Lee SH, Jun BH. Silver Nanoparticles: Synthesis and Application for Nanomedicine. Int J Mol Sci 2019; 20(4): 1-23.
[http://dx.doi.org/10.3390/ijms20040865] [PMID: 30781560]

[113]   Péter Bélteky ZK, Igaz Andrea rónavári Nóra, Tóth Bettina szerencsés Ildikó Y, Kiricsi Ilona Pfeiffer Mónika. Silver nanoparticles : Aggregation behavior in biorelevant conditions and its impact on biological activity. Int J Nanomedicine 2019; 14: 667-87.
[http://dx.doi.org/10.2147/IJN.S185965]

[114]   Saifuddin KN, Nian NCY, Zhang LW. Chitosan silver nano particles composite as point-of-use drinking water filtration system for household to remove pesticides in water. Academic journal 2011; 142-59.

[115]   Cheng X, Kan AT, Tomson MB. Naphthalene Adsorption and Desorption from Aqueous C 60 Fullerene. Chem Eng J 2004; 49: 675-83.

[116]   Tungittiplakorn W, Lion LW, Cohen C, Kim JY. Engineered polymeric nanoparticles for soil remediation. Environ Sci Technol 2004; 38(5): 1605-10.
[http://dx.doi.org/10.1021/es0348997] [PMID: 15046367]

[117]   Tungittiplakorn W, Cohen C, Lion LW. Engineered polymeric nanoparticles for bioremediation of hydrophobic contaminants. Environ Sci Technol 2005; 39(5): 1354-8.
[http://dx.doi.org/10.1021/es049031a] [PMID: 15787377]

[118]   Zhao H, Nagy KL. Dodecyl sulfate-hydrotalcite nanocomposites for trapping chlorinated organic pollutants in water. J Colloid Interface Sci 2004; 274(2): 613-24.
[http://dx.doi.org/10.1016/j.jcis.2004.03.055] [PMID: 15144837]

[119] Moreno N, Querol X, Ayora C, Pereira CF, Janssen-Jurkovicová M. Utilization of zeolites synthesized from coal fly ash for the purification of acid mine waters. Environ Sci Technol 2001; 35(17): 3526-34.
[http://dx.doi.org/10.1021/es0002924] [PMID: 11563657]

[120] Deliyanni E A, Bakoyannakis D N, Zouboulis A I, Matis K A. Sorption of As ( V ) ions by akagan e ite-type nanocrystals. Chemosph 50 2003; 50: 155-63.

[121] Alvarez-Ayuso E, García-Sánchez A, Querol X. Purification of metal electroplating waste waters using zeolites. Water Res 2003; 37(20): 4855-62.
[http://dx.doi.org/10.1016/j.watres.2003.08.009] [PMID: 14604631]

[122] Li Y, Ding J, Luan Z, *et al.* Competitive adsorption of $Pb^{2+}$, $Cu^{2+}$ and $Cd^{2+}$ ions from aqueous solutions by multiwalled carbon nanotubes. Carbon 2003; 41(14): 2787-92.
[http://dx.doi.org/10.1016/S0008-6223(03)00392-0]

[123] Qi L, Xu Z, Jiang X, Hu C, Zou X. Preparation and antibacterial activity of chitosan nanoparticles. Carbohydr Res 2004; 339(16): 2693-700.
[http://dx.doi.org/10.1016/j.carres.2004.09.007] [PMID: 15519328]

[124] Peng X, Luan Z. Carbon nanotubes-iron oxides magnetic composites as adsorbent for removal of Pb (II) and Cu (II) from water Carbon N Y 2005; 43(Ii): 855-94.
[http://dx.doi.org/10.1016/j.carbon.2004.11.009]

[125] Crini G. Recent developments in polysaccharide-based materials used as adsorbents in wastewater treatment. Prog Polym Sci 2005; 30(1): 38-70.
[http://dx.doi.org/10.1016/j.progpolymsci.2004.11.002]

[126] Peng X, Luan Z, Ding J, Di Z, Li Y, Tian B. Ceria nanoparticles supported on carbon nanotubes for the removal of arsenate from water. Mater Lett 2005; 59(4): 399-403.
[http://dx.doi.org/10.1016/j.matlet.2004.05.090]

[127] Lazaridis N K, Bakoyannakis D N, Deliyanni E A. Chromium (VI) sorptive removal from aqueous solutions ` ite by nanocrystalline akagane ite by nanocrystalline akagane 2005; 58: 65-73.

[128] Bhakat P B, Gupta A K, Ayoob S, Kundu S. Investigations on arsenic (V) removal by modified calcined bauxite 2006; 281: 237-45.
[http://dx.doi.org/10.1016/j.colsurfa.2006.02.045]

[129] Lu C, Liu C. Removal of nickel (II) from aqueous solution by carbon nanotubes. J Chem Technol Biotechnol 2006; 81(July): 1932-40.
[http://dx.doi.org/10.1002/jctb.1626]

[130] Wang H, Zhou A, Peng F, Yu H, Yang J. Mechanism study on adsorption of acidified multiwalled carbon nanotubes to Pb(II). J Colloid Interface Sci 2007; 316(2): 277-83.
[http://dx.doi.org/10.1016/j.jcis.2007.07.075] [PMID: 17868683]

[131] Kabbashi NA, Atieh MA, Al-Mamun A, Mirghami MES, Alam MD, Yahya N. Kinetic adsorption of application of carbon nanotubes for Pb(II) removal from aqueous solution. J Environ Sci (China) 2009; 21(4): 539-44.
[http://dx.doi.org/10.1016/S1001-0742(08)62305-0] [PMID: 19634432]

[132] Li Y, Liu F, Xia B, *et al.* Removal of copper from aqueous solution by carbon nanotube/calcium alginate composites. J Hazard Mater 2010; 177(1-3): 876-80.
[http://dx.doi.org/10.1016/j.jhazmat.2009.12.114] [PMID: 20083351]

[133] Dave PN, Chopda LV. Application of Iron Oxide Nanomaterials for the Removal of Heavy Metals. J Nanotechnol 2014; 2014: 14.
[http://dx.doi.org/10.1155/2014/398569]

[134] Nicomel N R, Leus K, Folens K, Van Der Voort P. Technologies for arsenic removal from water : Current status and future perspectives Env Res public Heal 2015; 13(1): 62.
[http://dx.doi.org/10.3390/ijerph13010062]

[135] Lee SH, Kim KW, Lee BT, *et al.* Enhanced Arsenate Removal Performance in Aqueous Solution by Yttrium-Based Adsorbents. Int J Environ Res Public Health 2015; 12(10): 13523-41.
[http://dx.doi.org/10.3390/ijerph121013523] [PMID: 26516879]

[136] Li X, Liu Y, Zhang C, Wen T, Zhuang L, Wang X. Porous $Fe_2O_3$ microcubes derived from metal organic frameworks for efficient elimination of organic pollutants and heavy metal ions. Chem Eng J 2017.

[137] Mangun CL, Yue Z, Economy J, Maloney S, Kemme P, Cropek D. Adsorption of organic contaminants from water using tailored ACFs Chem Mater 2001; 13: 2356-60.
[http://dx.doi.org/10.1021/cm000880g]

[138] Peng X, Li Y, Luan Z, *et al.* Adsorption of 1, 2-dichlorobenzene from water to carbon nanotubes. Chem Phys Lett 2003; 376(1-2): 154-8.
[http://dx.doi.org/10.1016/S0009-2614(03)00960-6]

[139] Zhang J, Zhuang J, Gao L, *et al.* Decomposing phenol by the hidden talent of ferromagnetic nanoparticles. Chemosphere 2008; 73(9): 1524-8.
[http://dx.doi.org/10.1016/j.chemosphere.2008.05.050] [PMID: 18804842]

[140] Zhang S, Zhao X, Niu H, Shi Y, Cai Y, Jiang G. Superparamagnetic $Fe_3O_4$ nanoparticles as catalysts for the catalytic oxidation of phenolic and aniline compounds. J Hazard Mater 2009; 167(1-3): 560-6.
[http://dx.doi.org/10.1016/j.jhazmat.2009.01.024] [PMID: 19201085]

[141] Afkhami A, Saber-tehrani M, Bagheri H. Modi fi ed maghemite nanoparticles as an ef fi cient adsorbent for removing some cationic dyes from aqueous solution. Desalination 2010; 263(1–3): 240-8.
[http://dx.doi.org/10.1016/j.desal.2010.06.065]

[142] Zhu H, Jiang R, Xiao L. Adsorption of an anionic azo dye by chitosan / kaolin / γ -$Fe_2O_3$ composites. Appl Clay Sci 2010; 48(3): 522-6.
[http://dx.doi.org/10.1016/j.clay.2010.02.003]

[143] Zhang Z, Kong J. Novel magnetic $Fe_3O_4$@C nanoparticles as adsorbents for removal of organic dyes from aqueous solution. J Hazard Mater 2011; 193: 325-9.
[http://dx.doi.org/10.1016/j.jhazmat.2011.07.033] [PMID: 21813238]

[144] Inbaraj BS, Chen BH. Dye adsorption characteristics of magnetite nanoparticles coated with a biopolymer poly(γ-glutamic acid). Bioresour Technol 2011; 102(19): 8868-76.
[http://dx.doi.org/10.1016/j.biortech.2011.06.079] [PMID: 21775135]

[145] Chaudhary S, Ram Ganga, Priya Saharan, Gagandeep Gagandeep, Mehta SK, Mor . Removal of Coomassie Brilliant Blue R-250 Dye from Water Using γ-$Fe_2O_3$ Nanoparticles. J Nanoeng Nanomanufacturing 2012; 2(13): 291-303.

[146] Panneerselvam P, *et al.* Removal of rhodamine B dye using activated carbon prepared from palm kernel shell and coated with iron oxide nanoparticles. Sep Sci Technol 2013; 47: 742-52.

[147] Yang X, Li J, Wen T, Ren X, Huang Y, Wang X. Adsorption of naphthalene and its derivatives on magnetic graphene composites and the mechanism investigation. Colloids Surf A Physicochem Eng Asp 2013; 422: 118-25.
[http://dx.doi.org/10.1016/j.colsurfa.2012.11.063]

[148] Le TH, Kim SJ, Bang SH, *et al.* Phenol degradation activity and reusability of Corynebacterium glutamicum coated with NH(2)-functionalized silica-encapsulated $Fe_3O_4$ nanoparticles. Bioresour Technol 2012, 104: 795-8.
[http://dx.doi.org/10.1016/j.biortech.2011.10.064] [PMID: 22093979]

[149] Kakavandi B, Jonidi A, Rezaei R, Nasseri S, Ameri A, Esrafily A. Synthesis and properties of $Fe_3O_4$-activated carbon magnetic nanoparticles for removal of aniline from aqueous solution: Equilibrium, kinetic and thermodynamic studies. Iran J Environ Health Sci Eng 2013; 10(1): 19.
[http://dx.doi.org/10.1186/1735-2746-10-19] [PMID: 23414171]

[150] Chaudhary G R, Saharan P, Kumar A, Mehta S K. Adsorption Studies of Cationic, Anionic and Azo-Dyes *via* Monodispersed Fe$_3$O$_4$ Nanoparticles J Nanosci Nanotechnol 2013; 13: 3240-5.
[http://dx.doi.org/10.1166/jnn.2013.7152]

[151] Jing J, Li J, Feng J, Li W, Yu WW. Photodegradation of quinoline in water over magnetically separable Fe$_3$O$_4$ / TiO$_2$ composite photocatalysts. Chem Eng J 2013; 219: 355-60.
[http://dx.doi.org/10.1016/j.cej.2012.12.058]

[152] Nanoparticles UO, Chaudhary GR, Saharan P, Mehta SK, Mor S, Umar A. Fast and Efficient Removal of Hazardous Congo Red from Its Aqueous Solution Using -Fe$_2$O$_3$ Nanoparticles. J Nanoeng Nanomanufacturing 2013; 3: 142-6.
[http://dx.doi.org/10.1166/jnan.2013.1120]

[153] Shahid Arshad G, Djinović Petar, Zavašnik Janez, Pintar lbin. Electron trapping energy states of TiO$_2$–WO$_3$ composites and their influence on photocatalytic degradation of bisphenol A Appl Catal B Environ 209: 273-84.

[154] Likun Gao J L, Gan Wentao, Qiu Zhe, Zhan Xianxu. Preparation of heterostructured WO$_3$/TiO$_2$ catalysts from wood fibers and its versatile photodegradation abilities Sci Rep 2017; 24

[155] Yantasee W, Lin Y, Fryxell GE, Busche BJ, Birnbaum JC. Removal of Heavy Metals from Aqueous Solution Using Novel Nanoengineered Sorbents : Self - Assembled Carbamoylphosphonic Acids on Mesoporous Silica. Sep Sci Technol 2012; 38(15): 3809-25.
[http://dx.doi.org/10.1081/SS-120024232]

[156] Kelly SD, Kemner KM, Fryxell GE, Liu J, Mattigod SV, Ferris KF. X-Ray-Absorption Fine-Structure Spectroscopy Study of the Interactions between Contaminant Tetrahedral Anions and Self-Assembled Monolayers on Mesoporous Supports. J Phys Chem B 2001; 105(27): 6337-46.
[http://dx.doi.org/10.1021/jp0045890]

[157] Yantasee W, Lin Y, Hongsirikarn K, Fryxell GE, Addleman R, Timchalk C. Electrochemical sensors for the detection of lead and other toxic heavy metals: The next generation of personal exposure biomonitors. Environ Health Perspect 2007; 115(12): 1683-90.
[http://dx.doi.org/10.1289/ehp.10190] [PMID: 18087583]

[158] Afkhami A, Saber-Tehrani M, Bagheri H. Simultaneous removal of heavy-metal ions in wastewater samples using nano-alumina modified with 2,4-dinitrophenylhydrazine. J Hazard Mater 2010; 181(1-3): 836-44.
[http://dx.doi.org/10.1016/j.jhazmat.2010.05.089] [PMID: 20542378]

[159] Alemayehu E, Thiele-bruhn S, Lennartz B. Author's personal copy Adsorption behaviour of Cr (VI) onto macro and micro-vesicular volcanic rocks from water. Separ Purif Tech 2011; 78(1): 55-61.
[http://dx.doi.org/10.1016/j.seppur.2011.01.020]

[160] Önnby L, Kumar PS, Sigfridsson KGV, Wendt OF, Carlson S, Kirsebom H. Improved arsenic(III) adsorption by Al$_2$O$_3$ nanoparticles and H$_2$O$_2$: Evidence of oxidation to arsenic(V) from X-ray absorption spectroscopy. Chemosphere 2014; 113: 151-7.
[http://dx.doi.org/10.1016/j.chemosphere.2014.04.097] [PMID: 25065803]

[161] Banerjee K, Amy GL, Prevost M, *et al.* Kinetic and thermodynamic aspects of adsorption of arsenic onto granular ferric hydroxide (GFH). Water Res 2008; 42(13): 3371-8.
[http://dx.doi.org/10.1016/j.watres.2008.04.019] [PMID: 18538818]

[162] Liu G, Zhang X, Talley JW, Neal CR, Wang H. Effect of NOM on arsenic adsorption by TiO($_2$) in simulated As(III)-contaminated raw waters. Water Res 2008; 42(8-9): 2309-19.
[http://dx.doi.org/10.1016/j.watres.2007.12.023] [PMID: 18316108]

[163] Pan JJ, Jiang J, Xu RK. Removal of Cr(VI) from aqueous solutions by Na$_2$SO$_3$/FeSO$_4$ combined with peanut straw biochar. Chemosphere 2014; 101: 71-6.
[http://dx.doi.org/10.1016/j.chemosphere.2013.12.026] [PMID: 24380440]

[164] Madadrang CJ, Kim HY, Gao G, *et al.* Adsorption behavior of EDTA-graphene oxide for Pb (II)

removal. ACS Appl Mater Interfaces 2012; 4(3): 1186-93.
[http://dx.doi.org/10.1021/am201645g] [PMID: 22304446]

[165]  Chakraborty A, Deva D, Sharma A, Verma N. Adsorbents based on carbon microfibers and carbon nanofibers for the removal of phenol and lead from water. J Colloid Interface Sci 2011; 359(1): 228-39.
[http://dx.doi.org/10.1016/j.jcis.2011.03.057] [PMID: 21507421]

[166]  Gao Z, Bandosz TJ, Zhao Z, Han M, Qiu J. Investigation of factors affecting adsorption of transition metals on oxidized carbon nanotubes. J Hazard Mater 2009; 167(1-3): 357-65.
[http://dx.doi.org/10.1016/j.jhazmat.2009.01.050] [PMID: 19264402]

[167]  Harry ID, Saha B, Cumming IW. Effect of electrochemical oxidation of activated carbon fiber on competitive and noncompetitive sorption of trace toxic metal ions from aqueous solution. J Colloid Interface Sci 2006; 304(1): 9-20.
[http://dx.doi.org/10.1016/j.jcis.2006.08.012] [PMID: 17011569]

[168]  Leyva-Ramos R, Berber-Mendoza MS, Salazar-Rabago J, Guerrero-Coronado RM, Mendoza-Barron J. Adsorption of lead(II) from aqueous solution onto several types of activated carbon fibers. Adsorption 2011; 17(3): 515-26.
[http://dx.doi.org/10.1007/s10450-010-9313-3]

[169]  Pan BC, Zhang QR, Zhang WM, *et al.* Highly effective removal of heavy metals by polymer-based zirconium phosphate: A case study of lead ion. J Colloid Interface Sci 2007; 310(1): 99-105.
[http://dx.doi.org/10.1016/j.jcis.2007.01.064] [PMID: 17336317]

[170]  Mouflih M, Aklil A, Sebti S. Removal of lead from aqueous solutions by activated phosphate. J Hazard Mater 2005; 119(1-3): 183-8.
[http://dx.doi.org/10.1016/j.jhazmat.2004.12.005] [PMID: 15752864]

[171]  Nah KYH, Hwang K-Y, Jeon C, Choi HB. In Wook, "Removal of Pb ion from water by magnetically modified zeolite. Miner Eng 2006; 19(14): 1452-5.
[http://dx.doi.org/10.1016/j.mineng.2005.12.006]

[172]  Zou Y, Wang X, Khan A, *et al.* Environmental Remediation and Application of Nanoscale Zero-Valent Iron and Its Composites for the Removal of Heavy Metal Ions: A Review. Environ Sci Technol 2016; 50(14): 7290-304.
[http://dx.doi.org/10.1021/acs.est.6b01897] [PMID: 27331413]

[173]  Wang C, Zhang W. Synthesizing Nanoscale Iron Particles for Rapid and Complete Dechlorination of TCE and PCBs. Environ Sci Technol 1997; 31(7): 2154-6.
[http://dx.doi.org/10.1021/es970039c]

[174]  Hoag GE, Collins JB, Holcomb JL, Hoag JR, Nadagouda MN, Varma RS. Degradation of bromothymol blue by 'greener' nano-scale zero-valent iron synthesized using tea polyphenols. J Mater Chem 2009; 19(45): 8671-7.
[http://dx.doi.org/10.1039/b909148c]

[175]  Chen SS, Der Hsu H, Li CW. A new method to produce nanoscale iron for nitrate removal. J Nanopart Res 2004; 6(6): 639-47.
[http://dx.doi.org/10.1007/s11051-004-6672-2]

[176]  Matheson LJ, Tratnyek PG. Reductive dehalogenation of chlorinated methanes by iron metal. Environ Sci Technol 1994; 28(12): 2045-53.
[http://dx.doi.org/10.1021/es00061a012] [PMID: 22191743]

[177]  Puls RW, Paul CJ, Powell RM. The application of *in situ* permeable reactive (zero-valent iron) barrier technology for the remediation of chromate-contaminated groundwater: A field test. Appl Geochem 1999; 14(8): 989-1000.
[http://dx.doi.org/10.1016/S0883-2927(99)00010-4]

[178]  Briley DA, Tucker-drob EM. Mechanism of ozone oxidation of polycyclic aromatic hydrocarbons

during the reduction of coking wastewater sludge. Clean (Weinh) 2015; 1-27.

[179] Cumbal L, Sengupta AK. Arsenic removal using polymer-supported hydrated iron(III) oxide nanoparticles: Role of donnan membrane effect. Environ Sci Technol 2005; 39(17): 6508-15.
[http://dx.doi.org/10.1021/es050175e] [PMID: 16190206]

[180] Hartono T, Wang S, Ma Q, Zhu Z. Layer structured graphite oxide as a novel adsorbent for humic acid removal from aqueous solution. J Colloid Interface Sci 2009; 333(1): 114-9.
[http://dx.doi.org/10.1016/j.jcis.2009.02.005] [PMID: 19233379]

[181] Noubactep C, Caré S, Crane R. Nanoscale metallic iron for environmental remediation: Prospects and limitations. Water Air Soil Pollut 2012; 223(3): 1363-82.
[http://dx.doi.org/10.1007/s11270-011-0951-1] [PMID: 22389536]

[182] Jan Filip O Š. Nanoscale Zerovalent Iron Particles for Treatment of Metalloids. springer 2019.

[183] Jan Filip MC, Karlicky Frantisek, ăk Zdenek Marus, Petr Lazar M O Erník, Zboriľ R. Anaerobic ics. J Phys Chem 2014; 118(6): 13817-25.

[184] O'Carroll BD, Krol CK, Magdalena HB. Nanoscale zero valent iron and bimetallic particles for contaminated site remediation. Adv Water Resour 2013; 51(january): 104-22.
[http://dx.doi.org/10.1016/j.advwatres.2012.02.005]

[185] Kang YS, Risbud S, Rabolt JF, Stroeve P. Synthesis and characterization of nanometer-size $Fe_3O_4$ and $\gamma$-$Fe_2O_3$ particles. Chem Mater 1996; 5(96): 2209-11.
[http://dx.doi.org/10.1021/cm960157j]

[186] Ai L, Jiang J. Removal of methylene blue from aqueous solution with self-assembled cylindrical graphene-carbon nanotube hybrid. Chem Eng J 2012; 192: 156-63.
[http://dx.doi.org/10.1016/j.cej.2012.03.056]

[187] Vijayakumar R, Koltypin Y, Felner I, Gedanken A. Sonochemical synthesis and characterization of pure nanometer-sized $Fe_3O_4$ particles. Mater Sci Eng A 2000; 286(1): 101-5.
[http://dx.doi.org/10.1016/S0921-5093(00)00647-X]

[188] Zhou ZH, Wang J, Liu X, Chan HSO. Synthesis of $Fe_3O_4$ nanoparticles from emulsions. J Mater Chem 2001; 11(6): 1704-9.
[http://dx.doi.org/10.1039/b100758k]

[189] Hao YM, Man C, Hu ZB. Effective removal of Cu (II) ions from aqueous solution by amino-functionalized magnetic nanoparticles. J Hazard Mater 2010; 184(1-3): 392-9.
[http://dx.doi.org/10.1016/j.jhazmat.2010.08.048] [PMID: 20837378]

[190] Me W, *et al.* Carbon-encapsulated magnetic nanoparticles as separable and mobile sorbents of heavy metal ions from aqueous solutions. Carbon N Y 2009; 7: 1201-4.

[191] Banerjee SS, Chen DH. Fast removal of copper ions by gum arabic modified magnetic nano-adsorbent. J Hazard Mater 2007; 147(3): 792-9.
[http://dx.doi.org/10.1016/j.jhazmat.2007.01.079] [PMID: 17321674]

[192] Xiaowang Liu BZC, Hu Q, Fang Z, Zhang X. Magnetic chitosan nanocomposites : A useful recyclable tool for heavy metalion removal. Am Chem Soc 2009; 2(5): 3-8.

[193] Shen HY, Zhu Y, Wen XE, Zhuang YM. Preparation of $Fe_3O_4$-C18 nano-magnetic composite materials and their cleanup properties for organophosphorous pesticides. Anal Bioanal Chem 2007; 387(6): 2227-37.
[http://dx.doi.org/10.1007/s00216-006-1082-1] [PMID: 17221237]

[194] Petrakis L, Ahner F, Science G, Company T. Use of high gradient magnetic separation techniques for the removal of oil and solids from water efflue, *IEEE Trans. Magn.*, vol. Mag 1976; 12(5): 486-8.
[http://dx.doi.org/10.1109/TMAG.1976.1059078]

[195] Kulkarni SJ, Goswami AK. A Review on Wastewater Treatment for Petroleum Industries and Refineries. Eng Technol A 2015; 1(3): 280-3.

[196]  Zhiwei Zhao LZ, Liu J, Cui F. One pot synthesis of tunable $Fe_3O_4$–$MnO_2$ core–shell nanoplates and their applications for water purification. J Mater Chem 2012; 22(18): 9052-7.
[http://dx.doi.org/10.1039/c2jm00153e]

[197]  Chen X, Mao SS. Titanium dioxide nanomaterials : Synthesis, properties, modifications, and applications chem rev 2007; 107(7): 2891-959.

[198]  Bessekhouad Y, Robert D, Weber JV. Synthesis of photocatalytic $TiO_2$ nanoparticles : Optimization of the preparation conditions. J Alloys Compd 2003; Vol. 157: 47-53.

[199]  Khoie MM, Marashi P, Rezaee M. Synthesis of $TiO_2$ nanoparticles *via* a novel mechanochemical method. J Alloys Compd 2009; 469(1–2): 386-90.

[200]  Chae SY, Park MK, Lee SK, Kim TY, Kim SK, Lee WI. Preparation of Size-Controlled $TiO_2$ Nanoparticles and Derivation of Optically Transparent Photocatalytic Films. Chem Mater 2003; 15(117): 3326-31.
[http://dx.doi.org/10.1021/cm030171d]

[201]  Kim C, Kee B, Park J, Tae S, Son S. Synthesis of nanocrystalline $TiO_2$ in toluene by a solvothermal route. J Cryst Growth 2003; 254(3–4): 405-10.
[http://dx.doi.org/10.1016/S0022-0248(03)01185-0]

[202]  Li X, Peng Q, Yi J, Wang X, Li Y. Near Monodisperse $TiO_2$ Nanoparticles and Nanorods 2006; 2383-91.
[http://dx.doi.org/10.1002/chem.200500893]

[203]  Seifried BS, Winterer M, Hahn H. Nanocrystalline Titania Films and Particles by Chemical Vapor Synthesis. Chem Vap Depos 2000; 6(5): 239-44.
[http://dx.doi.org/10.1002/1521-3862(200010)6:5<239::AID-CVDE239>3.0.CO;2-Q]

[204]  Dou Y, Zhang W, Kaiser A. Electrospinning of Metal-Organic Frameworks for Energy and Environmental Applications. Adv Sci (Weinh) 2019; 7(3): 1902590.
[http://dx.doi.org/10.1002/advs.201902590] [PMID: 32042570]

[205]  Dandan L. Metal-organic frameworks for catalysis: State of the art, challenges, and opportunities energy chem 2019; 100005

[206]  Efome JE, Rana D, Matsuura T, Lan CQ. Metal – organic frameworks supported on nano fibers to remove heavy metals. J Mater Chem A Mater Energy Sustain 2018; 6(10): 4550-5.
[http://dx.doi.org/10.1039/C7TA10428F]

[207]  Zhu B, *et al.* Iron and 1, 3, 5-Benzenetricarboxylic Metal − Organic Coordination Polymers Prepared by Solvothermal Method and Their Application in E ffi cient As (V) Removal from Aqueous Solutions. J Phys Chem 2012; 116(V): 8601-7.

[208]  Ding AY, Xu Y, Ding B, *et al.* Structure Induced Selective Adsorption Performance of ZIF-8 Nanocrystals in Water. Colloids Surf A Physicochem Eng Asp 2017; 520: 661-7.
[http://dx.doi.org/10.1016/j.colsurfa.2017.02.012]

[209]  Rahimi E, Mohaghegh N. New hybrid nanocomposite of copper terephthalate MOF-graphene oxide: Synthesis, characterization and application as adsorbents for toxic metal ion removal from Sungun acid mine drainage. Environ Sci Pollut Res Int 2017; 24(28): 22353-60.
[http://dx.doi.org/10.1007/s11356-017-9823-6] [PMID: 28801872]

[210]  Vu TA, Le GH, Dao CD, *et al.* Arsenic removal from aqueous solutions by adsorption using novel MIL-53 (Fe) as a highly e ffi cient adsorbent. RSC Advances 2014; 5(7): 5261-8.
[http://dx.doi.org/10.1039/C4RA12326C]

[211]  Li Z, Yang J, Sui K, Yin N. Facile synthesis of metal-organic framework MOF808 for arsenic removal. Mater Lett 2015.

[212]  Z. M., Jie Li F. L., Wu. Yi-nan, Li Zehua, Z. M., Z. M.. Characteristics of arsenate removal from water by metal-organic frameworks (MOFs). Water Sci Technol 2014; 1391-8.

[213] Wang C, Liu X, Keser Demir N, Chen JP, Li K. Applications of water stable metal-organic frameworks. Chem Soc Rev 2016; 45(18): 5107-34.
[http://dx.doi.org/10.1039/C6CS00362A] [PMID: 27406473]

[214] Xiong W, Zeng G, Yang Z, *et al.* Adsorption of tetracycline antibiotics from aqueous solutions on nanocomposite multi-walled carbon nanotube functionalized MIL-53(Fe) as new adsorbent. Sci Total Environ 2018; 627: 235-44.
[http://dx.doi.org/10.1016/j.scitotenv.2018.01.249] [PMID: 29426146]

[215] Chaohai Wang J, Cheng Ping, Yao Yiyuan, Yamauchi Yusuke, Yan Xin, Li Jiansheng. *In-situ* fabrication of nanoarchitectured MOF filter for water purification. J Hazard Mater 2020; 392(15): 122164.
[http://dx.doi.org/10.1016/j.jhazmat.2020.122164]

[216] Abid HR, Periasamya V, Sunc H, Wang S. Cascade applications of robust MIL-96 metal organic frameworks in environmental remediation: Proof of concept. Chem Eng J 2018; 341: 262-71.
[http://dx.doi.org/10.1016/j.cej.2018.02.030]

[217] Azhar MR, Vijay P, Tadé MO, Sun H, Wang S. Submicron sized water-stable metal organic framework (bio-MOF-11) for catalytic degradation of pharmaceuticals and personal care products. Chromosphere 2018; 196: 105-14.
[http://dx.doi.org/10.1016/j.chemosphere.2017.12.164] [PMID: 29294423]

[218] Azhar MR, Abid HR, Sun H, Periasamy V, Tadé MO, Wang S. One-pot synthesis of binary metal organic frameworks (HKUST-1 and UiO-66) for enhanced adsorptive removal of water contaminants. J Colloid Interface Sci 2017; 490(15): 685-94.
[http://dx.doi.org/10.1016/j.jcis.2016.11.100] [PMID: 27940035]

[219] Fang QR, Yuan DQ, Sculley J, Li JR, Han ZB, Zhou HC. Functional mesoporous metal-organic frameworks for the capture of heavy metal ions and size-selective catalysis. Inorg Chem 2010; 49(24): 11637-42.
[http://dx.doi.org/10.1021/ic101935f] [PMID: 21082837]

[220] Ke F, Qiu LG, Yuan YP, *et al.* Thiol-functionalization of metal-organic framework by a facile coordination-based postsynthetic strategy and enhanced removal of $Hg^{2+}$ from water. J Hazard Mater 2011; 196: 36-43.
[http://dx.doi.org/10.1016/j.jhazmat.2011.08.069] [PMID: 21924826]

[221] Zou F, Yu R, Li R, Li W. Microwave-assisted synthesis of hkust-1 and functionalized HKUST-1-@ $H_3PW_{12}O_{40}$ : Selective adsorption of heavy metal ions in water analyzed with synchrotron radiation chemphyschem 2013; 14: 2825-32.

[222] Sohrabi MR. Preconcentration of mercury (II) using a thiol-functionalized metal-organic framework nanocomposite as a sorbent. Mikrochim Acta 2014; 181(3-4): 435-44.
[http://dx.doi.org/10.1007/s00604-013-1133-1]

[223] Tahmasebi E, Masoomi MY, Yamini Y, Morsali A. Application of mechanosynthesized azine-decorated zinc(II) metal-organic frameworks for highly efficient removal and extraction of some heavy-metal ions from aqueous samples: A comparative study. Inorg Chem 2015; 54(2): 425-33.
[http://dx.doi.org/10.1021/ic5015384] [PMID: 25548873]

[224] Yang Wang GY. Functionalized Metal-Organic Framework as a New Platform for Efficient and Selective Removal of Cadmium (II) from Aqueous Solution. J Mater Chem A Mater Energy Sustain 2015; 3(29): 15292-8.
[http://dx.doi.org/10.1039/C5TA03201F]

[225] Saleem H, Rafique U, Davies RP. Investigations on post-synthetically modified UiO-66-NH₂ for the adsorptive removal of of heavy metal ions from aqueous solution. Microporous Mesoporous Mater 2015; 221: 238-44.
[http://dx.doi.org/10.1016/j.micromeso.2015.09.043]

[226]  Maleki A, Hayati B, Naghizadeh M, Joo SW. Adsorption of hexavalent chromium by metal organic frameworks from aqueous solution. J Ind Eng Chem 2015; 28: 211-6.
[http://dx.doi.org/10.1016/j.jiec.2015.02.016]

[227]  Zhang J, Xiong Z, Li C, Wu C. Exploring a thiol-functionalized MOF for elimination of lead and cadmium from aqueous solution. J Mol Liq 2016; 221: 43-50.
[http://dx.doi.org/10.1016/j.molliq.2016.05.054]

[228]  Abbasi A, Moradpour T, Van Hecke K. A new 3D cobalt (II) metal-organic framework nanostructure for heavy metal adsorption. Inorg Chim Acta 2015; 430: 261-7.
[http://dx.doi.org/10.1016/j.ica.2015.03.019]

[229]  Feng Luo JQ, Jing LC, Li LD, *et al.* High-performance $Hg^{2+}$ removal from ultra-lowconcentration aqueous solution using both acylamideand hydroxyl-functionalized metal-organic framework. J Mater Chem A Mater Energy Sustain 2015; 3(18): 9616-20.
[http://dx.doi.org/10.1039/C5TA01669J]

[230]  Huang BHL, Man He BC. Designable magnetic MOF composite and facile coordination-based post-synthetic strategy for enhanced removal of $Hg^{2+}$ from water. J Mater Chem A Mater Energy Sustain 2015; 3(21): 11587-95.
[http://dx.doi.org/10.1039/C5TA01484K]

[231]  Folens K, Leus K, Nicomel NR, *et al.* $Fe_3O_4$ @ MIL-101 – A Selective and Regenerable Adsorbent for the Removal of As Species from Water. Eur J Inorg Chem 2016; 2016(27): 4395-401.
[http://dx.doi.org/10.1002/ejic.201600160]

[232]  Chakraborty A, Bhattacharyya S, Hazra A, Ghosh AC, Maji TK. Post-synthetic metalation in an anionic MOF for efficient catalytic activity and removal of heavy metal ions from aqueous solution. Chem Commun (Camb) 2016; 52(13): 2831-4.
[http://dx.doi.org/10.1039/C5CC09814A] [PMID: 26776086]

[233]  Fe OM, *et al.* Fabrication of core-shell $Fe_3O_4$@ MIL-100(Fe) magnetic microspheres for the removal of Cr(VI) in aqueous solution. J Solid State Chem 2016; 244: 25-30.
[http://dx.doi.org/10.1016/j.jssc.2016.09.010]

[234]  Sánchez-Polo M, von Gunten U, Rivera-Utrilla J. Efficiency of activated carbon to transform ozone into *OH radicals: Influence of operational parameters. Water Res 2005; 39(14): 3189-98.
[http://dx.doi.org/10.1016/j.watres.2005.05.026] [PMID: 16005933]

[235]  Mon M, Lloret F, Ferrando-Soria J, Martí-Gastaldo C, Armentano D, Pardo E. Selective and Efficient Removal of Mercury from Aqueous Media with the Highly Flexible Arms of a BioMOF. Angew Chem Int Ed Engl 2016; 55(37): 11167-72.
[http://dx.doi.org/10.1002/anie.201606015] [PMID: 27529544]

[236]  Marta Mon A. a Xiaoni Qu, "Fine-tuning of the confined space in microporous metal-organic frameworks for efficient mercury removal. J Mater Chem A Mater Energy Sustain 2017; 5(38): 20120-5.
[http://dx.doi.org/10.1039/C7TA06199D]

[237]  Zheng TT, Zhao J, Fang Z-W, *et al.* A luminescent metal organic framework with high sensitivity for detecting and removing copper ions from simulated biological fluids. Dalton Trans 2017; 46(8): 2456-61.
[http://dx.doi.org/10.1039/C6DT04630D] [PMID: 28112321]

[238]  Huang >J Yan, Zeng Xiaofei, Guo Lingling, Zhang Liangliang. Heavy metal ion removal of wastewater by zeolite-imidazolate frameworks. Separ Purif Tech 2018; 194(3): 462-9.

[239]  Ayoub Abdullah E, Alqadami , Khan Moonis Ali, Raza Masoom , Siddiqui Z. Development of citric anhydride anchored mesoporous MOF through post synthesis modification to sequester potentially toxic lead (II) from water. Microporous Mesoporous Mater 2018; 261(1): 198-206.

[240]  Wen J, Fang Y, Zeng G. Progress and prospect of adsorptive removal of heavy metal ions from

aqueous solution using metal-organic frameworks: A review of studies from the last decade. Chemosphere 2018; 201: 627-43.
[http://dx.doi.org/10.1016/j.chemosphere.2018.03.047] [PMID: 29544217]

[241]   Aboutorabi L, Morsali A, Tahmasebi E, Büyükgüngor O. Metal-Organic Framework Based on Isonicotinate N-Oxide for Fast and Highly Efficient Aqueous Phase Cr(VI) Adsorption. Inorg Chem 2016; 55(11): 5507-13.
[http://dx.doi.org/10.1021/acs.inorgchem.6b00522] [PMID: 27195982]

[242]   Sofia Rapti AP. Rapid, green and inexpensive synthesis of high quality UiO-66 amino-functionalized materials with exceptional capability for removal of hexavalent chromium from industrial waste. R Soc Chem 2016; 3: 635-44.

[243]   Wang C, Du X, Li J, Guo X, Wang P, Zhang J. Environmental Photocatalytic Cr (VI) reduction in metal-organic frameworks : A. Appl Catal B 2016; 193: 198-216.
[http://dx.doi.org/10.1016/j.apcatb.2016.04.030]

[244]   Raffaele A Ricco, Konstas Kristina, Styles a Mark J. Lead(II) uptake by Aluminium Based Magnetic Framework Composites (MFCs) in Wate materials chemistry A. J Mater Chem A Mater Energy Sustain 2015; 3(39): 19822-31.
[http://dx.doi.org/10.1039/C5TA04154F]

[245]   Qin Q, Wang Q, Fu D, Ma J. An efficient approach for Pb (II) and Cd (II) removal using manganese dioxide formed *in situ.* Chem Eng J 2011; 172(1): 68-74.
[http://dx.doi.org/10.1016/j.cej.2011.05.066]

[246]   Rivera J M, Rincón S, Ben Youssef C, Zepeda A. Highly efficient adsorption of aqueous Pb (II) with mesoporous metal-organic framework-5 : An equilibrium and kinetic study. 2016.
[http://dx.doi.org/10.1155/2016/8095737]

[247]   Liang L, Chen Q, Jiang F, *et al. In situ* large-scale construction of sulfur-functionalized metal- organic framework and its efficient removal of Hg(II) from water. J Mater Chem A Mater Energy Sustain 2016; 4(40): 15370-4.
[http://dx.doi.org/10.1039/C6TA04927C]

[248]   Liu T, Che JX, Hu YZ, Dong XW, Liu XY, Che CM. Alkenyl/thiol-derived metal-organic frameworks (MOFs) by means of postsynthetic modification for effective mercury adsorption. Chemistry 2014; 20(43): 14090-5.
[http://dx.doi.org/10.1002/chem.201403382] [PMID: 25210002]

[249]   Zhou J, *et al.* Layered structures very important paper a two-dimensional zirconium carbide by selective etching of Al$_3$C$_3$ from nanolaminated Zr$_3$Al$_3$C$_5$ angewandte 2016; 14: 2825-32.

[250]   Zhen□Dong Lei ZL, Xue Y-C, Wen□Qian . The influence of carbon nitride nanosheets doping on the crystalline formation of MIL□88B(Fe) and the photocatalytic activities. wiley Online Libr 2018; 14(35)

[251]   Zhang Y, Zhou J, Feng Q, Chen X, Hu Z. Visible light photocatalytic degradation of MB using UiO-66/g-C$_3$N$_4$ heterojunction nanocatalyst. Chemoshhere 2018; 212(Dec): 523-32.
[http://dx.doi.org/10.1016/j.chemosphere.2018.08.117] [PMID: 30165279]

[252]   Xiaodong Zhang L, Yang Y, Huang W, *et al.* g-C$_3$N$_4$/UiO-66 nanohybrids with enhanced photocatalytic activities for the oxidation of dye under visible light irradiation Mater Res buttetin 2018; 99: 349-58.

[253]   Li Xiyi, Pi Yunhong, Wu Liqiong, Xia Qibin, liang Jun, Zhong Wu. Facilitation of the visible light-induced Fenton-like excitation of H$_2$O$_2$*via* heterojunction of g-C3N4/NH2-Iron terephthalate metal-organic framework for MB degradation. Appl Catal B Environ 2017; 202: 653-63.

[254]   Gong Y, Yang B, Zhang H, Zhao X. A g-C3N4/MIL-101(Fe) heterostructure composite for highly efficient BPA degradation with persulfate under visible light irradiation. J Mater Chem A Mater Energy Sustain 2018; 6(46): 23703-11.

[http://dx.doi.org/10.1039/C8TA07915C]

[255]  Giannakoudakis A, Nikolina A, Travlou , Secor Jeff, Bandosz T J. Oxidized g-$C_3N_4$ Nanospheres as Catalytically Photoactive Linkers in MOF/g-$C_3N_4$ Composite of Hierarchical Pore Structure Dimitrios A. Materialsview 2016; 13(1): 1601758.

[256]  Wang C, Wang P, Yi Xiao-Hong. Powerful combination of MOFs and $C_3N_4$ for enhanced photocatalytic performance," *Appl. Chem.* BEnvironmental 2019; 247: 24-48.

[257]  Hou Wang H, Yuan Xingzhong, Zeng YanWudGuangming, Chen Xiaohong, Leng Lijian. Synthesis and applications of novel graphitic carbon nitride/metal-organic frameworks mesoporous photocatalyst for dyes removal. Appl Catal B 2015; 174–175(september): 445-54.
[http://dx.doi.org/10.1016/j.apcatb.2015.03.037]

[258]  Luoa H, *et al.* Recent progress on metal-organic frameworks based and derivedphotocatalysts for water splitting. Chem Eng J 2020; 383: 123196.
[http://dx.doi.org/10.1016/j.cej.2019.123196]

[259]  Cheng X, Liu M, Zhang A, *et al.* Size-controlled silver nanoparticles stabilized on thiol-functionalized MIL-53(Al) frameworks. Nanoscale 2015; 7(21): 9738-45.
[http://dx.doi.org/10.1039/C5NR01292A] [PMID: 25963664]

[260]  Bakhtiari N, Azizian S. Adsorption of copper ion from aqueous solution by nanoporous MOF-5: A kinetic and equilibrium study. J Mol Liq 2015; 206: 114-8.
[http://dx.doi.org/10.1016/j.molliq.2015.02.009]

[261]  Conde-González JE, Peña-Méndez EM, Rybáková S, Pasán J, Ruiz-Pérez C, Havel J. Adsorption of silver nanoparticles from aqueous solution on copper-based metal organic frameworks (HKUST-1). Chemosphere 2016; 150: 659-66.
[http://dx.doi.org/10.1016/j.chemosphere.2016.02.005] [PMID: 26879292]

[262]  Jiang Q Y H. Oxidation or reduction state of au stabilized by an MOF: Active site identification for the three-component coupling reaction small methods 2018; 2(12): 180026.

[263]  Varma AJ, Deshpande SV, Kennedy JF. Metal complexation by chitosan and its derivatives : A review. Carbohydr Polym 2004; 55(1): 77-93.
[http://dx.doi.org/10.1016/j.carbpol.2003.08.005]

[264]  Higazy A, Hashem M, Elshafei A, Shaker N, Abdel M. Development of antimicrobial jute packaging using chitosan and chitosan – metal complex. Carbohydr Polym 2010; 79(4): 867-74.
[http://dx.doi.org/10.1016/j.carbpol.2009.10.011]

[265]  Agnihotri SA, *et al.* Recent advances on chitosan-based micro- and nanoparticles in drug delivery B. Biomaterials 2005; 45(February): 5-26.

[266]  Hesketh T, Li L, Ye X, Wang H, Jiang M, Tomkins A. HIV and syphilis in migrant workers in eastern China. Sex Transm Infect 2006; 82(1): 11-4.
[http://dx.doi.org/10.1136/sti.2004.014043] [PMID: 16461594]

[267]  Ritchie RH. Plasma Losses by Fast Electrons in Thin Films. 1956; 184

[268]  Wang X, Pang Y, Ku G, Xie X, Stoica G, Wang LV. Noninvasive laser-induced photoacoustic tomography for structural and functional *in vivo* imaging of the brain. Nat Biotechnol 2003; 21(7): 803-6.
[http://dx.doi.org/10.1038/nbt839] [PMID: 12808463]

[269]  He L, *et al.* Colloidal Au-Enhanced Surface Plasmon Resonance for Ultrasensitive Detection of DNA Hybridization J Am Chem Soc 2000; 122(25): 9071-7.
[http://dx.doi.org/10.1021/ja001215b]

[270]  Huber MM, Göbel A, Joss A, *et al.* Oxidation of pharmaceuticals during ozonation of municipal wastewater effluents: A pilot study. Environ Sci Technol 2005; 39(11): 4290-9.
[http://dx.doi.org/10.1021/es048396s] [PMID: 15984812]

[271] Hyeon T, Lee SS, Park J, Chung Y, Na HB. Synthesis of highly crystalline and monodisperse maghemite nanocrystallites without a size-selection process. J Am Chem Soc 2001; 123(51): 12798-801.
[http://dx.doi.org/10.1021/ja016812s] [PMID: 11749537]

[272] Pankhurst QA, Connolly J, Jones SK, Dobson J. Applications of magnetic nanoparticles in biomedicine. J Phys D Appl Phys 2003; 36(13): R167-81.
[http://dx.doi.org/10.1088/0022-3727/36/13/201]

[273] Lee S weissleder R. Ultrasmall with Iron Oxide : For Assessing rats. Radiology 1990; 175(2): 494-8.
[PMID: 2326475]

[274] Peng ZA, Peng X. Formation of high-quality CdTe, CdSe, and CdS nanocrystals using CdO as precursor. J Am Chem Soc 2001; 123(1): 183-4.
[http://dx.doi.org/10.1021/ja003633m] [PMID: 11273619]

[275] Graedel TE, Mandich ML, Weschler CJ. Kinetic model studies of atmospheric droplet chemistry: 2. Homogeneous transition metal chemistry in raindrops. J Geophys Res 1986; 91(D4): 5205-21.
[http://dx.doi.org/10.1029/JD091iD04p05205]

[276] Bastien courtY M D Se´, Bouzigues Ce´ Dric, Luccardini Camilla, Ehrensperger Marie□Virginie, Bonneau STE´ Phane. Tracking individual proteins in living cells using single quantum dot imaging. Methods Enzymol 2006; 414(06): 211-28.

[277] Han M, Gao X, Su JZ, Nie S. Quantum-dot-tagged microbeads for multiplexed optical coding of biomolecules. Nat Biotechnol 2001; 19(7): 631-5.
[http://dx.doi.org/10.1038/90228] [PMID: 11433273]

[278] Jaiswal JK, Mattoussi H, Mauro JM, Simon SM. Long-term multiple color imaging of live cells using quantum dot bioconjugates. Nat Biotechnol 2003; 21(1): 47-51.
[http://dx.doi.org/10.1038/nbt767] [PMID: 12459736]

[279] Gao X, Cui Y, Levenson RM, Chung LWK, Nie S. *In vivo* cancer targeting and imaging with semiconductor quantum dots. Nat Biotechnol 2004; 22(8): 969-76.
[http://dx.doi.org/10.1038/nbt994] [PMID: 15258594]

[280] Alivisatos AP, Gu W, Larabell C. Quantum dots as cellular probes. Annu Rev Biomed Eng 2005; 7: 55-76.
[http://dx.doi.org/10.1146/annurev.bioeng.7.060804.100432]

[281] Siegel SJ, Winey KI, Gur RE, *et al.* Surgically implantable long-term antipsychotic delivery systems for the treatment of schizophrenia. Neuropsychopharmacology 2002; 26(6): 817-23.
[http://dx.doi.org/10.1016/S0893-133X(01)00426-2] [PMID: 12007752]

[282] Qian J, Peng Z, Wu D, Fu X. Synthesis and Characterization of WO₃/S Core/Shell Nanoparticles by Thermal Evaporation. Key Eng Mater Vols 2014; 602–603: 51-4.
[http://dx.doi.org/10.4028/www.scientific.net/KEM.602-603.51]

[283] Nomoev AV, Bardakhanov SP, Schreiber M, *et al.* Structure and mechanism of the formation of core-shell nanoparticles obtained through a one-step gas-phase synthesis by electron beam evaporation. Beilstein J Nanotechnol 2015; 6: 874-80.
[http://dx.doi.org/10.3762/bjnano.6.89] [PMID: 25977857]

[284] Verma M, Kumar V, Katoch A. Sputtering based synthesis of CuO nanoparticles and their structural, thermal and optical studies Curr Opin Colloid Interface Sci 2018; 8: 127-33.
[http://dx.doi.org/10.1016/j.mssp.2017.12.018]

[285] Swihart MT. Vapor-phase synthesis of nanoparticles. Curr Opin Colloid Interface Sci 2003; 8(1): 127-33.
[http://dx.doi.org/10.1016/S1359-0294(03)00007-4]

[286] Geng XSQ, Karkyngul B, Sun C, Liang X, Yang C, Su X. In₂O₃ nanocubes derived from

monodisperse InOOH nanocubes: Synthesis and applications in gas sensors. J Mater Sci 2017; 52(9): 5097-105.
[http://dx.doi.org/10.1007/s10853-017-0747-9]

[287] Ghorbani HR. A Review of Methods for Synthesis of Al Nanoparticles. Orient J Chem 2014; 30(4): 1941-9.
[http://dx.doi.org/10.13005/ojc/300456]

[288] Valverde-Alva MA, García-Fernández T, Villagrán-Muniz M, *et al.* Synthesis of silver nanoparticles by laser ablation in ethanol: A pulsed photoacoustic study. Appl Surf Sci 2015; 355: 341-9.
[http://dx.doi.org/10.1016/j.apsusc.2015.07.133]

[289] Kim S, Yoo BK, Chun K, *et al.* Catalytic effect of laser ablated Ni nanoparticles in the oxidative addition reaction for a coupling reagent of benzylchloride and bromoacetonitrile. J Mol Catal Chem 2005; 226(2): 231-4.
[http://dx.doi.org/10.1016/j.molcata.2004.10.038]

[290] Tarasenko NV, Butsen AV, Nevar EA, Savastenko NA. Synthesis of nanosized particles during laser ablation of gold in water Appl Surf Sci 2006; 252: 4439-44.
[http://dx.doi.org/10.1016/j.apsusc.2005.07.150]

[291] Garcı FR. Adsorption of heavy metals onto sewage sludge-derived materials. 2008; 99: 6332-8.

[292] Hench LL, West NK. The Sol-Gel Process. Chem Rev 1990; 90(1): 33-72.
[http://dx.doi.org/10.1021/cr00099a003]

[293] Peña-Flores JI, Palomec-Garfias AF, Márquez-Beltrán C, Sánchez-Mora E, Gómez-Barojas E, Pérez-Rodríguez F. Fe effect on the optical properties of $TiO_2$:$Fe_2O_3$ nanostructured composites supported on $SiO_2$ microsphere assemblies. Nanoscale Res Lett 2014; 9(1): 499.
[http://dx.doi.org/10.1186/1556-276X-9-499] [PMID: 25276103]

[294] Aruna ST, Mukasyan AS. Combustion synthesis and nanomaterials. Curr Opin Solid State Mater Sci 2008; 12(3–4): 44-50.
[http://dx.doi.org/10.1016/j.cossms.2008.12.002]

[295] Agrafiotis C, Roeb M, Konstandopoulos AG, *et al.* Solar water splitting for hydrogen production with monolithic reactors. Sol Energy 2005; 79(4): 409-21.
[http://dx.doi.org/10.1016/j.solener.2005.02.026]

[296] Morales W, Cason M, Aina O, de Tacconi NR, Rajeshwar K. Combustion synthesis and characterization of nanocrystalline $WO_3$. J Am Chem Soc 2008; 130(20): 6318-9.
[http://dx.doi.org/10.1021/ja8012402] [PMID: 18439012]

[297] Vivekanandhan S, Venkateswarlu M, Satyanarayana N. Ammonium carboxylates assisted combustion process for the synthesis of nanocrystalline $LiCoO_2$ powders. Mater Chem Phys 2008; 109(2-3): 241-8.
[http://dx.doi.org/10.1016/j.matchemphys.2007.11.027]

[298] Kalimuthu K, Suresh Babu R, Venkataraman D, Bilal M, Gurunathan S. Biosynthesis of silver nanocrystals by Bacillus licheniformis. Colloids Surf B Biointerfaces 2008; 65(1): 150-3.
[http://dx.doi.org/10.1016/j.colsurfb.2008.02.018] [PMID: 18406112]

[299] Saifuddin N, Wong CW, Yasumira AANUR. Rapid biosynthesis of silver nanoparticles using culture supernatant of bacteria with microwave irradiation E-JChem 2009; 6(1): 61-70.
[http://dx.doi.org/10.1155/2009/734264]

[300] Shahverdi AR, Minaeian S, Reza H, Jamalifar H, Nohi A. Rapid synthesis of silver nanoparticles using culture supernatants of Enterobacteria : A novel biological approach. Process Biochem 2007; 42(5): 919-23.
[http://dx.doi.org/10.1016/j.procbio.2007.02.005]

[301] Ingle A, Gade A, Pierrat S, Sönnichsen C, Rai M. Mycosynthesis of Silver Nanoparticles Using the Fungus Fusarium acuminatum and its Activity Against Some Human Pathogenic Bacteria. Curr

Nanosci 2008; 4(2): 141-4.
[http://dx.doi.org/10.2174/157341308784340804]

[302] Xie J, Marijnissen JC, Wang CH. Microparticles developed by electrohydrodynamic atomization for the local delivery of anticancer drug to treat $C_6$ glioma *in vitro*. Biomaterials 2006; 27(17): 3321-32.
[http://dx.doi.org/10.1016/j.biomaterials.2006.01.034] [PMID: 16490248]

[303] Vakarelski IU, Toritani A, Nakayama M, Higashitani K. Deformation and Adhesion of Elastomer Microparticles Evaluated by AFM. Langmuir 2001; 17(16): 4739-45.
[http://dx.doi.org/10.1021/la001588q]

[304] Tagit O, Tomczak N, Vancso GJ. Probing the morphology and nanoscale mechanics of single poly(N-isopropylacrylamide) microgels across the lower-critical-solution temperature by atomic force microscopy. Small 2008; 4(1): 119-26.
[http://dx.doi.org/10.1002/smll.200700260] [PMID: 18098239]

[305] Lofrano G, Pedrazzani R, Libralato G, Carotenuto M. Advanced Oxidation Processes for Antibiotics Removal: A Review. Curr Org Chem 2017; 21(12): 1054-67.
[http://dx.doi.org/10.2174/1385272821666170103162813]

[306] Deshpande BD, Agrawal PS, Yenkie MKN. Prospective of nanotechnology in degradation of waste water: A new challenges. nano Struct nano objects 2020; 22: 100442.
[http://dx.doi.org/10.1016/j.nanoso.2020.100442]

[307] Jiang J, Oberdörster G, Biswas P. Characterization of size, surface charge, and agglomeration state of nanoparticle dispersions for toxicological studies. J Nanopart Res 2009; 11(1): 77-89.
[http://dx.doi.org/10.1007/s11051-008-9446-4]

[308] Kagan VE, Tyurina YY, Tyurin VA, *et al.* Direct and indirect effects of single walled carbon nanotubes on RAW 264.7 macrophages: Role of iron. Toxicol Lett 2006; 165(1): 88-100.
[http://dx.doi.org/10.1016/j.toxlet.2006.02.001] [PMID: 16527436]

[309] Monteiro-riviere NA, Nemanich RJ, Inman AO, Wang YY, Riviere JE. Multi-walled carbon nanotube interactions with human epidermal keratinocytes Toxicol Lett 2005; 155: 377-84.
[http://dx.doi.org/10.1016/j.toxlet.2004.11.004]

[310] Warheit DB, Laurence BR, Reed KL, Roach DH, Reynolds GAM, Webb TR. Comparative pulmonary toxicity assessment of single-wall carbon nanotubes in rats. Toxicol Sci 2004; 77(1): 117-25.
[http://dx.doi.org/10.1093/toxsci/kfg228] [PMID: 14514968]

[311] Oberdörster E. Manufactured nanomaterials (fullerenes, C60) induce oxidative stress in the brain of juvenile largemouth bass. Environ Health Perspect 2004; 112(10): 1058-62.
[http://dx.doi.org/10.1289/ehp.7021] [PMID: 15238277]

[312] Yin ZF, Wu L, Yang HG, Su YH. Recent progress in biomedical applications of titanium dioxide. Phys Chem Chem Phys 2013; 15(14): 4844-58.
[http://dx.doi.org/10.1039/c3cp43938k] [PMID: 23450160]

[313] Weir A, Westerhoff P, Fabricius L, Hristovski K, von Goetz N. Titanium dioxide nanoparticles in food and personal care products. Environ Sci Technol 2012; 46(4): 2242-50.
[http://dx.doi.org/10.1021/es204168d] [PMID: 22260395]

[314] Tang Y, Wang F, Jin C, Liang H, Zhong X, Yang Y. Mitochondrial injury induced by nanosized titanium dioxide in A549 cells and rats. Environ Toxicol Pharmacol 2013; 36(1): 66-72.
[http://dx.doi.org/10.1016/j.etap.2013.03.006] [PMID: 23598258]

[315] Petković J, Zegura B, Stevanović M, *et al.* DNA damage and alterations in expression of DNA damage responsive genes induced by $TiO_2$ nanoparticles in human hepatoma HepG2 cells. Nanotoxicology 2011; 5(3): 341-53.
[http://dx.doi.org/10.3109/17435390.2010.507316] [PMID: 21067279]

[316] Yaqoob AA, Parveen T, Umar K, Ibrahim MN. Role of nanomaterials in the treatment of wastewater: A review. Water 2020; 12(2): 495.

[http://dx.doi.org/10.3390/w12020495]

[317]  Gerloff K, Fenoglio I, Carella E, *et al.* Distinctive toxicity of TiO$_2$ rutile/anatase mixed phase nanoparticles on Caco-2 cells. Chem Res Toxicol 2012; 25(3): 646-55.
[http://dx.doi.org/10.1021/tx200334k] [PMID: 22263745]

[318]  Valos AMD. TiO$_2$ nanoparticles induce dysfunction and activation of human endothelial cells. 2012; 25: 920-30.

[319]  Wang C, Cheng K, Zhou L, *et al.* Evaluation of Long-Term Toxicity of Oral Zinc Oxide Nanoparticles and Zinc Sulfate in Mice. Biol Trace Elem Res 2017; 178(2): 276-82.
[http://dx.doi.org/10.1007/s12011-017-0934-1] [PMID: 28120304]

[320]  Zoroddu MA, Medici S, Ledda A, Nurchi VM, Lachowicz JI, Peana M. Toxicity of nanoparticles. Curr Med Chem 2014; 21(33): 3837-53.
[http://dx.doi.org/10.2174/0929867321666140601162314] [PMID: 25306903]

[321]  Chen Z, Meng H, Xing G, *et al.* Acute toxicological effects of copper nanoparticles *in vivo.* Toxicol Lett 2006; 163(2): 109-20.
[http://dx.doi.org/10.1016/j.toxlet.2005.10.003] [PMID: 16289865]

[322]  Ballestri M, Baraldi A, Gatti AM, *et al.* Liver and kidney foreign bodies granulomatosis in a patient with malocclusion, bruxism, and worn dental prostheses. Gastroenterology 2001; 121(5): 1234-8.
[http://dx.doi.org/10.1053/gast.2001.29333] [PMID: 11677217]

[323]  Gehrke I, Geiser A, Somborn-Schulz A. Innovations in nanotechnology for water treatment. Nanotechnol Sci Appl 2015; 8: 1-17.
[http://dx.doi.org/10.2147/NSA.S43773] [PMID: 25609931]

[324]  Lee SH, Lee HR, Kim Y, Kim M. Toxic Response of Zinc Oxide Nanoparticles in Human Epidermal Keratinocyte HaCaT Cells. Toxicol Environ Health Sci 2012; 4(1): 14-8.
[http://dx.doi.org/10.1007/s13530-012-0112-y]

[325]  Brame J, Li Q, Alvarez PJJ. Nanotechnology- enabled water treatment and reuse : Emerging opportunities and challenges for developing countries. Trends Food Sci Technol 2011; 22(11): 618-24.
[http://dx.doi.org/10.1016/j.tifs.2011.01.004]

[326]  Telenko C, Rourke JMO, Webber ME. A Compilation of Design for Environment Guidelines. J Mech Des 2016; 138(March): 031102-1, 11.
[http://dx.doi.org/10.1115/1.4032095]

[327]  Abrahama MA, Nguyenb N. Green Engineering :Defining the Principles, Sandestin Conference. Environ Prog 2003; 22(4): 233-6.

# CHAPTER 7

# Utilization of Water: Environmental Impact and Health Issues

**Bhavna D. Deshpande[1,*], Pratibha S. Agrawal[1], M.K.N. Yenkie[1] and S.J. Dhoble[2]**

[1] *Department of Applied Chemistry, Laxminarayan Institute of Technology, R.T.M. Nagpur University, Nagpur–440010, India*

[2] *Department of Physics, R.T.M. Nagpur University, Nagpur-440033, India*

**Abstract:** The availability of pure drinking water to individuals reflects the progress of any region, which is linked directly with the quality of life across the globe. Variations in the quality and quantity of water systems control all aspects of human life. Both its shortage and excess affect the growth and development of the community. The utilization and conservation of our water world must be an integral part of sustainable development and should be appealed to by all sectors. While effective wastewater treatment has the tendency to recover the water, integration of all policies with periodic improvement using research outcomes is still essential. In order to tackle the challenges in the coming decades, it is important that all stakeholders are sensitized about the current scenario, future needs, and the need for a proper scientific and rational approach to moderate the issues and challenges collectively. Integrated Water Conservation Techniques are acknowledged as the only sustainable solution to water scarcity.

**Keywords:** Adsorbent, Advanced oxidation process, Photo-catalyst, Nano-materials, Non-biodegradable, Oxidation, Water treatment.

## INTRODUCTION

Water is a basic supporting system for the sustainable development of society as well as the country. The importance of water remains as there exists no alternative or substitute for it. In spite of large water bodies surrounding the earth, water scarcity has engulfed large areas, affecting the normal life of humans. Changing lifestyles, climatic conditions, population explosions, urbanization, and industrialization have changed the quality as well as the availability of this indispensable resource worldwide. Pollutants emerging from factories, industries,

---

* **Corresponding author Bhavna D. Deshpande:** Department of Applied Chemistry, Laxminarayan Institute of Technology, R.T.M. Nagpur University, Nagpur–440010, India; Tel: +91-9881370448; E-mail: dnd.bhavna@gmail.com

**R. M. Belekar, Renu Nayar, Pratibha Agrawal and S. J. Dhoble (Eds)**
**All rights reserved-© 2022 Bentham Science Publishers**

and institutions, agricultural runoff, and domestic sewage directly or indirectly enter aquatic bodies, disturbing the entire ecosystem.

The pollutants present in water can be categorized as biological, organic, inorganic, radioactive, and thermal pollutants. Biological contaminants include viruses, worms, planktons, fungi, bacteria, and protozoa, which are responsible for spreading diseases in humans and animals. Organic pollutants are highly toxic. Their presence in low concentrations leads to severe genetic disorders. These include dues, detergents, sewage, phenols, pharmaceuticals, pesticides, oils, *etc*. Inorganic pollutants determine the quality of water. It includes mineral acids, inorganic salts, metal compounds, trace elements, and organometallic compounds. Radioactive contaminants emit radiations that affect humans and the environment adversely. Their presence in nature may be due to mining activities and processing, nuclear weapon manufacturing, isotopes used in medical fields and industrial and research applications. Generally, water is used as a coolant in many industries. Differences in temperatures of coolants and large water bodies affect the aquatic ecosystem, causing thermal pollution. The industrial boom has led to an increase in water pollutants and the depletion of freshwater resources. It is imperative that pollutants adversely affect the environment and cause a scarcity of water.

To cater to the extended human needs, this problem may arise in the future. There are various reasons for water scarcity, most likely contamination through organic and inorganic chemicals, natural calamities, excess and unjudicial use, population explosion, underdeveloped management systems, *etc*. Climatic changes are supposed to aggravate these problems more in the near future. The unavailability of safe freshwater has led to severe health issues because of lack of sanitation and exposure to microbes through the food chain, affecting people worldwide. Pollution of freshwater resources exposes humans to waterborne diseases in various ways [1]. This paper emphasizes problems faced by the planet today in terms of quantity and quality of freshwater. Integrated techniques, along with traditional techniques, can be a viable tool to increase storage capacity. Water conservation and management can be made effective with good governance, policies, participation of stakeholders, spreading awareness, public-private partnership, and advanced techniques.

## HUMAN DESIRE AND EARTHS LIMIT

History has witnessed that all the civilizations in the past have flourished near the rivers. About 5000 years ago, humans used water for irrigation purposes. They controlled and diverted river water to their fields; hence, water was primarily used for irrigation. Water remains an inevitable source for domestic as well as economic use then and even now. The flow chart below (Fig. **1**) shows the water usage timeline of humans.

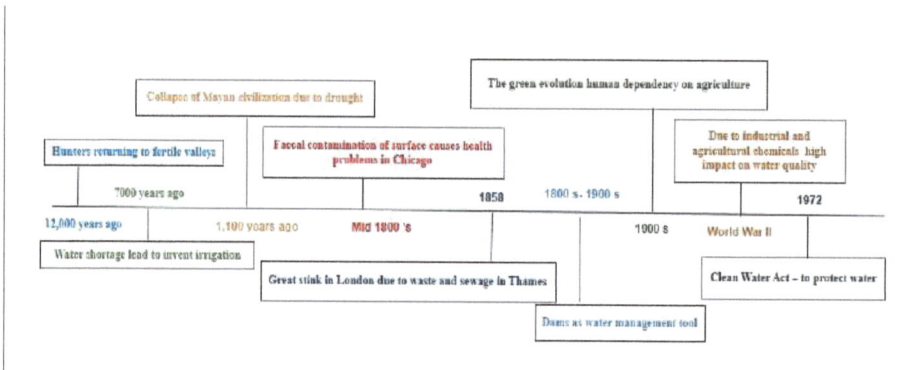

**Fig. (1).** Flow chart showing water usage timeline of humans.

The use of natural resources and urban development has brought environmental changes. Our actions have changed the landscapes and disrupted river flow, reallocating forests, hence disturbing wildlife, for our survival, which has affected our environment [2]. With 70% of oceans that cover our planet, the real usable water remains only 2.5%, of this, only approximately 1% remains accessible for human usage which is found in reservoirs, groundwater, lakes, ponds, and rivers. These resources are replenished by rainwater and snowfall; hence, these are the only sources of our sustenance. Today, the demand for freshwater has increased largely for domestic purposes, irrigation, and municipal and industrial usage [3]. Largely, water is used for household water (washing machine, sink, shower, cooking), communities (fire fighting, public areas, hotels, clubs, shops, public drinking, schools, colleges, and libraries, public gardens and parks, malls, *etc.*) farming (irrigation, spraying fertilizer, herbicides, and pesticides, dairy, vegetables, and grain crops), generating electricity, industries recreation, and transportation (water parks, fountains, golf courses, *etc.*).

For years, we have enjoyed the fruits of abundant resources without actually considering what might be the ill effects of their unjudicial use. Today, nearly 40% of the world's population is living in water stress; this count is expected to rise to 50-65% by 2025, and about 90% of the population by 2050 will grow in the water-scarce areas [4].

In the last few years, global warming has led to frequent droughts affecting the world economy [5]. The stress on the water is further added due to unplanned agricultural and industrial usage along with improper and mismanaged use of groundwater resources. Rivers and rain are the major sources of surface freshwater. India receives about 3 trillion M3 of water from rainfall and 1869 BMC from various river basins. In spite of having a wide-ranging river network in India, the availability of freshwater remains an enigma (Fig. **2**). A large amount of water (1150 km$^3$) flows as surface runoff, and the same amount percolates as groundwater. Groundwater resources known as aquifers are profusely found in the northern and coastal plains. These are continuously replenished through seepage, infiltration, evaporation, and transpiration. Only 25% of the groundwater used by humans is profoundly used for agricultural practices. Data reveals a decrease in the per capita water availability in India. Hence, we are already under water stress and by 2050, the water availability will further reduce to 1140 m$^3$ as the population growth is expected to reach 1.64 billion. A shortage of water leads to alarming food security, raising social and economic problems.

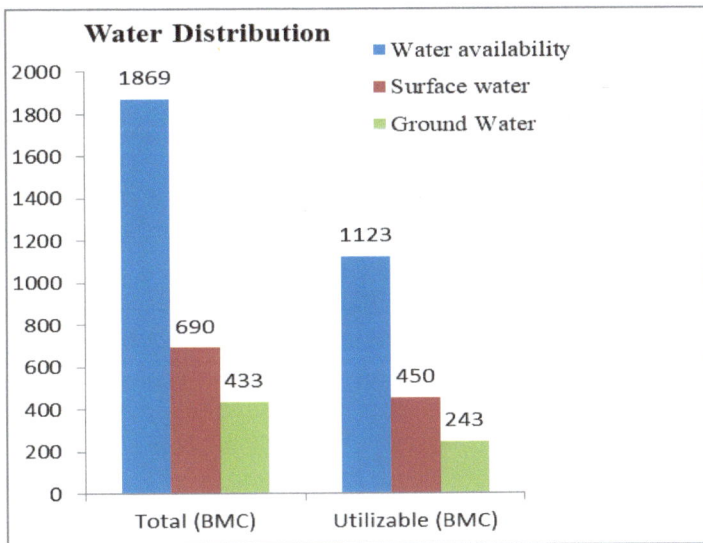

**Fig. (2).** Water Distribution – Water availability on Surface and Ground.

The alarming water crises can be controlled by recovering wastewater, upgrading existing storage and management of the freshwater resources, and harvesting water from several means that can sustain the environment for the future (Fig. **3**). Water treatment and reuse alone cannot solve the scarcity problem. More emphasis should be given to improving the quality of available water. Reused water often adds to the health risks and has limitations. Hence, it should be judicially used.

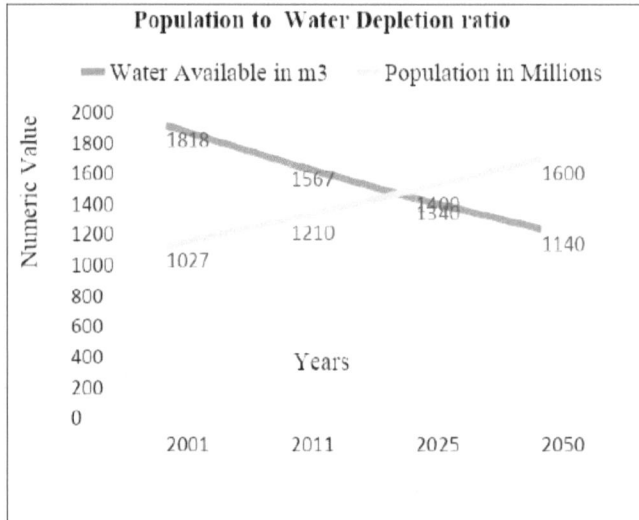

**Fig. (3).** Depletion of water resources with the increase in Population.

Government authorities can only control the severity of water pollution and its conversation and management with strict governance and stringent rules. Strict laws should be enforced in every industry, organization, and institution for proper waste management and disposal of effluent. An increase in contamination levels deteriorates the quality and makes the water unfit for consumption [6]. Hence, the WHO and the Directive 2015/1787 have amended the quality of water for human consumption for drinking, recreation, and recycling purposes. Many developed countries, such as Australia, have adopted ('Australian Guidelines for Water Recycling') their own directives and the guidance for recycling of water, management of surface water, aquifer recharge, and irrigation. United States Environmental Protection Agency (USEPA) in 2012, which included wide reuse applications of wastewater. In 2015, the International Organization for Standardization (ISO) have laid out guidelines for the use of treated wastewater for irrigation projects and agricultural use.

On December 7, 2007, National Water Council, Spain's communities, and local authorities laid their own specifications for carrying out tests, mainly physio-chemical characterization and microorganism growth level. However, in 2019 European Parliament legislative have laid minimum requirements that treated water should be reused for agricultural irrigation. This includes risk analysis and the safe use of treated water for various applications [7]. Therefore, the choice of treatment for wastewater treatment depends upon its later usage. A considerable amount of chemicals should be used for treatment to prevent excess contamination and overdose of salts in residue that is leached and may affect freshwater resources. Such processes that do not harm human health and ecology are preferred [8].

The absolute quantities of freshwater sources remain nearly constant on Earth, yet their accessibility and availability have created a problem due to increased human desires and an uneven distribution of water. Uncertainty in water availability is also a challenge faced by most countries, hampering economic growth. According to IEC, certain guidelines for creating awareness among the stakeholders on safe drinking water, sanitation, and hygiene, have been laid that include, the judicial use of water, avoiding wastage, water harvesting, the importance of reuse and recycling of water, water quality, its handling, its operation, planning, and management, waterborne diseases, the importance of hygiene, and low-cost water treatment technologies, *etc*.

It is a well-established fact that the need and demand for water will continue to increase. Hence, extreme weather conditions, water unavailability, groundwater depletion, and disturbed hydrological cycles are expected in the time to come. We, humans, need to limit our desires until the mother earth limits them. Otherwise, effective, efficient, and limited use of natural resources can help to replenish these resources. Excessive and unplanned use has produced immense stress on water resources. Furthermore, anthropological means have led to climatic changes that have altered water availability.

## EFFECTS OF CLIMATE AND ENVIRONMENTAL CHANGE ON WATER AVAILABILITY

Rivers and lakes, though tiny, constitute the most available sources for human use and consumption. Water scarcity results when the amount of water withdrawn from natural resources exceeds its replenishment. Hence, the water supply remains inadequate to fulfil human needs. The supply of total global freshwater remains constant, though the local supply remains subjected to an ambiguity of geography, rainfall, humidity, and temperature, resulting in climatic changes [9]. The simultaneous increase in economic activity, population density, urbanization, and industrialization has altered the environment and forest ecosystems, leading to climatic changes worldwide.

The world's water sources have largely been affected by climatic changes. The water cycle below (Fig. **4**) depicts the changes caused due to global warming affecting fresh water supply, and food production, with the property evaluation worldwide.

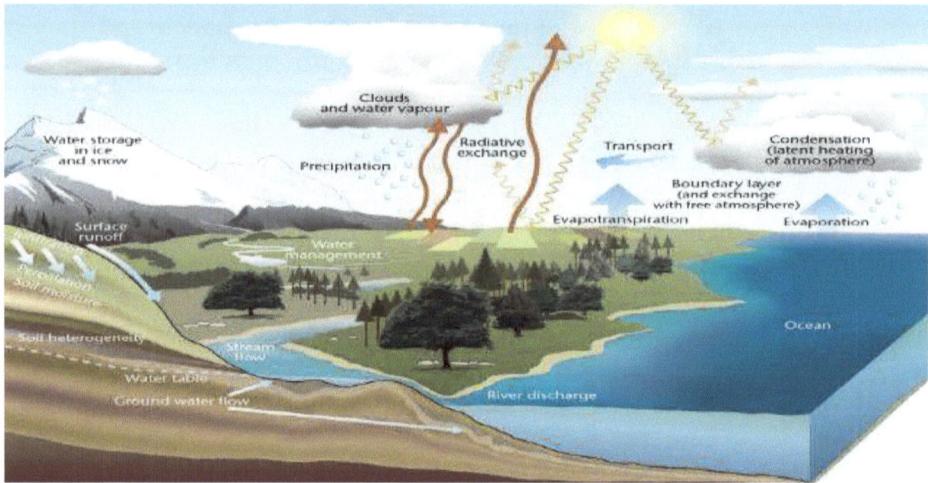

**Fig. (4).** The water cycle depicts the changes caused due to global warming, Source: UK Met Office.

Climatic changes can bring the following shifts: a rise in temperature increases evaporation losses, warm air takes up more moisture from water bodies, unlike dry air. In addition, the high humidity in the air can be dangerous to humans, as it would increase the temperature, thus blocking the cooling effect due to sweating. An increase in evaporation leads to rains and snowfall. Changes in air temperature and its circulation pattern may cause heavy precipitation as well as intense and prolonged drought-like conditions. These dry conditions lead to wildfires that can disturb the habitats of wildlife.

Rainfall patterns disturb the natural ecosystem along with agricultural practices. Rainstorms lead to surface runoff, which takes away the essential nutrients and humus from the soil. In addition, it carries with it undesirables garbage (dirt, polybags, sags, pollutants), pouring them into nearby water bodies. The dumping from runoff affects the aqua ecosystem. Pollutants like fertilizer suffocate the water dwellers and insufficient oxygen affects fisheries. Such water bodies are abandoned by people who like to swim and use them for recreational activities. Flood destroys the infrastructure of the affected area, damages crops, and endangers lives as a result, affecting the economy.

Ocean dwellers are equally affected by climatic changes. The food chain has been disturbed due to an increase in acidity and temperature. Fishing industries are facing immense losses as the fish are traveling to lower temperature regions. Fish lovers are, therefore, paying high prices for their meals. Global warming has resulted in the melting of glaciers, thereby, results in a rise at the sea level, causing potential risk to coastal villages and towns.

The slow melting of glaciers in springs helps replenish drinking water supplies in rivers and streams. In places where the rainfall exceeds the snowfall, it means that less water is stored for later use as snowpacks. Moreover, rain melts the snow that is present on the ground. The regions that depend on melted snow to refill drinking water resources are experiencing problems due to a lack of snowpack. Winter tourism and wildlife are affected by changes in snowpack [10].

Hence, the change in the atmospheric concentration of CO and $CO_2$ has led to global warming, unpredicted temperatures, and precipitation. The rise in sea level increases the risk of flooding and extreme weather conditions, which force the population from the coastal part to migrate, which has led to stress and resource competition in a new area on one hand and the depletion of fresh water and soil on the other. This stress is more prevalent in underdeveloped countries where the resources are few. This high population to resource ratio at times leads to communal tension. The developed countries, with economic, political prowess, and a stable social fabric, are better equipped to handle the threats emerging from environmental factors. Three major implications expected to follow as per United Nations Intergovernmental Panel on Climate Change (IPCC) 2007, are degradation of cropland (decreased infertility), population displacement (migration from extreme places to safe places; IPPC, 2001:36), and depletion of freshwater resources (increasing scarcity). A climatic change has a major impact on food production capacity. A slight rise in temperature in the temperate areas increases the crop yield whereas; a higher temperature may reduce the yield [11]. In tropical areas, minimal fluctuations in temperature may reduce the output change. Extreme changes in temperature and precipitation directly increase the degradation of soil and water resources. However, these impacts can be moderated by adapting to better land use and management ways (IPCC, 2001:32) [12].

Construction of large reservoir is important to store water for unpredicted climatic changes. To sustain the odd changes in the climate, infrastructure that can absorb the shocks, is required. Climatic changes alter the hydrological cycle directly or indirectly, affecting developing countries globally [13]. These changes alter the periodic seasons and show their might at unexpected times, which brings uncertainty and chaos for the entire humanity. Extreme climatic changes will have a catastrophic effect on food security; creating water scarcity, drought, or deluge, thereby affecting the livelihood of most of the people on earth. An increase in the precipitation and its magnitude may flood the regional water bodies, affecting the local ecosystem [14]. On the other hand, in the absence of adequate precipitation, drought-like situations can prevail. Apart from quantitative changes (floods, droughts, and deluge), the quality of surface water and groundwater is affected largely by climatic changes. Land use, urban scattering, deforestation, and areas of waterproofing can alter the quality of water. Quality parameters that show modification can be physicochemical (pH, temperature, DO, dissolved organic matter), nutrients (minerals N2, S, K, P), micro-pollutants (metals, organics, inorganics, pesticides, pharmaceuticals), and biological (microorganisms) [15]. Therefore, climatic changes alter air, food, and water quality. If the disruption occurs in a similar pattern for a longer duration, it can lead to vector and water-borne diseases.

## ENVIRONMENTAL IMPACT AND HEALTH ISSUE

Pollution and health hazards are imperative. Pollution leads to numerous health problems. Ground or surface water may be contaminated due to industrial, agricultural, or domestic water outflow into the nearby water resources, making it unsafe for drinking purposes. Waterlogging in low line areas, stagnant water, improper sanitation habits, faecal disposal in open [16], garbage dumping, and disposal of dead bodies of animals and humans may lead to bacteriological contamination; the runoff water may give rise to water-borne diseases like malaria, diarrhea, cholera, typhoid, *etc.* Diarrhea is a very common bacterial disease in rural areas worldwide. It is usually caused by faecal contamination of water and consumption of untreated drinking water. Cholera is a digestive tract disease caused due to contaminated water, which leads to dehydration and renal failure. Shigellosis, a disease, which affects the digestive tract and damages the intestinal lining, leading to vomiting, abdominal cramps, and watery and bloody diarrhea, can be treated by antibiotics and proper sanitation [17, 18].

Studies show that chemical contamination in drinking or usable water has hostile effects on the human body. Chemicals such as sulphate with magnesium cause diarrhea, fluoride causes fluorosis, deformities in bones and joint problems, discoloration, damage to teeth, cadmium affects bones, boron, and mercury affects the nervous system, sodium with chloride increases blood pressure [19, 20]. Iron aggravates stomach disorders in humans and adversely affects fish gills, which get clogged due to excess iron in the water.It might be fatal for the fish as well as those who have consumed such fish, and it may lead to major health problems [21, 22]. Heavy metals such as lead retards mental growth in children cause anemia, affects the central nervous system, and damages kidneys in adults, arsenic causes skin cancer and skin-related diseases [23, 24]. Some metals lead to hair loss, liver cirrhosis, and neural disorders [25, 26]. Nitrates cause blue baby syndrome and cancer [27]. In comparison to urban areas, the mortality rate due to cancer is more in rural areas because of the unavailability of treated and safe drinking water. The unavailability of safe water, inadequate sanitation facilities, and a careless attitude towards self-hygene are the factors that make rural poor more prone to diseases. A large population consumes food and vegetables grown in contaminated water. Hence, the toxic chemicals find their way into the food chain [28]. Crop production is hampered due to poor water quality [29]. Such water has a negative effect on pregnant females, resulting in low birth weight, retarded growth, and respiratory problems [30].

Amongst viral diseases, the most common disease caused due to contaminated water is hepatitis, which infects the liver, causing loss of appetite. If undetected for a longer time, it may be fatal. Vaccines and proper hygiene can cure it. The Culex mosquito breeds in contaminated water and its bite causes encephalitis. It may result in a coma and paralysis. Until now, vaccines were not available for a complete cure. Poliomyelitis is another dreadful disease that may lead to paralysis, vaccines and polio drops can prevent it. Gastroenteritis is caused by various viruses that can be fatal in infants [31].

Parasitic diseases are caused due to cyst, which enters humans through contaminated water. Most of these parasites affect the stomach or interstitial lining causing watery diarrhea, bloating, and weight loss [32]. Cryptosporidiosis and Giardiasis are known parasitic diseases.

Pesticides, usually from the factory outlet or agricultural runoff cause cancer and damage the nervous system, reproductive system, and immune system [33]. Hence, knowing the effect of excess chemicals on the human body is imperative to impart knowledge to rural as well as the urban population on health issues [34]. Several heavy metals, organic and inorganics, enter the food chain through polluted water. These pollutants have an adverse effect on the metabolism, and if the concentration exceeds the permissible limit, may mutate the cell's organelles [35]. Minerals are important for sustenance up to a certain limit after which it becomes slow poison, which may lead to permanent disability and sometimes-genetic disorders too. Entry of contaminants may be through different means. In many cases, the working conditions or places are not conducive; the people working with asbestos, glass, minerals, mining, salt production, rare metals, heavy metals, *etc.* are exposed constantly and hence are always at high risk of contamination [36, 37]. These occupational hazards can be lessened by providing them with the proper kit and educating them about disaster management. The workplaces should have all the necessities (sanitation, airy, safe water, sick room, complaint box, *etc.*), upgraded machines, and the proper hierarchy should be known, and maintained to fill the gap between management and labor.

All living beings on earth are affected by polluted water and can be subjected to dreadful diseases, including cancer, neurological and heart-related ailments, and respiratory problems [38].

## Preventive Measures

Proper management of water resources can help to overcome waterborne illnesses and reduce the health hazards caused by the intake of contaminated water. As said, "Prevention is better than cure". People should be educated about self-hygiene, safe drinking water, water quality standards, waste management, and judicial use of locally available resources. Water pipes need to be checked at regular intervals for leakages and cracks, water should be boiled and filtered before drinking.

Moreover, public awareness is a must. Various plans, schemes, and investment proposals should reach the masses. Upgradation of such plans has to be done regularly by the local governing body. All such programmes should be monitored from time to time by the authorities to check whether they are benefiting the targeted population. Hence, certain objectives as specified by the IEC campaign should be imbibed, which include, creating awareness, motivation, initiating community participation, adoption of hygienic practices, changing mindsets, promoting personal accountability, and taking responsibility, for a better future. Awareness in the masses can be promoted by audio-visual (means as TV, record players), audio spots/jingles (radio), talk shows, discussions, lectures, street theatre (plays, Folk dance, Katha Kathan, nritraya Natak, and Nautanki). Local magazines, newspapers, brochures, pamphlets, flips, charts, and leaflets can be distributed amongst the masses at regular intervals [39].

## WATER CONSERVATION AND MANAGEMENT: TRADITIONAL METHODS

Science and technology have been part of our culture for ages. History has witnessed that humans have always come up with innovative ideas to overcome natural troubles. Water conservation and management sound a relatively new term, but history has witnessed a variety of storing and extracting processes in past. Water harvesting remains a very old and traditional method of collecting rainwater, diverting and storing it from streams springs, and from sub-surface water. Water harvesting methods are typically based on local climate and geographical areas. India's prime occupation being agricultural, farmers' dependency remained on annual rainfall; Hence, trapping the rainfall was done by various means in different parts of India.

Most of the well-known civilizations, such as the Indus Valley, which flourished near the Indus River 5000 years ago, shows magnificent urban water management and sewage systems, which remain unparalleled to date. Another well-planned city in the Rann of Gujarat was Dholavira. Furthermore, an old water harvesting system was found along Naneghat in the Western Ghats, 130 km from Pune. In ancient times, houses were such built that rainwater directly reached underground tanks. Such systems can be witnessed today in forts, palaces, and houses in Western Rajasthan. The pipes and tunnels used to transport and maintain water flow were made of baked earthen soil; such water management systems are still functional at Burnpur in Madhya Pradesh, Golkunda in AP, Aurangabad in Maharashtra, Bijapur in Karnataka, and temple tanks in Tamil Nadu. In Rajasthan, sloping courtyards were constructed in available space. If enough space was available, a kund was made in the center, lined in a way that no seepage and contamination were possible. Hence, they were the sustainable source of drinking water. Rajasthan was known for the tank and khadins, baories were common in Gujarat (Fig. **5**). Integrated surface and groundwater systems in forts of Jodhpur and Chittor, lakes of Udaipur, Ahar Pynes in Bihar (Fig. **6c**), Zings of Ladakh (Fig. **6a - c**) Panhala Fort of Shivaji Maharaj in Maharashtra, and Sanchi in Madhya Pradesh, are excellent examples of rainwater harvesting where baolis and wells were constructed to trap groundwater. Pat System in Madhya Pradesh (Jhabua) and Johads in Karnataka, Odisha were the traditional water management systems from Central Highlands (Fig. **7**). Similarly, traditional water Management systems were observed in Bhandara Phad in Maharashtra (Dhule), Nasik, and Ramtek Model in Maharashtra (Fig. **8**) from the Deccan plateau.

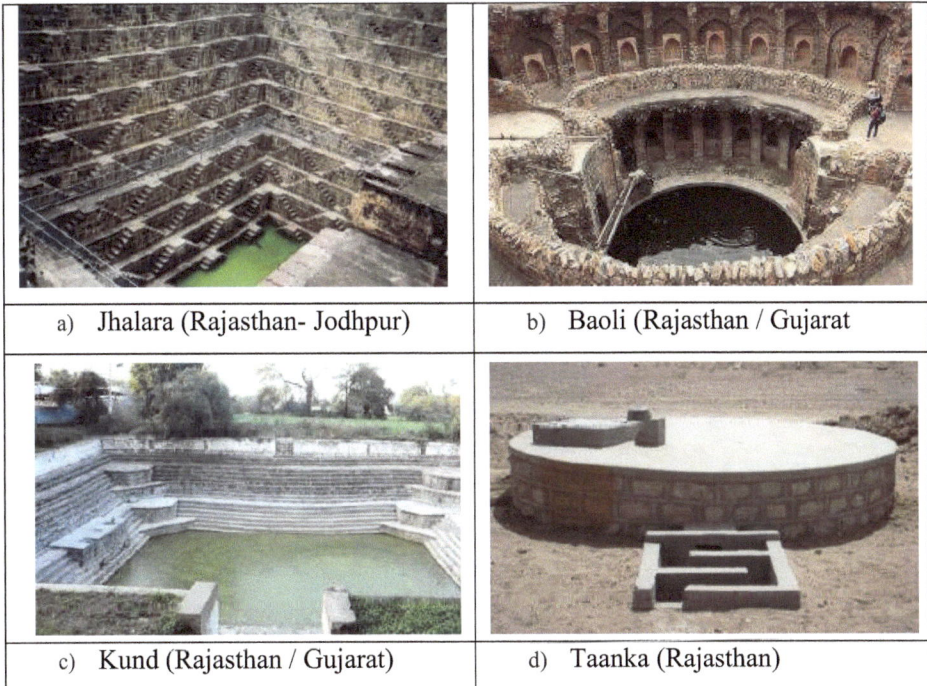

| | |
|---|---|
| a)  Jhalara (Rajasthan- Jodhpur) | b)  Baoli (Rajasthan / Gujarat |
| c)  Kund (Rajasthan / Gujarat) | d)  Taanka (Rajasthan) |

*(Fig. 8) contd.....*

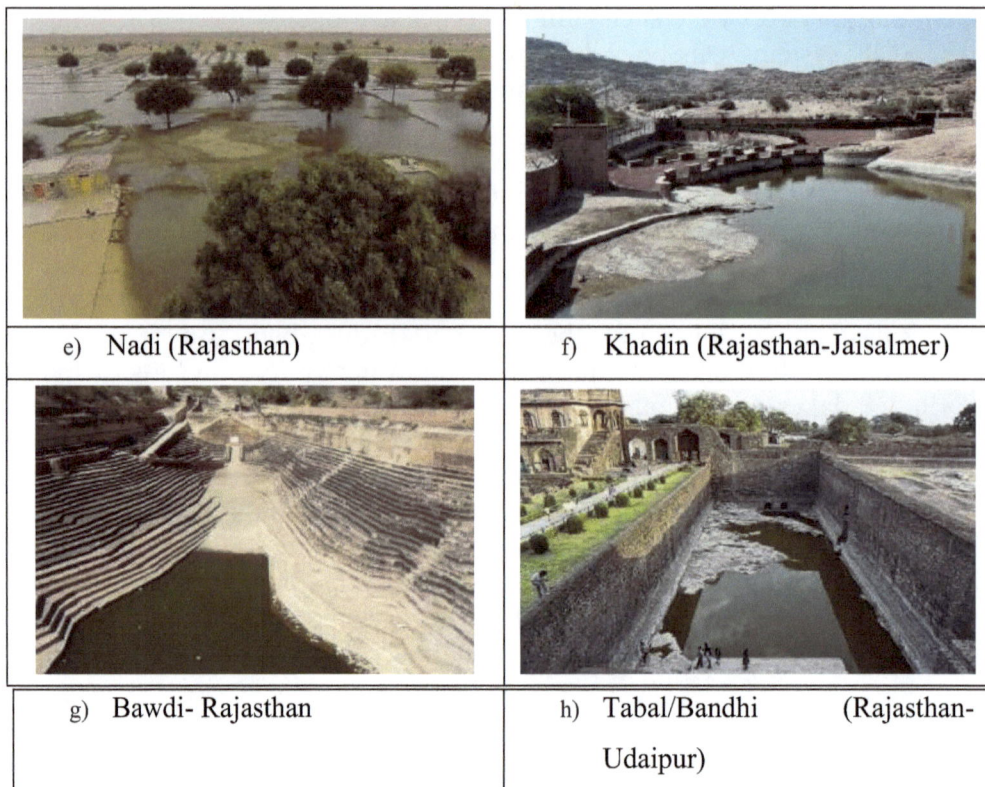

| | |
|---|---|
| e)  Nadi (Rajasthan) | f)  Khadin (Rajasthan-Jaisalmer) |
| g)  Bawdi- Rajasthan | h)  Tabal/Bandhi  (Rajasthan-Udaipur) |

**Fig. (5).** Traditional water Management System from The Thar Desert Ecological.

| | |
|---|---|
| a)  Zing (Ladakh) | b)  Kuhls (Himachal Pradesh) |
| c)  Ahar-pynes (Bihar) | |

**Fig. (6).** Traditional water Management System from **a)** Trans- Himalayan Region **b)** Western Himalaya and **c)** Indo-Gangetic Plains.

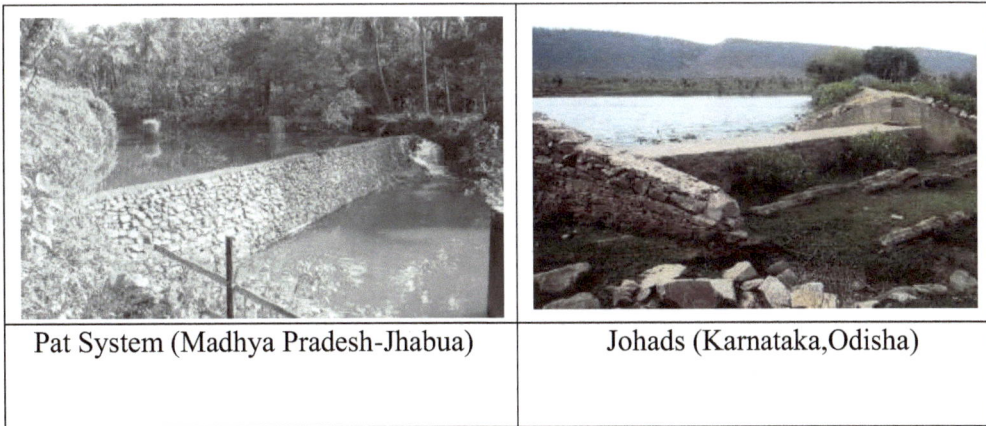

| Pat System (Madhya Pradesh-Jhabua) | Johads (Karnataka,Odisha) |
|---|---|

**Fig. (7).** Traditional water Management System from Central Highlands.

| a)  Bhandara Phad(Maharashtra-Dhule, Nasik) | b)  Ramtek Model (Maharashtra) |
|---|---|

**Fig. (8).** Traditional water Management System from Deccan plateau.

The main source of water on earth is rain. The rainwater travels many folds before entering the sea. As the rain touches the earth, it is taken up by trees, plants, some part is evaporated; some are retained by ground, and the remaining water reaches the sea through the drainage system. Rainwater harvesting is an old-age remedy for water conservation. People from diverse parts of the country have practiced this method depending upon the geography, soil nature, climate, and contours for storing water. Rainwater is collected from rooftops or the ground and stored in large tanks for later use. Rainwater harvesting was popular in Rajasthan (Fig. **5**), West Bengal, Meghalaya, and Nagaland (Fig. **10a - b**), Kerala (Fig. **9a**), and Tamil Nadu (Fig. **9b**) as shown in pictures.

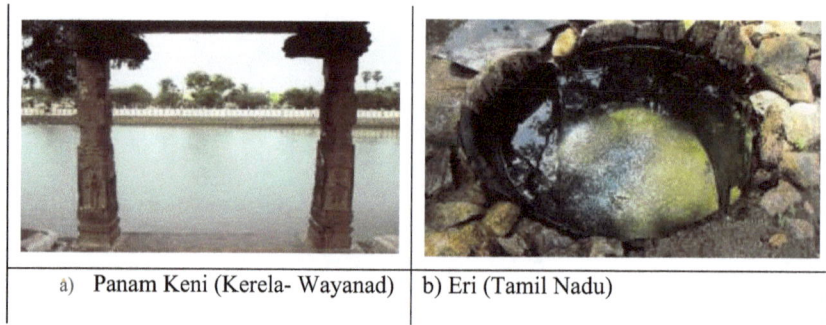

| a) Panam Keni (Kerela- Wayanad) | b) Eri (Tamil Nadu) |

**Fig. (9).** Traditional water Management System from Eastern Coast.

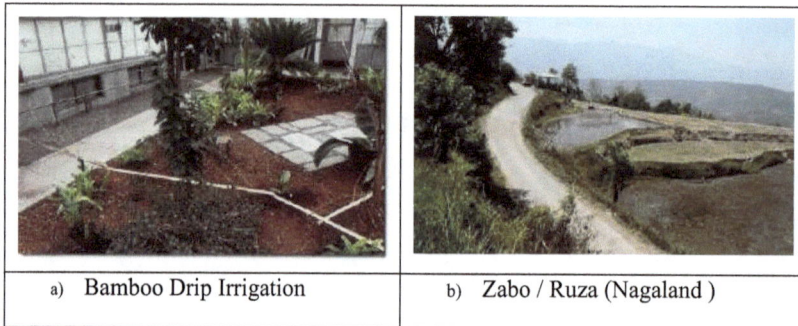

| a) Bamboo Drip Irrigation | b) Zabo / Ruza (Nagaland ) |

**Fig. (10).** Traditional water Management System from North East Hill Ranges. Sources: https://www.the betterindia.com/61757/traditional-water-conservation-syste and Water – its Conservation, Management and Governance (Central Water commission pdf. 2017).

The traditional methods used in Rajasthan for trapping runoff farming water to increase crop productivity were Khadin and Nadis, which were constructed to meet the needs of drinking water (Fig. **5e** and **f**). These Nadis showed high sedimentation, high evaporation, seepage losses, and water pollution. Percolation tanks built artificially were popular in Maharashtra, Madhya Pradesh, Tamil Nadu, Karnataka, and Gujarat. These tanks facilitate the percolation of stored surface runoff to recharge groundwater. Another way to conserve water used way back, which is still relevant today, is to check dams. They can be seen near Bhabar, Kandi, and other areas of Uttar Pradesh, Punjab, and Maharashtra. Check dams are constructed on land with gentle slopes with small streams entering them; they work efficiently in hard rock and alluvial soil. A series of check dams can be constructed to trap the maximum ruff from the streams. Along with check dams, a series of bunds or weirs are constructed on some nalas to resist the water flow and retain it on the previous soil/ rock surface for a longer duration. Nala bund behaves like a mini percolation tank. Underground bandharas also known as dykes, are constructed to conserve groundwater by arresting sub-surface flow. There are no evaporation losses, no siltation occurs, disasters such as dam collapse are out of the question, dykes can be used for perennial streams. The quality of water in dykes is better than in other water storing processes as the water stored is filtered by sandy soil; however, the seepage risk of surface contaminants cannot be neglected. Farm ponds or dug well are often constructed to collect the agricultural runoff. A series of such ponds can result in active groundwater recharge and help to dilute contaminants considerably [40].

Some methods adopted in entire Himalayan regions to store natural water for irrigation and drinking were constructed by Zings, Kul, Kulh, and Guhl. Zabo in Nagaland (Fig. **10a - b**), dongs by Assam tribes, jampois in West Bengal, were different ways to harvest water. North-East states were experts in using bamboo split pipes for harvesting purposes. In southern India, people stored water in ooranis, cheuvus, and temple tanks. The brick and ring wells were introduced by the Satvahanas. Later dynasties constructed large-scale advanced irrigation systems, which brought happiness and prosperity to the Deccan region. The areas where the recharge of groundwater becomes ineffective or shows very low efficiency are phreatic aquifers, which are partly covered with a poorly permeable layer. Such a situation is also evident in ponds where siltation resists recharge. To overcome this, a recharge shaft or an artificial recharge structure provides an alternate solution by penetrating the impervious zone and helping surface water recharge the phreatic aquifer. These shafts are preferably used in deep water levels.

Contour bundling is the reverse of bund management, where trenches are excavated at varying contour levels to conserve the runoff and aid percolation. This watershed method helps to build up soil moisture storage. In hilly and sloping areas, a series of bunds/trenches are interconnected for the easy flow of runoff water from the top to the bottom of the hill base by masonry chutes. These are known as half-moon trenches, and that help in slope stabilization and effective percolation of rainwater into the ground.

All above traditional methods of water conservation and management were operated at small scales using local resources that catered for a limited population using the local geographic conditions. Various traditional methods used in different parts of the country could not support the famine, floods, and droughts that occurred then. But the demand for food production kept on increasing. Hence, methods that could change the economy and were self-reliant during such calamities were required [41].

# WATER CONSERVATION AND MANAGEMENT: CONTEMPORARY PRACTICES

Traditional methods have contributed to the fullest of water conservation for years. However, with certain lifestyle changes, growing materialistic demands, and climatic changes, these methods fall short of making up for the scarcity of water, To evaluate the water sector effectively and efficiently current challenges need attention.

- The water availability has suffered due to uncertain climatic changes. Uneven and unpredicted precipitation, non-seasonal rainfall has led to drought-flood situations.
- With a growing population, the per capita availability of water is declining rapidly. With such a constantly rising population, it is predicted that by 2050, 17% of the population will experience total scarcity of water.
- Rising demand in the food sector, industrialization, and energy production might cause a tug of war between water availability and water demand in near future.
- Low irrigation efficiency might be due to lack of channels and field drainage, improper levelling of fields, unlined channel systems, dilapidated irrigation systems, inadequate water distribution, waterlogging, and salinity.
- There is a number of governance issues like lack of coordination of various agencies, poor management practices, problems in land acquisition, rehabilitation, resettlement, environmental and forest clearance, weak planning, and narrow approach.
- Water quality is affected due to the discharge of effluents from domestic and industrial sources. Runoff water from mining, agriculture, and various land activities, also leads to the entry of organic pollutants, heavy metals into water bodies. The natural purification capacity of rivers and lakes has been reduced due to higher doses of pollutants or due to the reduced natural flow. In addition, floods reduce the water quality, thereby affecting exploitable water. Polluted water gives rise to water-borne diseases and epidemics.
- Overuse of groundwater in arid and hard rock regions has resulted in affecting water quality immensely. In coastal areas, unattended and unscientific development may lead to seawater flowing toward the land, affecting the quality of land and water. Human interference and natural means are responsible for the presence of fluoride, arsenic, and iron decreasing the groundwater quality.

Seeing that the traditional methods could not turn the tables, hence, in the post-independence period, the Planning Commission decided to implement the Soviet Model of Industrialisation. This model was based on food security obtained through creating large reservoirs and canal systems for irrigation, cultivation of hybrid seeds and chemical fertilisers, and mechanisation of farming systems. Conventional methods helped India to become self-reliant in the food sector and develop the overall economy. However, it failed to bridge the gap between supply and demand.

The reasons for the above challenges that came up unexpectedly were that the conventional systems were centralized, stakeholders participation was minimal, and hence, ownership was missing. There existed a communication gap between the central and state governments, this stress deprived state governments to execute the projects with efficacy. The potential sites are placed in remote areas that they are difficult to approach. There were tremendous losses while distributing water to end-users. Hence, it was not a sustainable and integrated approach. So, there needed to be a solution to the above challenges. However, journey until the date was valued, but building the gaps was required on an urgent basis. Considering the present data on demand and supply, a long-term, well-planned, integrated approach to water management under stress conditions should be practiced.

For the planning and design of water resources management and structures, data collection, monitoring, processing, storage, retrieval, and dissemination are important aspects. In addition to these, inputs to the decision-makers, knowledge sharing, people's participation, mass communication, and capacity building are equally important and essential for effective water resources management. Thus, there is an urgent need to make a critical assessment of the combined strategy of traditional & conventional methods and to updated available technologies for better, precise, judicious, and scientific water conservation and management in the near future.

## BIOETHICAL, SOCIAL, AND LEGAL CONSIDERATIONS: PREVENTION AND REGULATION LAWS

Environmental degradation due to unplanned urbanization, intensive agriculture, population, and technology explosion, unthoughtful use of resources is ultimately enhancing rural poverty. These factors affect the soil fertility, air quality, quality, and quantity of water, causing deforestation, endangering wildlife, and affecting fisheries and agro cycles. Hence, farmers are directly affected by climatic cycles and environmental degradation. The rural population, especially tribes, is dependent on natural resources and is self-reliant. The collection of forest resources by natives is time-consuming and tedious, snatching away precious time, which refrains them from developing earning skills and educating children. Sudden changes in climate turn their distress. The loss of natural resources might not affect the urban population directly nor affect the economy, but the rural folks' breakdown might. For urban clans environmental degradation means lack of sanitation, water treatment, industrial pollution, and the impact of air, water, and soil quality on health. These factors affect their employment, attending school, and the economic decisions on gender inequality. Hence, social disparities affect different strata differently. It is evident that human health is adversely affected by environmental quality [42].

Climatic changes have affected the environment immensely worldwide. Large societies have been exposed to environmental hazards, unknowingly causing health imbalances. Quality of life depends on health, which remains a fundamental human right. The trio of community, individual, and environment, remains responsible for public health. Hence, to protect the planet, immediate action should be taken to promote a healthy ecosystem. For the population struggling with malnutrition, poverty, and the effects of natural calamities (floods and drought), existing health disparities add up as an additional obstacle in their lives [43]. The population is most vulnerable to climatic changes and health issues are the world's poor, and not those who are actually responsible for carrying out these changes. However, the use of new technologies to reduce environmental issues may indirectly affect the poor labour class. Should they be made the scapegoats for industrialization? In this changing climatic era, WHO, several health organizations, and agencies have taken human health as a core issue. These agencies are conducting health impact assessments and collecting data on risk as well as solutions to climatic changes. Enormous health problems can be reduced by improving sanitation (individual and community), checking pollution (air and water), protecting safe drinking water and soil contaminations, and providing better public health facilities. It can be inferred that education regarding environmental awareness in public can save the planet [44].

Legal failures, unjudicial access to environmental resources result in their degradation. In rural areas, common water sources, grazing lands, local forest products, fisheries, and entertainment zones are governed by local bodies. They protect them from over-exploitation under legal forms and penalize law-breakerssuch as overgrazing of lands on account of others, without considering the damage to the owner. In some cases, these norms, decided by the local community, are altered by the development processes, including urbanization, reduced mortality resulting in population growth, and state action giving importance to individual strengthening over community regulations and rights. Such manipulation in forms leads to the overuse of resources that affect the community's livelihood.

Policy fiascos can adversely affect environmental wellbeing (can cause environmental imbalance). Policy failure may occur due to explicit and implicit subsidies given for the use of resources. Improper use of these policies exceeds the use of natural resources and leads to altering the entire system. Anthropogenic climatic changes have affected precipitation patterns, ecosystems, agricultural potential, forest produce, water resources, coastal and marine resources, and have enhanced a wide variety of diseases. Climatic changes, stratospheric ozone depletion, and biodiversity losses are major concerns emerging globally. The responsibility should be shouldered by all countries in a global forum. Further, global natural resources must be shared on an equal sharing per-capita basis across the globe [41]. Economic growth to results in environmental degradation due to excessive use of natural resources that generate pollution aggravated by formal (legal rights) and informal (community norms) institutions. The following are the water management techniques practices by the EPA.

*Meter/Measure/Manage:* These facilities help to determine water-saving opportunities. It assures proper function of equipment, its proper maintenance, and checks if there are no leakages associated.

*Optimize Cooling Towers:* Cooling towers require a large amount of water. To obtain maximum benefits, the same water should be recycled at least six or more times. The ratio of water discharged to water evaporated needs to be controlled to minimize the total cost. The meter installed identifies the water leakage and checks malfunctions of the equipment. Condensate water from air-conditioning units can be reused for cooling towers.

*Replace Restroom Fixtures*: The old sanitary fixture should be replaced by those that are efficient and water-saving. The total water consumption can be reduced to half with new toilets and showerheads with low flush as well as flowing rates.

*Eliminate Single-Pass Cooling*: Single-pass cooling should be replaced by recirculating chilled water in laboratories and offices before it goes into drain this saves 80% of water consumption.

*Use Water-Smart Landscaping and Irrigation*: Supplementary water supply can be lowered or curtailed using hybrid, native, and drought-tolerant plant species. Crops should be weather-based; soil moisture sensors can control the excessive flow of water of irrigation. Proper landscaping reduces 10 to 20% of water consumption.

*Recover Rainwater*: Recovered rainwater from the rooftop is stored in tanks this water is used for cooling towers, irrigation of parks, golf courses, public places maintenance, and flushing toilets (Fig. **11**).

**Fig. (11).** A complete figure showing rainwater harvest and ground water recharge. Source: https://jalam-jeevam.telangana.gov.in/rain-water-harvesting/how-to-construcrt/

Besides the above remedies given by EPA some water management strategies used worldwide are as follows:

***Recycling, treating, and reuse of wastewater*-** Domestic, sewage, and industrial wastewater can be treated and reused for various purposes (public areas beautification such as parks, landscaping, gardening, car washing, *etc.*) to reduce the stress on freshwater resources. Special ternary treatments (UV and O3) can turn the wastewater drinkable.

***Irrigation techniques*-**Controlled irrigation, sensory devices, and smart agricultural practices (crop cycle, quality of seeds, hybrid methods) can reduce water usage. The technique adopted depends upon the geography, contour, soil, and climate of the local area. Drip irrigation and sprinklers are effective and tested ways to reduce water flow.

***Love for natural resources*-** Every individual should feel for Mother Nature. All the water resources, including streams, lakes, rivers, ponds, glaciers, ice, and seas, have their own important places. Water and marine ecosystems are niches for different organisms. Hence, water is a priceless possession; avoid polluting it, care for it, and preserve it. Individuals must ensure that they conserve enough clean, fresh, and safe water for generations to come [45].

Groundwater pollution is associated with improper and excessive use of agrochemicals (pesticides, insecticides, herbicides, and other agro practices) and their pricing policies that neglect the potential effect on the environment. Several organic pollutants are eliminated while traveling through various soil layers; however, chemical pesticides are leached into the groundwater, therefore, polluting it. Contamination of groundwater may be due to geogenic causes, such as fluorides and arsenic being leached from natural resources [46, 47]. Several health problems are reported due to groundwater contamination, as it is also a major source of drinking water in many parts of the world. Overexploitation of groundwater in coastal areas leads to inadequate recharge followed by serious salinity ingress, affecting health and land fertility [48].

***Partnerships and Stakeholder Involvement-*** Environmental conservation can be effectively processed if all the stakeholders work hand in hand. Policymaking and implementing government agencies of Central, State, Municipal, Panchayat level; public and private sector; academicians, researchers, industrialists, professional experts, youth clubs, voluntary and community organizations may play important role in not only formulation but also in practical application and promote environmental conservation techniques. Local self-government plays an important role in managing natural resources and conserving them. The 73 and 74 constitutional amendments provide the framework for their empowerment. To actually realize the role, certain amendments in policy and the legislature are required [49].

## ENVIRONMENTAL AWARENESS, EDUCATION, AND INFORMATION

Enhancing environmental consciousness is vital for environmental conservation. It needs the sinking of individual behaviour with the necessities for conservation, which probably would lessen the demands of observing and implementing systems. Awareness needs to be given to the public, youth, adolescents, urban dwellers, industrial and construction workers, municipal and other employees, *etc*. Awareness displays responsible behaviour towards the environment and the impacts of irresponsible actions that can harm the ecosystem, which includes public health, living conditions, sanitation, and livelihood prospects. Environmental awareness can be created for the general public or for a focused group, it can be formal as well as informal or a combination of both. The print, electronic, or live media, formal and informal settings are various means for spreading awareness to the public. According to IEC, various awareness activities at state, district, and block level includes:

***Mass Media:*** Audio Visual spots, jingles.

***Print Media***: Distribution of relevant materials to anganwadis, schools, religious groups, health workers, social workers, marketplaces, and government offices.

***Outdoor Publicity***: Hoardings and banners near the most visited places. Wall painting at schools, colleges, offices, railway and bus stations, hospitals, government offices, poster making, painting, and slogan competitions at all levels in schools and colleges.

***Other activities***: Health walks for all, conducting group discussions in communities, rallies, pad yatras, street plays that can fascinate people in villages and towns, interactive games, and exhibitions.

People from various sections of society need to be aware of water management processes so that they can be a part of policymaking and hence contribute to protecting the environment. Mass communization programs using modern means can be effective to educate people about not only water utilization, conservation, and management but also to spread awareness regarding their responsibility in the sustainable management of water resources. Capacity building in the community makes them active, confident, and well-informed in decision-making. Hence, joint efforts of people and government administration are a promising key. The Supreme Court has made it mandatory to impart environmental education to all informal systems. However, existing programs should be made more interactive and inclusive. To invoke the compliance with respect to the regulatory and legal provisions related to environmental protection, a just and punctual structure must be in place. Certain compulsory enforced actions must be taken to inspire the masses; their feedback is important for effective participation and to ensure the public impact on the developmental projects and environmental assessments.

## FRAMEWORK FOR LEGAL ACTION

Irrespective of the technological advancements today, the human component makes it difficult to achieve our water management goals. Policies and procedures set by the government are so complex and tedious that it takes time to get the results, which in turn slows down the process of meeting water demand worldwide. A wide range of issues needs to be considered for the purpose of achieving effective water governance. One such initiative to improve water management is the application of integrated water resources management (IWRM). This offers corresponding progress and management of land, water, and associated resources to maximize social and economic well-being. IWRM is goal-oriented and believes in the concept of sustainability. The proper functioning of IWRM requires a cyclic response from targeted groups and a knowledgeable and negotiable management system, along with a comprehensive governance framework. This would include communication, participation of the local public, governing technological, educational, and scientific issues, along with sustainable water policy.

Today, humans are dealing with water problems globally. This is probably due to a very slow approach to such an important and inevitable resource. It was only since 1998 when water was considered a major constituent for sustainable development. It was when the UN Commission adopted "Strategic Approaches to

Freshwater Management" [UN, 1998, p. 2] in their directives. Lately, the thought of effective management, conservation, governance, and policies, has widened the water agenda. The decision on policies and regulations are laid by considering social, political institutions and various stakeholders. International and national transboundary water bodies come as a great challenge to water managers. For countries sharing rivers and aquifers, an agreement to share the water at the time of stress should be predetermined.The various parameters that are required for proper water management are as follows:

***Strategies and Actions***: All levels of government - central, state/UT, and local should be encouraged to formulate action plans and strategies consistent with NEP. Consent from Panchayats and the urban local bodies on different issues relating to funds, officials, and capacities will help to make major provisions during policymaking. Environmental concerns need to be integrated with all development processes with diligence at all levels of government. The action plan must have an outline of strategic themes, monitoring of activities, functions, and roles that are ongoing, and seize new initiatives coming in the way. Strategic themes listed may not follow the same features in each case.

## PROCESS RELATED REFORMS

***Approach:*** For the proper regulation and functioning of reforms, the *Committee of Reforming Investment Approval and Implementation Procedures* (The Govindarajan Committee) was set up, which monitors and identifies delays in projects, reviews fund clearances, and various approvals required for the development. The aim of such committees is to ensure accountability, transparency, quick decision making and access to information, and reduce delays.

***Framework for Legal Action:***The present-day environmental protection and redressal structure is based on principles laid down under the criminal liability doctrines. However, it has proved ineffective as adeterrence and hence needs review. The prevailing civil liability doctrine as prevalent in western and the European states could be an effective deterrence to restrict environmentally harmful actions and would pave the way to commensurately compensate the affected.

The approach to fix civil liability regime for environmentally harmful actions, could be derived by laying down the following parameters:-

***Fault based Liability:*** The violators would be held liable for breach legal duties, which are mandatory to maintain environmental standards.

*Strict Liability:* This doctrine is based on the obligation to compensate the affected parties for the harm inflicted on them by the act of omission and commission, which may not be illegal action per say.

## Substantive Reforms

*Environment and Forests Clearances*: Environmental Impact Assessment (EIA) remains the systematic theoretical analysis for evaluating and reviewing new projects. The assessment is in accordance with the Govindarajan committee recommendation. Decentralisation of powers would be effective with the proper development of human and institutional capacities.

*Coastal Areas:* Coastal Regulation Zone notifications and Integrated Coastal Zone Management (ICZM) are responsible for the development of coastal areas. These authorities ensure the protection of valuable coastal resources, economic activities, and infrastructure development without hampering the livelihoods of local residents. The uniqueness of the island has to be preserved at any cost. Environmental safeguards should be the priority while building development projects in the islands. Moreover, high-value agricultural crops, deep-sea fishing, prospecting for oil and natural gas, and tourism should be endorsed for more revenue generation. Preparing for ICZM plans state requires technical and financial support.

*Living Modified Organism (LMO's)*: LMOs can be used as bio-techno agents to escalate the issues arising due to environmental pollution and have immense potential to contribute to economic development worldwide. Its scientific applicability should be assessed to ensure that LMOs do not have an adverse effect on the environment, ecology, health, and economy. An ecologically sustainable protocol needs to be evolved for the trans-border movement consistent with multilateral bio-safety.

*Environmentally Sensitive Zones:* The areas with rich biodiversity should be identified as 'Incomparable Values'. A local institution should be set up to ensure development plans for such areas.

*Monitoring of Compliance:* An effective mechanism to monitor the compliance would ensure a compliant regime. Local governing bodies and private-public partnerships can work hand in hand with the environmental management plans.

*Use of Economic Principles in Environmental Decision-Making:* The rapid degradation and depletion of natural resources, which is in fact, is an outcome of capitalist cronyism, has forced policymakers, watchdog bodies, and to revamp the existing policy framework. Since the Industrial Revolution, natural resources were considered as 'free goods', and its exploitation resultant in the form of population and degradation was passed on to the next generation. The supreme court of India in its 2G spectrum judgment has laid down the maxim, to ascertain the royalty fees towards usage of airwaves, and has firmly eroded the tendency to use the natural resources as "free goods".

The development *versus* destruction debate is now firmly encircled around sustainable development. Existing policy and regulatory frameworks have now firmly laid down the maxims to determine the "Net Presentable Value" for the utilization of natural resources. Robust compliance and monitoring compliances would always entail huge economic costs. Effective utilization of electronic surveillance systems like remote sensing would ensure a better compliance regime. This is now evident by the utilization of eco tagging, being effectively implemented by the Indian Bureau of Mines.

***Regulatory Reforms:*** To minimize environmental degradation and further curtail the cost of developmental projects, a robust legislative framework and regulatory institutions are imperative. Along with the legislation, sectoral and cross-sectoral laws and policies also influence environmental quality.

***Revisiting the Policy and Legislative Framework:*** The present legislative framework broadly comprises the Environment Protection Act 1986, Water (Prevention and Control of Pollution) Act, 1974 Water Cess Act, 1977 Air (Prevention and Control of Pollution) Act, 1981. The law regarding forest and biodiversity is contained in the Indian Forest Act, 1927; the Forest (Conservation) Act, 1980; the Wildlife (Protection) Act, 1972 and Biodiversity Act, 2002.

Environmental standards refer to both ambients as well as emission standards. The studies show that the quality parameters are location specific, hence, they should be according to the nation's priority, objectives, and resources. The standard may change as per the development of a country, its access to technologies, financial as well as natural resources. Within the country, local bodics may implement standards considering the local problem, they would require agreement from central government to ensure commitment to the policy [41, 50].

## RESEARCH NEEDS AND DIRECTIONS: EMERGING ISSUES

A good society and a healthy environment should be the objectives in the coming years. Effective resource management and its judicial use can help in meeting the objective. The processes involved in manufacture and recycling are the key elements in the long run. To sum up, it's our social obligation to identify, through research and development, the full potential in engineering and management, political and economic measures, and implementation by governance that is needed to sustain and safeguard our future. The rapid change in technology, with the help of scientific research, can help in sustainable development. A more objective-oriented approach will be beneficial in the process.

Research and development help us understand the optimization of water with minimum wastage. Researchers more intensively understand the water requirement, of the ecosystem, which can assist us to reduce the wastage of water from numerous stages of the supply chain. The human tradition has deeper relations with water usage. We need to utilize energy to the fullest. The basic need is to reuse water in different aspects of life. New ideas in research in this area are important for the sustainability of the ecosystem.

To sum up, it is imperative that universities and institutes relentlessly pursue research and development activities within the context of proactive engagement with the industrial fraternity; this will ensure contemporary research activity stays relevant. A strong industry, institute, the academic collaboration will lead to a better product development outcome. This will ensure the curriculum being talked about remains relevant to the need of society [51].

***Facing Challenges:*** Humans have always adapted to the changing environment. History stands as a witness as to how humans have survived all odds using scientific approach and intellect. For the past few decades, the environment has been subjected to global warming, causing drastic changes, affecting humankind globally. Our pace and efforts are not in agreement in dealing with global warming. Global warming has resulted in climatic changes and water availability to a greater extent than has affected agriculture, natural habitats, economic systems, and the water cycle. Scientists from all over the world expressed their concern at international climatic meetings about the increasing emission of greenhouse gases. However, these meetings have slightly persuaded concrete action and political decisions.

The scientific approach and results are difficult to accept and implement by the public, politicians, and economic systems. Hence, to bring behavioural changes and readiness for adaptation, new ideas are needed to bring changes worldwide for a desirable future. Therefore, novel thinking, a fresh approach, innovative research and scientific attitude, political and social acceptance can lead to new water management and conservation methods. Besides the conventional infrastructure, new infrastructure can be built with low environmental risk, sustain the growing need, and for desirable livelihood. Water storage can be done through wetlands and large dams. While considering any infrastructure its environmental and social impact should be considered. To avoid the previous errors World Commission on Large Dams [52] has stated guidelines.

***Water Scarcity- A Problem of the future:*** Every life on this planet desires a healthy ecosystem. It is impossible to think about the future without water as an elixir. With a growing population by 2050, it is expected that developed (China) and developing countries (India) will have scanty water resources. Therefore, it is imperative to take urgent measures to control water losses, conservation of water resources, and discontinuation of poisoning of existing resources. Intensive research should be carried out to meet these goals.

A nation like India, which is among the fastest developing nations, has to ensure adequate water for better economic and financial growth in the near future to have a healthier society. There should be accountability for every drop of water used for the goods and services we provide. On the other hand, the backward trend currently being seen has to be reversed to meet the demands of the increasing population. Coping with fresh and marine water to the citizens is yet another challenge. Recognition and problem-solving approaches will surely be a topic of research [53].

We need innovative, modern, and improved ideas to avoid water crises. For long-term sustainability, an objective-oriented approach is required. Today's approach to handling situations may be irrelevant in future scenarios or maybe have a worse impact on them. The government has to take strict measures to adhere to new laws and support financial aid to the researchers. The researcher should get a free hand to identify, establish, and implement his ideas, taking cognizance of environmental laws to face future challenges. Local people, government, environmentalists, and politicians should put the water challenge into top priority and help to tackle the issue to the fullest [54].

Financial gaps between the wealthy and poor are increasing day by day [55, 56]. Millions of people across different strata of society are deprived of access to nutritious food and safe drinking water. Neglect of proper sanitization has added to the miseries of people, especially the urban poor. Data shows that though food production and supply have remarkably increased in the past 15 years with respect to the population, malnutrition in children has shown a notable rise instead. A reason for the unavailability of food to the needy may be because of overeating by a large population in some countries. A study conducted shows that 1.5 billion people across the globe have been found to be obese due to overeating at the age of 20 or above. Another reason could be the wastage of food before and after reaching the dining hall due to inadequate transportation and storage and overfilling during meals [57, 58 - 60]. Therefore, an increase in production is not a solution to the issue. Moreover, management of resources (water, crops, food) has to be done "on every level by everybody" responsibly.

Resources and commodities are also hit by these imperfections. There exists a bond between crop, food, water, energy, humans, and the environment. Over use, unjudicial use will surely affect the health of the ecosystem and the mother earth. Hence, water crises can be minimized by reducing insufficiencies and managing the food supply chain with respect to water consumption [61].

***Importance of Technology:*** To meet the growing demands in a short duration technology plays an important role. Various technological trends and advances can benefit the water sector effectively. To name a few: artificial intelligence, nanotechnology, renewable technology, geoengineering, remote sensing, mapping groundwater resources, biotech methods for pest control, desalinization, and habitat preservation techniques [49].

Computer-operated simulators and optimizers with graphical and audio interpreters can help in planning, decision-making, designing, and policymaking that will maximize the impact and reduce the undesired process.

Innovations are high-risk and uncertain endeavors. However, financial support from the government and public-private enterprises can help sustain them for future development [62].

***The Way Forward:*** Global monitoring can fetch the comprehensive information required for water management; further, the progress of the targeted project can be studied. This could provide local, national, as well as global information, that can be used timely for decisions and policymaking. Data monitoring (figures, pictures) and document transparency could help to knit, bind, and empower society to meet collective goals.

The prior data on climate, soil content, constituents, water quality in the local area, and streamflow on mobile can assist farmers in taking decisions required in remote villages. Integration of monitoring network through satellites, improving ground-based network and knowledge from variable sources can improve water management remarkably [63].

For an advanced understanding of environmental issues, extensive research work from reputed and competent institutions, a continuous exchange of thoughts with the scientific community, academicians from government and public institutions, negotiators, and policymakers need to come on one forum frequently. A significant area of research (may change as per the need) are as follows:

- Nomenclatures of living natural resources
- Research for better understanding of ecological processes
- Research influencing policymaking
- R & D for management and clean environment

The need of the hour is to identify the area for research, have a detailed literature survey, generate a report and establish a research program within the government, with the outputs clearly stated. With the necessary financial aid, research in priority areas outside the Government can also be perused. There is an urgent need for techniques that can address the emerging challenges, which should be result orientated and could, bridge the gap between the conventional, contemporary, and new approaches of water management. Hence, integrated techniques should be augmented to meet the required water demand, thereby releasing the stress on water availability [41, 62].

## INTEGRATED WATER CONSERVATION TECHNIQUE: A FRAMEWORK FOR EFFECTIVE MANAGEMENT

Present-day challenges can be met by using some modern and innovative methods for water conservation. Water saved is water conserved. Hence, water stored in tanks, soil, ground aquifers, and surface reservoirs, if used judicially, can improve water availability. Conservation demands enough space and time to meet water requirements. Various economic, community-based, and administrative measures can assist water conservation. Population control can bring down the stress on all the natural resources as about 70% of water resources are used for irrigation purposes, 10% efficiency in irrigation methods can conserve a large amount of water. Several ways can be administered to increase the water efficiency and productivity by increasing more crop per drop motto. Some methods used include:

*Canal lining and Automation:* Canals are lined with geo-membranes and concrete canvas, which reduce the water loss due to percolation and sufficient water reaches the tail end farmer. Mechanized devices are used for the lining process. Canal automation can control the quantum of water supply to the crops and will restrict the over-irrigation of fields. Automation can plan a systematic supply of water throughout the season (Fig. **12a**).

*Pipe/Pressurized irrigation:* Evaporation and seepage are the means of major water loss during irrigation. Pipe irrigation can help overcome these losses. The initial cost of laying pipelines remains a costly affair, but in the long term, it has proved cost-effective. Pipes are laid underground to avoid theft, or on the ground as well. Water can flow under gravity (elevation) or pressure pumps can facilitate the flow as required. The pressure pumps are connected to sprinklers and drip pipes. Underground closed pipes hold water to the farmer's end, which can be used as and when required (Fig. **13 a - b**).

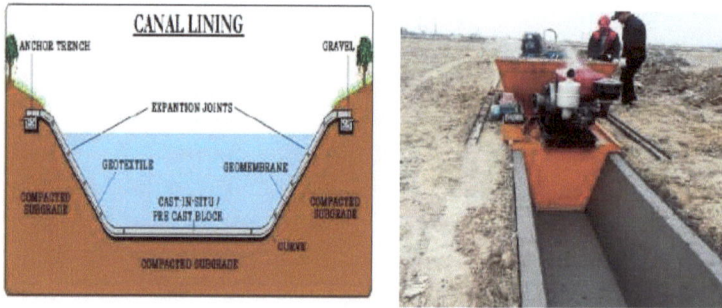

**Fig. (12).** (**a**). Detailed layout of canal lining, Source: https://climaxindia.com/canal-linning.html, (**b**) Machine laying concrete, Source: http://www.keluesolutions.com/ water-canal-making-machine/automatic-water-canal-maki

**Fig. (13).** Underground Pipes and underground sprinklers. Source: https://twitter.com/OdishaWater/status/1058250358323986432, https://in.pinterest.com/pin/263742121910356961/

***Micro Irrigation:*** Surface irrigation methods commonly used for a long time cater to large evaporation and percolation losses. Micro-irrigation has proved an efficient way to use water on farms. Sprinklers and drip irrigation are more efficient ways of irrigation than flood irrigation. Besides irrigation, these methods are used for the application of fertilizer and pesticides to the crop. In the case of drip irrigation, water reaches the roots directly, the flow rate can be controlled at any point. This saves water and increases the yield. Sub-surface irrigation further reduces evaporation losses but is costlier than drip and sprinkler systems (Fig. **14a - b**).

**Fig. (14).** Pressurised sprinklers and Detailed layout of drip irrigation. Source: http://74.52.53.155/sites/all/themes/ncpah/images/Drip_irrigation.jpghttps://modernfarmer.com/2015/07/irrigation-101/

***Lift Irrigation Technique:*** A system in which water is pumped from groundwater, canals, ponds, rivers, lakes, and reservoirs to the main delivery chamber, from where it is distributed to farmers by the proper delivery network. A constant water supply is required at the site, and pumps to draw water to the main chamber. This type of irrigation is benefits area where the target land is higher. It has a minimum water loss. Lift irrigation has stabilized agriculture production in drought situations. Crop production has increased due to abundant water availability, hence prosperous farmers. Lift irrigation schemes require the participation of beneficiaries, planning, designing, technique, and execution have to be done by a technically knowledgeable person. Unplanned lift irrigation can adversely affect the groundwater table. Since water needs to be lifted continuously, the method turns out to be costly. Usually, a cooperative society or group of farmers owns such schemes; few individuals having huge farming land can afford this scheme (Fig. **15**).

**Fig. (15).** Lift Irrigation Technique. Source: http://www.irrigation.kerala.gov.in/index.php/infrastructure/lift-irrigation

***Land management:*** Proper land management that includes integrated practices can acquire the ultimate yield with minimum expenditure and resources. Such practices include preparation of land according to the crop, soil-water conservation and moisture storage in soil, levelling of the surface for proper germination and yield, proper tillage for better water infiltration, reducing runoff, mixed cropping and alternate crop systems to retain fertility and nutritive content in the soil, rainwater harvesting, and recycling of agro wastewater.

The soil gets saturated and saline due to over-irrigation. The runoff washes away important nutrients, causing eutrophication of rivers and lakes. The proper drainage system can avoid such runoff. In developed countries, excess irrigated water is trapped in perforated pipes placed underground. Treated trapped water(primary process) can be redirected to another field. This helps to save nutrients and water, that would otherwise enter into water bodies. This drainage system requires proper administration until the process runs efficiently.

***Installations of Overhead Solar Panels:*** One of the innovative ways to reduce evaporation loss and provide free power to nearby villagers is the installation of solar panels on the canals. Floating solar panels can be installed on reservoirs and lakes. Such innovation can cut the cost and need for land acquisition from the government or private firms (Fig. **16**).

**Fig. (16).**   Overhead Solar Panels. Source: https://www.researchgate.net/figure/SPVS-system-over-irrigat-on-canals-India

***Automated Green houses:*** In automated green houses, environmental parameters such as humidity, temperature, and soil moisture can be controlled and monitored through the year. Drip irrigation is quite common in green houses with a controlled supply of water near the root zones (Fig. **17**).

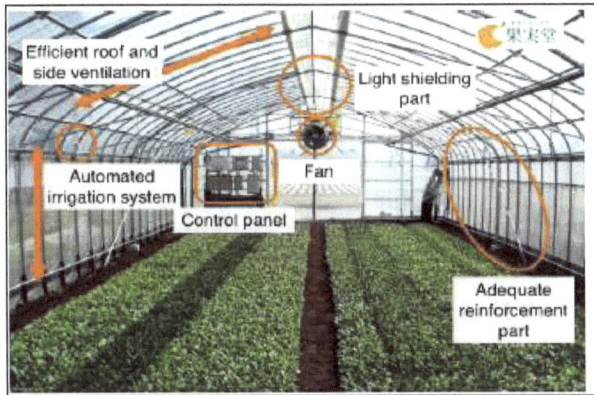

**Fig. (17).** Automated Green houses. Source:https://sensprout.com/wp-content/uploads/2017/02/kajitsudo.png

***Use of Information Technology:*** Information technology benefits farmers the most. They are updated about the weather forecast and floods through mobiles and the internet. Farmers can secure information regarding topography, soil quality, and weather conditions. The best variety of seeds, fertilizers, and pesticides available can be procured at cost-effective rates. Information and knowledge of crop patterns would yield more with minimal resources. The use of remote sensing (RS) can help to access whether the farmers are following the cropping calendar, quality of crops, and water requirements, RS can save the crops due to river meandering. It can identify the areas with high soil salinity and can help classify wetlands. A Geographical Information System (GIS) can help to estimate weather forecast climatic changes in water availability. It can also detect flood-prone areas, thereby assisting to reduce the disaster (Fig. **18**).

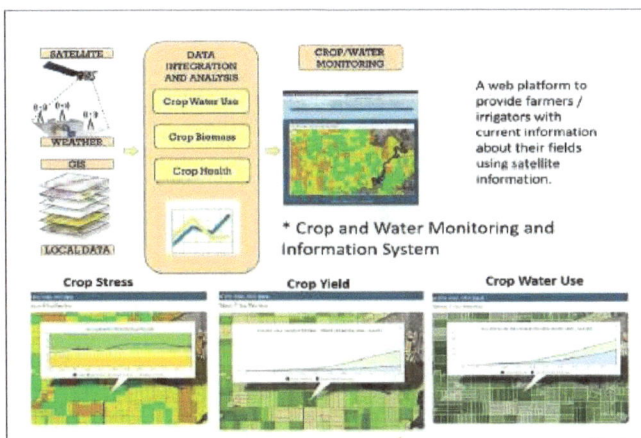

**Fig. (18).** Remote Sensing and Geographical Information System. Source: https://uwrl.usu.edu/water-resources/projects/remote-sensing

***Genetically Modified Crops:*** Scientists all over the world are engaged in developing genetically engineered crops that can sustain in less available water. Such crops can be made to grow on less fertile soil yet still produce a large yield (Fig. **19**).

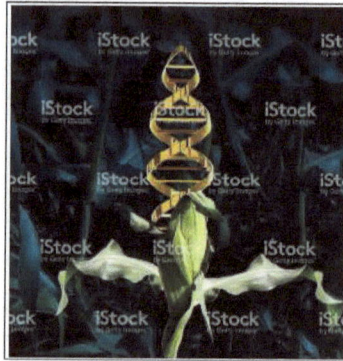

**Fig. (19).**   Genetically Modified Crops. Source:https://www.istockphoto.com/in/photo/genetically-modifie--crops-gm481377878-69345835.

***Magnetic water irrigation:*** Magnetic water irrigation is a promising technique to improve crop productivity and water efficiency. Water is magnetized by passing through a permanent magnet placed on the feed pipeline used for irrigation. It lasts a long time with no energy and no maintenance. Due to this, magnetization decreases surface tension and viscosity, which leads to increase fluidity and wet ability of water, allowing easy flow into the cell membrane. Hence, a high yield with low irrigation is achieved. The germination of seeds and water retention in the soil near roots increases substantially as compared to non-magnetic water (Fig. **20**).

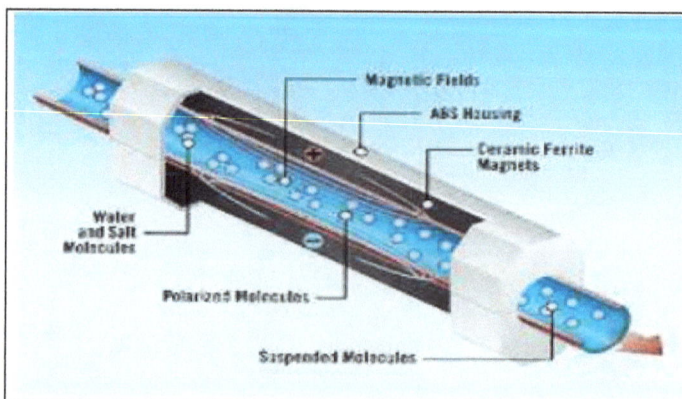

**Fig. (20).**   Genetically Modified Crops. Source: http://www.p3international.com/products/p7920.html

***Water Budgeting:*** Water budgeting defines an association between the input and output of water in a region. A comprehensive process where available water resources, causes of losses, consumption patterns, and demand from various sectors, where water requirements persists are monitored. Therefore, the cost of water usage is calculated.

Despite the above ways of water conservation, and its management there are new technologies that are helping urban areas save large amounts of water. To name a few water savers: automatic taps, water-less urinals, low flow showers, tap aerators, and dual flush. Rainwater harvesting has become mandatory in many countries; awareness regarding its importance has helped many societies and communities to overcome water scarcity. Excess rainwater can be directed through permeable pavements and footpaths directly to increase groundwater table. Both rainwater harvesting and ground recharge have made the urban population less dependent on external water resources [64].

Apart from the above methods, analytical techniques can be improved and integrated with existing techniques. Analytical methodologies can detect pharmaceuticals (ppb) in water sources that could have environmental significance. Integration of agency and public activities can be enhanced by responding to public inquiries, maintaining and expanding website, technical presentations, inviting articles for reviewed journals, and media press conferences. Forensic techniques can be used to detect emerging pollutants in water bodies. A sensitive and powerful method such as advanced mass spectrometric and chromatographic techniques can be used to detect emerging pollutants. Musks (communal sewage effluent) can be extracted and detected by an analytical method. Study of the presence of personal care products, incombustible organic compounds from the direct piping of small engine exhaust in Lake Tahoe, and lake deposition of airborne pollutants from industrial activity [65].

Most of the big cities in India are facing water scarcity. To mention a few, they are Chennai, Bengaluru, Hyderabad, and Pune. The rivers all over have low water levels. Dams, canals, groundwater tables, wells, and tanks are gradually turning dry. However, many individuals, groups, and communities are working for water conservation and its management. Many innovative low-cost ideas have come from students, researchers, academicians, and environmentalists [66 - 71]. Some enthusiasts have developed innovative and interesting methods using low-cost materials, which are attractive and friendly to use, make and apply, such as Ferro-cement tanks, cycle run water pumps, joy pumps, rainwater syringe, and water wheels. These are made by local people to meet their requirements. The contribution of women conservationists cannot be neglected; they are bringing immense change in society by educating village folks about the importance of water management and its conservation. Together, they are turning water scarcity into abundance in many small towns and villages [72]. Women from different parts of the country have contributed to managing, safeguarding, and protecting water resources; they are raising their voices and demanding water rights. Women from Odisha (Jagatsinghpur district) took up renovation of water bodies to overcome the acute water shortage in their locality. Sabarmati Tiki, a water conservationist founder of an NGO, that runs an organic farming training center near Bhubaneshwar (March 2018) was conferred with the Nari Shakti Award from the President of India, Ramnath Kovind, on international women's day [73]. Similarly, The Women's Alliance of Ladakh has taken the initiative to protect and conserve their culture and environment. They are pursuing farmers for organic farming using traditional water harvesting methods (in winters) as the region faces water scarcity due to low snowfall [74]. The Godavari from Marathwada (Maharashtra) has changed the farming pattern around her village from cash crops to food crops. She has moved to an innovative model of one-acre and 25 crops [75]. A Group of 98 women activists from 32 countries from 7 to 9 March 2019 gathered in Nepal for the first International Women and River Congress [76, 77].

## CONCLUSION

Water has a significant role in social and economic development. Anthropological, manmade, and artificial means have led to a scarcity of natural resources; hence, before the resources are completely wiped off and become expensive commodities, conservation is imperative. Various methodologies having wider applicability were used to solve water scarcity problems as well as issues related to food security, environmental protection, and disaster management. These methodologies included traditional, conventional, and modern technology-oriented techniques. Traditional practices would only solve limited water issues for a small population. They were insufficient to fulfill water demands for large-scale food production, industrial production, and energy requirements for a developing country like India, where escalating population along with urbanization has put immense stress on this precious resource.

Therefore, conventional methods were put under scrutiny for water resource development and extension of projects. These methods were developed in rivers and the natural depressions that turned into reservoirs. Due to the lack of feasible and suitable storage spaces and other affiliated issues like rehabilitation and resettlement, environmental and forest clearances and land acquisition, the total amount of utilizable water in the country has remained untapped. Interestingly, new complexities and challenges like environmental degradation, over-exploitation of ground and surface resources, climate change, operation and maintenance issues, *etc.* could not be handled by conventional methods alone. The above-mentioned problems led to the demand for modern methods that included wastewater treatment and reuse, micro-irrigation, piped irrigation, and the use of water-saving appliances. Presently, abundant innovative techniques and management practices at small and large scales are pouring in to tackle worldwide water scarcity issues such as desalination of water, unique irrigation projects, sharing irrigation management, water budgeting, *etc.* Invariably, the application of such methods is erratic, location-specific, and disintegrated. Due to further complicated challenges like water conflicts, social and environmental activism, strained external relationships with bordering countries, and frequent extremities (floods and droughts), all age-old water conservation and management techniques have resulted in poor results. This does not infer that these methods have outlived their impact, but rather is indicative of an inadequately planned strategy. In addition, attention needs to be given to the selection of the method as it might have conflicting terms with other techniques. Hence, an urgent need for integrated river basin planning and management has been proposed, which aims to combine traditional, contemporary, and modern practices to tackle problems effectively.

The only way to cater effectively to the future challenges of the water sector lies in an efficient master plan for a river basin that must be strong enough to adjust to the more likely spatial and temporal variations of rainfall that have been severely affected by climate change. Additionally, new scenario-based modelling needs to be developed for the river basin plan, along with relevant data collection, interpretation, dissemination, and collection needs to be accurate and dynamic. For the smooth and sustainable running of such machinery, constant development and research are required to aid the process. To enhance the quality of life globally, civil engineers serve as a backbone in designing, constructing, and planning the infrastructure. They help with policymaking, risk management, and way-outs during natural calamities. A researcher from a different stream can join hands and bring change in the future. Research can lead us to an alternative path for a given target. Hence, for the progress and development of any nation; scientific attitude building should be taken up by schools and colleges. Education institutes and governments should promote new ideas and innovation by conducting science competitions and exhibitions. State and central governments should allot separate budgets for research activities. Active participation of the private-public sector needs to be emphasized; together they hold tremendous potential and promise in the water sector. Simultaneously, the significance of hydrological, demographical, meteorological, and other relevant data for modelling purposes at the river basin level cannot be neglected as it helps in the prediction of precipitation, hence giving sufficient time for preparation towards disaster management like floods and droughts.

To impede the ever-increasing gap between demand and supply of freshwater, integrated water resource management (IWRM) provides an effective solution to this dire situation as a practical strategy for water conservation and management. The amalgamation of different master plans integrated with traditional, conventional, and modern innovative methods can be applied to various river basins depending on the geographic area and demography. Well-motivated and strong attempts on a global scale in this direction can possibly achieve water security over the current increasing water scarcity issue.

## CONSENT FOR PUBLICATION

Not applicable.

## CONFLICT OF INTEREST

The authors declare no conflict of interest, financial or otherwise.

## ACKNOWLEDGEMENTS

Declared none.

## REFERENCES

[1]     Schwarzenbach P, Egli T, Hofstetter TB, Von Gunten U, Wehrli B. Global water pollution and human health. Annu Rev Environ Resour 2010; 35(1): 109-36.
        [http://dx.doi.org/10.1146/annurev-environ-100809-125342]

[2]     Cosgrove WJ, Loucks DP. Water management: Current and future challenges and research directions. Water Resour Res 2015; 51(6): 4823-39.
        [http://dx.doi.org/10.1002/2014WR016869]

[3]     Imeson A. Desertification, land degradation and sustainability.
        [http://dx.doi.org/10.1002/9781119977759]

[4]     T O, Kana S. Global hydrological cycles and world water resources. Fresh Water 2020; 313(5790): 1068-72.

[5]     Tang WZ. Physicochemical treatment of hazardous wastes lewis publications 2016; p. 608. Available from: https://books.google.com/books

[6]     Jayaswal K, Sahu V, Gurjar B R. Water pollution, human health and remediation. 11-27.
        [http://dx.doi.org/10.1007/978-981-10-7551-3_2]

[7]     Garrido-cardenas J A, Agüera A. Wastewater treatment by advanced oxidation process and their worldwide research trends. 2020.

[8]     Bertinelli L, Strobl E, Zou B. Economic development and environmental quality : A reassessment in light of nature's self-regeneration capacity. 2007; Vol. 6: pp. 2-9.

[9]     Beck L, Bernauer T, Beck L, Bernauer T. How will combined changes in water demand and climate affect water availability in the Zambezi river basin ? How will combined changes in water demand and climate affect water availability in the Zambezi river basin ? Glob Environ Chang 2011.

[10]    Fecht S. How climate change impacts our water Clim week NYC 2019.

[11]    Wisser D, Frolking S, Douglas EM, Fekete BM, Vo CJ. Global irrigation water demand : Variability

and uncertainties arising from agricultural and climate data sets. 2008; 35: pp. 1-5.

[12]    Raleigh C, Urdal H. Climate change, environmental degradation and armed conflict. Polit Geogr 2007; 26(6): 674-94.
[http://dx.doi.org/10.1016/j.polgeo.2007.06.005]

[13]    Gleick PH, *et al.* Improving understanding of the global hydrologic cycle observation and analysis of the climate system The Global Water Cycle 2013.

[14]    Alcamo J, Flörke M, Märker M, Alcamo J, Flörke M, Märker M. Future long-term changes in global water resources driven by socio-economic and climatic changes Future long-term changes in global water resources driven by socio-economic and climatic changes. 2010; 6667.

[15]    Delpla I, Jung AV, Baures E, Clement M, Thomas O. Impacts of climate change on surface water quality in relation to drinking water production. Environ Int 2009; 35(8): 1225-33.
[http://dx.doi.org/10.1016/j.envint.2009.07.001] [PMID: 19640587]

[16]    Nel MW. New and emerging waterborne infectious diseases. Encycl life Support Syst. 2009; 1: pp. 1-10.

[17]    Raghav S, Painuli R, Kumar D. Threats to water : Issues and challenges related to ground water and drinking water. Springer International Publishing 2019.

[18]    Available from: https://www.nps.gov

[19]    Chetia M, Sarma HP, Banerjee S, Singh L, Dutta J. Use of surface water for drinking purpose in Golaghat district of Assam, India 2010; 2(1): 269-77.

[20]    Ramesh M, Narasimhan M, Krishnan R, Chalakkal P, Aruna RM, Kuruvilah S. The prevalence of dental fluorosis and its associated factors in Salem district. 2019; pp. 5-10.

[21]    Soetan KO, Olaiya CO, Oyewole OE. The importance of mineral elements for humans, domestic animals and plants. RE:view 2010; 4: 200-22.

[22]    Ahmed T, Scholz F, Al-Faraj W, *et al.* Water-related impacts of climate change on agriculture and subsequently on public health: A review for generalists with particular reference to Pakistan. Int J Environ Res public Heal 2013; 13: 1-16.

[23]    Muñoz O, Zamorano P, Garcia O, Bastías JM. Arsenic, cadmium, mercury, sodium, and potassium concentrations in common foods and estimated daily intake of the population in Valdivia (Chile) using a total diet study. Food Chem Toxicol 2017; 109(Pt 2): 1125-34.
[http://dx.doi.org/10.1016/j.fct.2017.03.027] [PMID: 28322969]

[24]    Rosborg I. S Precautions Drinking water minerals and mineral balance.

[25]    Farag A, Salem HM, Eweida EA. "Heavy metals in drinking water and their environmental impact on human health. ICEHM 2000; pp. 542-56.

[26]    Zimmermann KF, Chowdhury S, Annabelle K. Arsenic contamination of drinking water and mental health. 2015; pp. 1 28.

[27]    Krishnan S, Indu R. Groundwater contamination in India: Discussing physical processes, health and sociobehavioral dimensions. Anand, India: IWMI-Tata, Water Policy Research Programmes 2006.

[28]    Corcoran EC, Nellemann C. Sick water? The central role of wastewater management in sustainable development A Rapid Response Assessment. United Nations Environment Programme 2010.

[29]    Khan MA, Ghouri AM. "Environmental Pollution: Its effects on life and its remedies.," J arts, Sci Commer 2011; 2(2): 276-85.

[30]    Currie J, Zivin JG, Meckel K, Neidell M, Schlenker W. Something in the water: contaminated drinking water and infant health. Can J Econ 2013; 46(3): 791-810.
[http://dx.doi.org/10.1111/caje.12039] [PMID: 27134285]

[31]    Available from: http://www.in.gov/isdh/22963.htm

[32]   Andersson I. Environment and human health. Agency Eur Environ 2003; 250-71.

[33]   Pen G, Hincapie M. Degradation of pesticides in water using solar advanced oxidation processes. 2006; Vol. 64: pp. 272-81.

[34]   Fatta-Kassinos D, Kalavrouziotis IK, Koukoulakis PH, Vasquez MI. The risks associated with wastewater reuse and xenobiotics in the agroecological environment. Sci Total Environ 2011; 409(19): 3555-63.
[http://dx.doi.org/10.1016/j.scitotenv.2010.03.036] [PMID: 20435343]

[35]   Whelton A, Dietrich A M, Burlingame G A, Duncan S E. Minerals in drinking water : Impacts on taste and importance to consumer health 2007.
[http://dx.doi.org/10.2166/wst.2007.190]

[36]   Mishra P, Sahu HB, Patel RK. Environmental pollution status as a result of limestone and dolomite mining- a case study 2004; 23(3): 427-32.

[37]   Fu F, Wang Q. Removal of heavy metal ions from wastewaters: A review. J Environ Manage 2011; 92(3): 407-18.
[http://dx.doi.org/10.1016/j.jenvman.2010.11.011] [PMID: 21138785]

[38]   Ullah S, Javed WM. An integrated approach for quality assessment of drinking water using GIS: A case study of Lower Dir. J Himal Earth Sci 2014; 47(2): 163-74.

[39]   IECGuidelines ref.pdf. 2014.

[40]   Bringing sustainability to drinking water systems in rural India. 2007.

[41]   Policy N E. National Environment Policy. 2006.

[42]   Lee L M. A bridge back to the future : Public health ethics, bioethics, and environmental ethics a bridge back to the future : Public health ethics, bioethics, and environmental ethics. 2017; 5161

[43]   Patz JA, Gibbs HK, Foley JA, Rogers JV, Smith KR. Climate change and global health : Quantifying a growing ethical crisis. 2007.

[44]   Resnik DB, Portier CJ. Environment, Ethics, and Human Health Hast Cent

[45]   I. and M. Water Management – Meaning. Available from: https://www.importantindia.com/25072/water-management-meaning-importance-and-methods/

[46]   Jacks G. Controls on the genesis of some high-fluoride ground waters in India. 2018; 2927: 241-4.

[47]   Maheshwari RC. Fluoride in drinking water and its removal 2006; 137: 456-63.

[48]   Kousa M, Moltchanova A, Viik-Kajander E, *et al.* Geochemistry of ground water and the incidence of acute myocardial infarction in Finland. Community Health. J Epidemiol 2004; 58(2): 136-9.

[49]   Daniel E V B, Loucks P. Water resource, systems problems planning and management.

[50]   Human rights to water and Sanitation. Legislative regulatory and policy frame works. 2014.

[51]   Gray N. Water technology an introduction for environmental scientists and engineers.

[52]   UNEP. Dams and development: A new framework for decision-making The report of the World Commission on Dams. London, U. K.: Earthscan 2000.

[53]   Falkenmark M, J R. Balancing water for humans and nature: The new approach in ecohydrology. London: Earthscan 2004.

[54]   Gulbenkian . Water and the future of humanity (GTT) water and the future of humanity: Revisiting water security. Lisbon.: Calouste Gulbenkian Found. 2014.

[55]   Blanchard DCO. Washington, International Monetary Fund (IMF) the world economy in the twentieth century: Striking developments and policy lessons. World Econ Outlook 2000; 149-80.

[56]   Shah A. Global Illues Social. Political, economic and environmental issues that affect Us All 2010.

[57]    Beddington J. Achieving food security in the face of climate change: Final report from the commission on sustainable agriculture and climate change. Copenhagen: CGIAR Res. Program on Clim. Change, Agric. and Food Security 2012.

[58]    Lundqvist DM, De Fraiture J C. Saving water: From field to fork—curbing losses and wastage in the food chain, SIWI Policy Brief, Stockholm Int. Stockholm, Sweden: Water Inst. 2008.

[59]    Lundqvist J. Producing more or wasting less. Bracing the food security challenge of unpredictable rainfall, in Fourth Bot. ın Foundation Water Workshop on Re-thinking Water and Food Security.

[60]    Peter Alexander D M. Rethinking food waste for a healthier planet Lacncet Planet Heal 2017; 1(5)
[http://dx.doi.org/10.1016/S2542-5196(17)30077-3]

[61]    Hoff H. Understanding the Nexus. Background Paper for the Bonn2011 Conference: The Water, Energy and Food Security Nexus Stockholm Environ. Inst., Stockholm..

[62]    Pradesh H, Loucks DP. Water resources and environmental management : issues, challenges, opportunities and options. 2007; pp. 1-10.

[63]    L. S Grayman, Loucks DP. "Toward a sustainable water future: Visions for 2050. Reston, Va.: ASCE Press 2012.

[64]    Feldman D L. Water policy for sustainable development.

[65]    Grange A H. A new high resolution mass spectromey technique for identifying pharmaceuticals and potential endocrine disruptors in drinking water sources 2000.

[66]    Belekar RM, Dhoble SJ. Activated Alumina Granules with nanoscale porosity for water defluoridation. Nano-Structures & Nano-Objects 2018; 16: 322-8.
[http://dx.doi.org/10.1016/j.nanoso.2018.09.007]

[67]    Gedekar KA, Wankhede SP, Moharil SV, Belekar RM. Synthesis, crystal structure and luminescence in $Ca_3 Al_2 O_6$. J Mater Sci Mater Electron 2018; 29(8): 6260-5.
[http://dx.doi.org/10.1007/s10854-018-8603-5]

[68]    Gedekar KA, Wankhede SP, Moharil SV, Belekar RM. $Ce_3^+$ and $Eu_2^+$ luminescence in calcium and strontium aluminates. J Mater Sci Mater Electron 2018; 29(6): 4466-77.
[http://dx.doi.org/10.1007/s10854-017-8394-0]

[69]    Wani MA, Dhoble SJ, Belekar RM. Synthesis, characterization and spectroscopic properties of some rare earth activated LiAlO2 phosphor. Optik (Stuttg) 2021; 226(1): 165938.
[http://dx.doi.org/10.1016/j.ijleo.2020.165938]

[70]    Belekar RM, Athawale SA, Gedekar KA, Dhote AV. Various techniques for water defluoridation by alumina: Development, challenges and future prospects. AIP Conf Proc. 2104(1): 03004.
[http://dx.doi.org/10.1063/1.5100431]

[71]    Belekar RM. Suppression of coke formation during reverse water-gas shift reaction for $CO_2$ conversion using highly active Ni/Al2O3-CeO2 catalyst material. Phys Lett A 2021; 395: 127206.
[http://dx.doi.org/10.1016/j.physleta.2021.127206]

[72]    World water day 2019. Available from: https://www.worldwaterday.org/theme/

[73]    Nari Shakti Odissa. Available from: http://odishasuntimes.com/woman-conservationist-from-odisha-receives-nari-shakti-award/

[74]    Available from· https://scroll.in/article/892992/a-womens-group-in-ladakh-is-fighting-to-save-the-regions-environment-help-conserve-water

[75]    Available from: https://www.news18.com/news/buzz/when-a-bunch-of-women-took-over-farmlands -in-drought-prone-maharashtra-1745223.html

[76]    Available from: https://www.womenandrivers.com/statement

[77]    Available from: https://www.asiatimes.com/2019/03/opinion/womens-rights-and-river-protection/

# SUBJECT INDEX

## A

Abdominal cramps 204
Acetylcholinesterase 153
Acid 3, 14, 17, 18, 21, 34, 36, 50, 59, 60, 61,
    65, 75, 78, 80, 82, 110, 114, 138, 140,
    150, 154, 155, 158, 161
  alginic 158
  aliphatic 17
  aromatic 78
  benzoin 114
  corrosive 36
  dichlorophenoxyacetic 17
  glycolic 161
  haloacetic 3, 138
  hydrofluoric 50
  hydroxyl benzoic 80
  lactic 161
  linoleic 154
  nitrilotriacetic 18
  organic 21, 75, 80
  salicylic 82, 138, 140
  sulfuric 34
  sulphuric 80
Actions, photocatalytic 143
Activated 6, 7, 147
    carbon fibers (ACFs) 147
    sludge process 6, 7
Activated alumina 28, 30, 34, 36, 37, 40, 41,
    42, 43, 44, 53, 54, 55, 56, 60, 63, 64, 65,
    66
  alum-impregnated 41
  calcium-magnesium-coated 37
  copper-coated 42, 43
  impregnated 64, 65, 66
  prepared immobilized 40
Activation 10, 49, 63, 155, 165
  energy 49
  mechanical 165
  transition metal 10

Adsorbent(s) 21, 30, 31, 32, 33, 34, 43, 45, 46,
    47, 51, 52, 54, 55, 56, 59, 60, 61, 63, 64,
    66, 143
  activity 34
  alumina-based 21, 30
  material 61
Adsorption 1, 20, 21, 28, 29, 30, 31, 32, 33,
    34, 38, 39, 40, 42, 44, 47, 48, 49, 51, 52,
    53, 54, 55, 58, 59, 60, 63, 143, 144, 148,
    155, 158
  behaviour 55
  efficiency 40
  energy 51
  isotherms 28, 44, 54, 58, 155
  method 20, 31
  naphthalene 148
  process 21, 31, 32, 39, 47, 48, 49, 52, 53,
    55, 59
  properties 32
  techniques 29, 30, 31, 42
Advanced oxidation processes (AOPs) 2, 9,
    11, 75, 78, 79, 82, 83, 84, 91, 114, 115,
    116, 137, 154, 216
Agrochemicals 216
Air conditioning 91
Air pollutants 154
Alkaline sodium hydroxide solution 52
Alum-impregnated activated alumina (AIAA)
    41
Alumina 34, 35, 36, 40, 42, 43, 44, 45, 46, 49,
    50, 51, 55, 56, 59, 60, 61, 65, 66
  amorphous 49, 50
  hydrated 34
  impregnating 42
  mesoporous 43
  metallurgical grade 56
  nanotube-based 66
  oxide-coated 44
  surface 34, 35, 36, 40, 42, 66
  virgin-activated 55
Aluminium 19, 28, 35, 36, 40, 42, 46, 52, 53,
    58, 59, 61, 62, 63, 65, 66, 119, 145

**R. M. Belekar, Renu Nayar, Pratibha Agrawal and S. J. Dhoble (Eds)**
**All rights reserved-© 2022 Bentham Science Publishers**

anti-fouling 145, 160
lanthanum hydroxide 54
Complexes, toxic aluminum-fluoride 36
Components, inorganic 11
Computer-operated simulators 224
Concentration 8, 9, 11, 14, 16, 29, 47, 48, 49,
    63, 64, 79, 82, 106, 107, 108, 150, 203
    atmospheric 203
    catalyst 82
    ionic 106
    phosphate 108
Conservation 77, 171, 196, 210, 217, 218,
    219, 221, 223, 228, 232
    environmental 217
    soil-water 228
Constants, equilibrium binding 33
Consumption 36, 76, 116, 118, 135, 144, 172,
    200, 201, 204
    electricity 135
    low energy 116, 144
Contaminants 2, 6, 7, 15, 78, 138, 140, 144,
    173, 205, 211
    carbon-based 144
    organic 6, 7
Contaminated water 14, 29, 136, 143, 152,
    154, 204, 205, 206
Contamination 77, 95, 110, 197, 201, 204,
    205, 207, 216
    bacteriological 204
    faecal 204
Copper 2, 42, 163, 170
    nanoparticles 163, 170
    oxide-coated alumina (COCA) 42
    synthesised 42
Corrosion inhibitor 118
Crohn's disease 170
Crop(s) 202, 203, 205, 210, 215, 224, 225,
    226, 227, 228, 229, 230, 232
    damages 202
    engineered 230
    production 205, 227
    productivity 210, 230
Cross-linked poly microparticles 133
Cryostat bath 80
Cryptosporidiosis 205
Cyanotoxins microcystins 138

**D**

Damage 19, 29, 117, 169, 170, 204, 205, 214

chromosomal 170
    oxidative 170
Damages kidneys 204
Dechlorination electrochemical 18
Defluoridation 28, 29, 36, 40, 42, 52, 54, 60,
    61, 62, 66
    capacity 52, 54
    efficiency 60, 66
Degradation 7, 8, 9, 17, 75, 78, 79, 80, 82, 83,
    84, 114, 116, 118, 119, 120, 123, 124,
    125, 127, 138, 140, 141, 143, 203, 213,
    214, 215, 221, 232
    environmental 213, 214, 215, 221, 232
    of benzoic acid 114, 118, 119, 120, 123,
    124, 127
    organic 17
    oxidative 78, 84
    pathways 116
    photocatalytic 75, 138
    photochemical 75
    photo-oxidative 75
    processes 83
Degradative oxidation 80
Denitrification, electrochemical 18
Destruction 9, 17, 22, 115, 137, 167
    electrochemical 17
    immense 167
Deteriorates 82, 200
    contamination levels 200
Development 196, 218, 221, 222, 232
    sustainable 196, 218, 221, 222
    water resource 232
Diarrhea 91, 204, 205
    bloody 204
    watery 205
Differential scanning calorimetry (DSC) 167
Diffuse double-layer theory 15
Diseases 3, 4, 167, 170, 197, 201, 204, 205
    common bacterial 204
    digestive tract 204
    heart 4
    lung 167
    neurological 170
    skin-related 204
    viral 205
    waterborne 3, 197, 201
Disinfectant 4, 78, 138
    chemicals 4
Disorders 36, 118, 197, 204
    autism spectrum 36

# Z

www.ingramcontent.com/pod-product-compliance
Lightning Source LLC
Chambersburg PA
CBHW050822220326

41598CB00006B/292